Elements of Phase Transitions
and Critical Phenomena

Elements of Phase Transitions and Critical Phenomena

Hidetoshi Nishimori

Tokyo Institute of Technology

Gerardo Ortiz

Indiana University

OXFORD
UNIVERSITY PRESS

Great Clarendon Street, Oxford OX2 6DP

Oxford University Press is a department of the University of Oxford.
It furthers the University's objective of excellence in research, scholarship,
and education by publishing worldwide in

Oxford New York

Auckland Cape Town Dar es Salaam Hong Kong Karachi
Kuala Lumpur Madrid Melbourne Mexico City Nairobi
New Delhi Shanghai Taipei Toronto

With offices in

Argentina Austria Brazil Chile Czech Republic France Greece
Guatemala Hungary Italy Japan Poland Portugal Singapore
South Korea Switzerland Thailand Turkey Ukraine Vietnam

Oxford is a registered trade mark of Oxford University Press
in the UK and in certain other countries

Published in the United States
by Oxford University Press Inc., New York

First published 2011
Reprinted 2012, 2014

British Library Cataloguing in Publication Data

Data available

Library of Congress Cataloging in Publication Data

Nishimori, Hidetoshi.
Elements of phase transitions and critical phenomena /
Hidetoshi Nishimori, Gerardo Ortiz.
p. cm.
ISBN 978–0–19–957722–4
1. Phase transformations (Statistical physics)
I. Ortiz, Gerardo. II. Title.
QC175.16.P5N57 2011
530.4'74–dc22   2010032104

Typeset by SPI Publisher Services, Pondicherry, India
Printed in Great Britain
on acid-free paper by
Clays Ltd, St Ives plc

ISBN 978–0–19–957722–4

3 5 7 9 10 8 6 4

Dedicated to Sandra and Masae

Preface

As we enter the twenty-first century, techniques borrowed from equilibrium and non-equilibrium statistical physics have become widely applied to disciplines never imagined by their founders. Statistical physics is turning into an essential discipline and a fundamental framework for understanding and making quantitative predictions on diverse phenomena involving a large number of interacting degrees of freedom. These degrees of freedom may represent fundamental particles, such as electrons or quarks, or neurons carrying information through synapses, or even speculative agents trading in a competitive financial market. This holistic precept, that the whole is not necessarily equal to the sum of its parts, finds in statistical physics its most beloved tool.

Phase transitions and critical phenomena have consistently been among the principal subjects of active studies in statistical physics. The simple act of transforming one state of matter or phase into another, for instance by changing the temperature, has always captivated the curious mind. In that way, one can convert an almost uninteresting state of matter into a superconducting material with tremendous implications and applications. The Large Hadron Collider at the European Organization for Nuclear Research (CERN), which is currently exploring the nature of fundamental interactions at high energies, relies on the use of superconducting magnets, electromagnets built out of coils of superconducting niobium-tin wire cooled by liquid helium. Those magnets not only consume less power but most importantly can achieve an order of magnitude stronger fields than ordinary magnets, a fact that is crucial to reach such high energies.

The unusual set of physical properties known today as critical phenomena were discovered and apparently first reported in the *Annales de Chimie et de Physique* (1822–1823) by the Baron Charles Cagniard de la Tour. He performed experiments on liquids (water, alcohol, and ether) sealed in a glass cell under pressure, and observed the remarkable fact that above a certain temperature, that itself depends on the particular substance, the surface tension between the liquid and vapor disappeared, thus discovering what is known today as the supercritical fluid phase. Trying to prove that beyond a certain temperature the liquid gasifies regardless of pressure, he also noticed that near particular pressure and temperature values something unusual happened. In the neighborhood of this point, known as the critical point, the liquid becomes increasingly milky, indicating that visible light is being strongly scattered. The term critical point was coined later in 1869 by Thomas Andrews who observed that carbon dioxide at 31 degrees Celsius and 73 atmospheres pressure displayed the phenomenon of critical opalescence, that turbid and milky state previously observed by Cagniard de la Tour in other substances. The underlying universality of critical phenomena escaped the attention of their founders. It was Pierre Curie around 1895 who realized the similarity between the critical behaviors of a liquid–gas phase transition and that of the ferromagnetic transition in iron. The formal connection and derived analogies between unrelated physical materials behaving in a similar, universal, way near a continuous phase transition constitutes one of the landmarks of critical phenomena. Since the discovery of the renormalization group method in the early 1970s, the realm of applications of the concepts of scale invariance and criticality has pervaded several fields in the natural and social sciences. Thus, in perspective, it is of no surprise that these concepts, and the methods used to study them, can be applied to disciplines as diverse as the ones indicated in our introductory paragraph.

This book provides an introductory account of the theory of phase transitions and critical phenomena. The basic knowledge of the theory of phase transitions and critical phenomena is now recognized to be indispensable for students and researchers from many fields of physics and related disciplines. The book has been written having in mind an advanced undergraduate or graduate student in science or mathematics. It has been assumed that the reader has finished introductory courses of statistical mechanics in addition to elementary courses in calculus, Fourier analysis, and probability theory. Very basic undergraduate knowledge of quantum mechanics is required to understand the very few extensions of the classical theory. Clarity and detailed user-friendly derivations of usually accepted, as elementary and not so elementary, concepts have been our guiding principle. We preferred this style of presentation to what is sometimes known as rigorous, where at the expense of making the argument so sharp one loses track of the main idea.

One of our goals in writing this book is to provide the mathematical tools necessary for students to compute properties of critical systems in diverse contexts and disciplines, such as biophysics or complex systems. Almost all parts are written in a self-contained manner and all new concepts and calculations are explained in much detail without assuming prior knowledge of phase transitions and critical phenomena. We have avoided historical presentations of various topics allowing us to present compact derivations of the concepts without hiding details. For example, it is typical to first introduce the scaling hypothesis and then the renormalization group method as a way of justifying that hypothesis. Rather, we preferred to derive the scaling laws directly once the concept of a renormalization transformation is introduced, which, in our opinion, is a more natural and pedagogical way of presenting the material.

Another of the goals of this book is to prepare the reader to start reading more advanced books and research papers, in which basic accounts of common knowledge are often omitted and consequently beginners are trapped in the jungle of undefined jargon and complicated manipulations. Serious attempts have been directed toward a self-contained modular approach so that the reader does not have to refer to other sources for supplementary information. Accordingly, most of the concepts and calculations are described in detail, sometimes with additional/auxiliary descriptions given in appendices and exercises. It is, of course, impossible to cover all of the topics related to phase transitions and critical phenomena in a single volume of this introductory nature. One main omission is the general subject of quantum phase transitions, which happen at zero temperature as a result of changes in the parameters of the Hamiltonian representing the physical system. Although by itself a topic for a second volume, we have explained a few extensions of classical concepts to the quantum realm when appropriate and not in danger of jeopardizing the main ideas. Most of these extensions are written in the appendices. The bibliography at the end of the book will guide the reader to other topics not covered in this book and also to more advanced references.

A number of important concepts and methods have been developed, such as mean-field theory, scaling theory, the renormalization group method, exact solutions, series expansions, and Monte Carlo simulations, most of which have turned out to be valuable tools not only in statistical physics but also in other fields of physics. The present book also contains pedagogical presentations of statistical field theory methods, including a chapter on conformal field theory, random systems, percolation, the important use of dualities, and various modern developments hard to find in a single textbook on phase transitions. Moreover, as mentioned above, a series of appendices expand and clarify several issues not developed in the main text. It has been done in this way to avoid getting stuck in details and thereby losing the main flow of ideas. We would like to invite the reader, however, to seriously explore those appendices in a second reading since they are very useful to understand the depth and extensions of a particular topic.

In the first half of this book, standard topics such as mean-field theory, the renormalization group, and statistical field theory methods are explained. Then, slightly more advanced, but commonly encountered, concepts and methods follow, including the conformal field theory, the Kosterlitz–Thouless transition, effects of randomness, exact solutions, duality, and numerical techniques. Special emphasis has been placed on providing a physically intuitive description, sometimes with certain sacrifice of mathematical rigor, except in the chapters that discuss exact solutions and duality. The first five chapters are very basic and quintessential, followed by several chapters that can be read independently of each other, provided that the first five chapters have been finished. The important role played by symmetry and topology in understanding the competition between phases and the resulting emergent collective behavior, giving rise to rigidity and soft elementary excitations, is stressed throughout the book. Most importantly, in accordance with Sophocles' advice,[1] exercises are presented as the topics develop with solutions found at the end of the book, thus giving the text a self-learning character. It is strongly recommended that the reader solves (or at least tries to solve) the exercises as one proceeds in reading, since they often contain vital information to understand the logic developed in the main text.

The book reflects lectures given by the authors at their Universities to graduate students on the same topics and is thus classroom tested for its usefulness for beginners to this field. Students attending those courses contributed significantly to the improvement of presentation and material selection and the authors are very grateful to them. We would like to express our special thanks to Matthew Dean Jones and Zsolt Bertalan for proofreading and providing insightful remarks. We are also indebted to John Cardy, Pierluigi Contucci, Michael Fisher, Cristian Giardina, Norio Kawakami, Makoto Oka, Andrea Pelissetto, and David Sherrington for their crucial suggestions and comments on the draft. Shu Tanaka kindly drew the impressive picture on the cover of this book.

---

[1] "One learns by doing a thing; for though you think you know it, you have no certainty until you try."

Following the convention of many textbooks, we did not directly refer to original research papers for almost all topics in this book. However, we don't mean to claim priority for the materials presented. On the contrary, virtually almost all concepts, methods, and conclusions are well-established, standard ones. The book simply reflects the authors' interpretation of what constitutes a concise, consistent, coherent, and clear manner of presenting a wide range of topics. Correspondingly, we tried to avoid attributing each result to a specific person, except for a limited number of very common names including (but not limited to) the Ising model, Heisenberg model, Landau theory, Virasoro algebra, Kosterlitz–Thouless transition, Sherrington–Kirkpatrick model, and Lee–Yang zeros. The reader is referred to the bibliography at the end of the book for more detailed sources of information on the original references. We, nevertheless, would like to express our sincere apologies to those who contributed to the developments of the field for leaving out their names, with the expectation that our approach is understood and accepted.

We hope this book will help anyone interested in this fascinating subject and, moreover, inspire young scientists to continue developing this profound and far-reaching field of science.[2]

Hidetoshi Nishimori and Gerardo Ortiz
Tokyo and Bloomington
March 2010

---

[2] Updates, amendments and addenda will be posted on a dedicated web page at http://mypage. iu.edu/~ortizg/bookP.htm

Contents

1
Phase transitions and critical phenomena

As an introduction to the physics of phase transitions and critical phenomena, we explain in this chapter a number of basic ideas such as phases, phase transitions and critical phenomena. Intuitive accounts are given to the concepts of scaling and renormalization, which are powerful, systematic tools to analyze critical behavior of macroscopic systems. Also explained are several model systems, on the basis of which phase transitions and critical phenomena have been studied.

1.1 Phase and phase diagram

We are surrounded by a number of substances in different states. A *phase* is a state of matter in which the macroscopic physical properties of the substance are uniform on a macroscopic length scale, e.g. 1 mm. Familiar examples are ice, liquid water, and water vapor, each of which is a phase of water as a collection of macroscopic numbers of H_2O molecules. Roughly speaking, we call the length scale that we encounter in our daily life the *macroscopic* scale, which is to be contrasted with the *microscopic* scale as the standard of length in the atomic world. The goal of statistical mechanics is to elucidate physical phenomena occurring on the macroscopic scale as a result of the interactions among microscopic constituents.

A phase is characterized by a thermodynamic function, typically the free energy. A thermodynamic function is a function of a few macroscopic parameters such as the temperature and the pressure. Thus, the phase of a macroscopic substance is determined by the values of these parameters. A *phase diagram* is a graph with those parameters as the axes, on which the phase is specified for each point. An example of a phase diagram is given in Fig. 1.1. A typical phase diagram has several specific features including *phase boundaries*, a *critical point* (point C in Fig. 1.1), and a *triple point* (point TP). A phase boundary separates different phases. A change in parameters such as the temperature across a phase boundary causes a sudden change in the phase of a substance. For example, a solid phase changes into a liquid phase at the melting temperature. This is a *phase transition*. A phase boundary sometimes disappears at a critical point, where the two phases become indistinguishable and the substance shows anomalous behavior. The theory of *critical phenomena* explains this anomalous behavior.

Three different phases coexist at the triple point. Consider the example of water. Suppose that we confine some amount of water and ice in a container, seal it, and

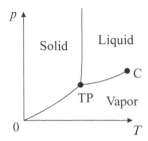

Fig. 1.1 Typical phase diagram. The phase of a substance is determined by the values of the control parameters such as temperature T and pressure p. C denotes the critical point and TP stands for the triple point.

evacuate the remaining air by using a vacuum pump. Then, the space above water and ice will be filled by vapor, realizing the triple point where ice, water and vapor coexist. The temperature and pressure of the triple point of water are $T = 273.16$ K and $p = 0.61$ kPa, respectively.

A phase can be characterized by various physical quantities. Especially important is the *order parameter*, which measures how microscopic elements constituting the macroscopic phase are ordered or in a similar state. As detailed in the following chapters, the order parameter is associated with the breaking of a symmetry of the system under consideration. The order parameter measures the degree of asymmetry in the *broken symmetry* phase (which is the ordered phase), i.e. it is non-zero in the ordered phase (lower-symmetry state) and vanishes in the disordered phase (symmetric phase).

In magnetic materials, for example, *magnetization* is a characteristic order parameter. Magnetization is the strength of a magnet, roughly speaking. The alignment of microscopic electronic spins gives rise to macroscopic magnetism. The symmetry that spontaneously gets broken is associated with the rotation of the spins. In solids, atoms or molecules occupy periodic positions. In this case, the spatial periodicity of molecules/atoms is the order parameter. A more abstract example is the quantum-mechanical phase of superconductors. A superconductor is characterized by a macroscopic quantum-mechanical wave function. The phenomenon of superconductivity is observed when the phase of this wave function has a constant value in a macroscopically extended region. It is not always an easy task to determine the order parameter. Indeed, some phases do not even have a local order parameter to characterize them. Moreover, some order parameters couple to external physical probes (e.g. the magnetization couples to an externally applied magnetic field), while others do not (e.g. the phase of the macroscopic wave function, which is the superconducting order parameter, does not couple to any physical external probe).

1.2 Phase transitions

A phase transition is a phenomenon in which a drastic change between thermodynamic phases occurs as the system parameters such as the temperature and pressure are

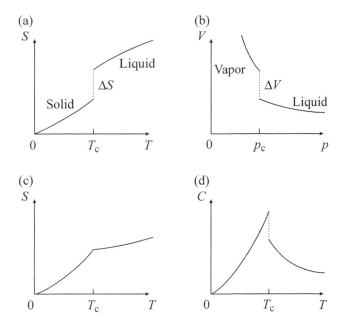

Fig. 1.2 Singularities in physical quantities at transition points. S is the entropy and C is the specific heat. (a) and (b) are first-order transitions, and (c) and (d) are second order.

varied. A familiar example is the melting of ice at $0\,°\mathrm{C}$ near 1 atm. The characterization of a phase transition as a drastic change of macroscopic properties is described theoretically as the emergence of singularities (non-analyticities) in functions representing physical quantities. As shown in Fig. 1.2, quantities such as the entropy S, the volume V and the specific heat C show such singularities as a discontinuity (jump), a *cusp* or a divergence. An example is the melting of ice, in which latent heat must be supplied to the system and consequently the entropy jumps as illustrated in Fig. 1.2(a). When water boils and changes to vapor, the volume changes discontinuously. From a physics standpoint the reason behind the occurrence of a phase transition is the competition between the (internal) energy E and the entropy S of the system, which together determine its free energy $F = E - TS$. While the first term (E) favors order, the second (S) privileges disorder, and depending on the value of the external parameters (such as T), one of the two terms dominates.

According to the conventional classification, phase transitions are roughly divided into two types by the degree of singularity in physical quantities. When the first-order derivative of the free energy F shows a discontinuity, the transition is of *first order*. The transition is called *continuous* if the second- or higher-order derivatives of the free energy show a discontinuity or a divergence. It is also common to name phase transitions by the order of the derivative that first shows a discontinuity or divergence, e.g. it is called *second order* if it is the second-order derivative of the free energy that first displays the discontinuity or divergence. For instance, the transition of ice to water accompanies latent heat and consequently a jump in entropy ($\Delta S > 0$). Since the entropy is the derivative of the free energy $S = -(\partial F/\partial T)_V$, such a transition is of first order. A transition with continuous entropy but a discontinuity in the specific heat C, which is the derivative of the entropy, is of second order (Figs. 1.2(c) and (d)). In many second-order transitions, the specific heat diverges at the transition temperature. Examples include the λ transition, i.e. the superfluid transition in liquid helium 4, and the paramagnetic–ferromagnetic transition in magnetic materials. A particularly interesting and common transition in systems of low space dimensionality is the Kosterlitz–Thouless transition (see Chapter 7), where all derivatives of the free energy are continuous, nonetheless, the free energy has a singularity, known in mathematics as an essential singularity.

Notice that from the statistical mechanics viewpoint, thermodynamics arises from the free energy, which is determined by the partition function Z,

$$Z = \mathrm{e}^{-F/k_{\mathrm{B}}T} = \mathrm{Tr}\ \mathrm{e}^{-H/k_{\mathrm{B}}T}, \tag{1.1}$$

where k_{B} is Boltzmann's constant, and Tr (trace) represents a sum over all the degrees of freedom that enter the Hamiltonian H of the system under study. Since Z is a sum of exponentials of $-H/(k_{\mathrm{B}}T)$, non-analyticities of the free energy can only happen in the *thermodynamic limit*, where the volume of the system V and number of degrees of freedom (e.g. spins in magnetic materials) N grow to infinity, such that its ratio remains constant, i.e. $N/V \to \mathrm{const}$.

A material may show both first- and second-order transitions depending on the conditions. Figure 1.3(a) illustrates the phase diagram of a magnetic material placed in an external magnetic field h. If the temperature T is lower than some T_{c} (*critical temperature, critical point, criticality, transition point*), the sign of the magnetization m jumps from minus to plus as the external magnetic field h is scanned from the negative direction to the positive direction as realized by the path (b) of Fig. 1.3(a), thus a first-order transition. For negative h, the spins in the magnetic material align with that negative direction on the macroscopic scale. They suddenly change the direction as the

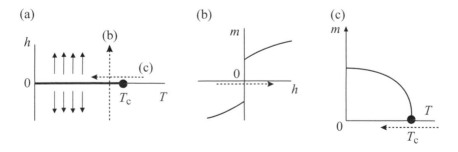

Fig. 1.3 (a) Phase diagram of a magnetic material, (b) first-order transition, and (c) second-order transition. The dotted arrows marked (b) and (c) in panel (a) correspond to the changes in parameters shown in panels (b) and (c), respectively.

external field becomes positive.[1] Thus, for $T < T_c$, a finite magnetization m remains even after we take the zero-field limit $h \to 0+$ as depicted in Fig. 1.3(b). The sign is of course negative, $m < 0$, when $h \to 0-$. This is called *spontaneous magnetization*, a typical example of an order parameter. When the temperature is high, $T > T_c$, the magnetization changes smoothly at $h = 0$ without any singularities. On the other hand, if we keep the external magnetic field infinitesimally small, $h = 0+$, and lower the temperature across T_c, then the spontaneous magnetization changes continuously from 0 to a positive value (Fig. 1.3(c)), thus defining a second-order transition.

1.3 Critical phenomena

Continuous phase transitions are often synonymous with critical phenomena, i.e. anomalous phenomena that appear around the critical point (C in Fig. 1.1) where two or more phases become indistinguishable. The main goal of the present book is to present the basic theory to understand critical phenomena.

Let us explain the idea of critical phenomena observed in magnetic materials. Suppose that we decrease the temperature T toward the critical temperature (critical point) T_c from above as in Fig. 1.3(c). The magnetization m keeps its vanishing value throughout this process. Nevertheless, as a precursor to finite spontaneous magnetization $m > 0$, $h \to 0+$ below T_c, the magnetization increases very rapidly if we apply a small but finite external magnetic field h at temperatures slightly above the critical temperature. Thus, according to the definition of *magnetic susceptibility* χ

$$m = \chi h + \mathcal{O}(h^3), \tag{1.2}$$

this χ assumes a very large value near the critical temperature T_c (Fig. 1.4).

The magnetization is proportional to the external field $m \propto h$ for $T > T_c$. When the temperature is adjusted to be exactly at the critical point ($T = T_c$), the magnetization grows more rapidly as a function of the external field, $m \propto h^{1/\delta}$ ($1/\delta < 1$). If we define the magnetic susceptibility as the first-order coefficient of h as in eqn (1.2), then the susceptibility at the critical point behaves as $\chi \approx m/h \propto h^{1/\delta - 1}$ and diverges as $h \to 0$. See Fig. 1.4. Spins do not spontaneously align on the macroscopic scale in the high-temperature region $T > T_c$ but they, nevertheless, tend to have a similar direction within fairly large regions for T close to T_c. These clustered spins respond coherently to the external field, and consequently the magnetization increases very rapidly as the field is applied. The degree of alignment of spins fluctuates significantly in space and time. The singularities in physical quantities reflect these *fluctuations*.

Essentially, the same phenomena are observed around the liquid–vapor critical point shown in Fig. 1.1. Suppose that we increase the temperature and pressure so that the system stays on the coexistence curve (phase boundary) of liquid and vapor along the curve between TP and C in Fig. 1.1. Since (low-density) vapor and (high-density) liquid become indistinguishable beyond the critical point, there exist fairly large regions in the liquid where the density is significantly lower than the average at temperatures sufficiently close to (but below) the critical temperature. Similarly,

---

[1] This is an idealized picture. The change in the direction of spins is actually much more complex due to the magnetic domain structure.

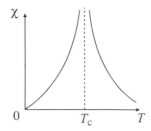

Fig. 1.4 The magnetic susceptibility diverges at the critical point.

large clusters of high density will show up in the vapor. These phenomena may be described as fluctuations in the density, the difference of which between vapor and liquid represents the order parameter. The length scale of such fluctuations ranges from microscopic to quasi-macroscopic near the critical point. Consequently, fluctuations of length scales close to the wavelength of visible light exist. Then, there appear white cloud-like regions in the transparent liquid/vapor due to the reflection of light from such clusters. This phenomenon is termed *critical opalescence*. One of the physical quantities that show a singularity in such a case is the variance of density per unit volume, which diverges at the critical temperature.

The degree of singularity or divergence of physical quantities near the critical point is described by *critical exponents* or *critical indices* $(\alpha, \alpha', \beta, \gamma, \gamma', \delta, \eta, \nu, \cdots)$. Experiments show that physical quantities generally have *power-law* singularities as functions of the difference between the control parameters (such as temperature) and their critical values. Let us denote this difference by t and take it as a dimensionless quantity. For example, $t = (T - T_c)/T_c$, where T_c is the critical temperature. Critical exponents of simple magnetic materials are defined as follows

$$\chi \propto |t|^{-\gamma} \quad (T > T_c), \quad |t|^{-\gamma'} \quad (T < T_c) \tag{1.3}$$

$$C \propto |t|^{-\alpha} \quad (T > T_c), \quad |t|^{-\alpha'} \quad (T < T_c) \tag{1.4}$$

$$m \propto |t|^{\beta} \quad (T < T_c) \tag{1.5}$$

$$m \propto |h|^{1/\delta} \quad (T = T_c) \tag{1.6}$$

$$G(r) \propto r^{-\tau} e^{-r/\xi} \quad (T \neq T_c) \tag{1.7}$$

$$\xi \propto |t|^{-\nu} \quad (T > T_c), \quad |t|^{-\nu'} \quad (T < T_c) \tag{1.8}$$

$$G(r) \propto r^{-d+2-\eta} \quad (T = T_c), \tag{1.9}$$

where χ is the susceptibility, C the specific heat, m the magnetization, $G(r)$ the connected two-point *correlation function* $G(r) = \langle S_i S_{i+r} \rangle - \langle S_i \rangle \langle S_{i+r} \rangle$ with two spins, S_i, S_{i+r}, separated by a distance r, and d is the space dimensionality of the system.

As already mentioned, the magnetic susceptibility χ diverges at the critical point, and the rate of divergence is described by eqn (1.3) using the critical exponents γ and γ'. The symbol $\propto$ expresses the most singular contribution among the singularities in the function on the left-hand side. There actually exist additional weaker singularities and non-singular (i.e. regular) terms and we have omitted the proportionality constants. Therefore, a more accurate expression for χ should look like

$$\chi = A|t|^{-\gamma} + B|t|^{-\gamma+1} + \cdots + \text{const} + t + t^2 + \cdots. \tag{1.10}$$

The critical exponents for the higher-temperature side $(t > 0)$ and the lower-temperature side $(t < 0)$ usually assume the same value. This is a non-trivial fact and should be confirmed in each case.

Similar remarks apply to the other critical exponents. The index of singularity in the specific heat C is α. The index β describes how the magnetization (order parameter) m approaches zero as the temperature increases toward the critical point. The magnetization m at T_c is a non-linear function of the external field h, a fact that is expressed by the exponent δ. The correlation function $G(r)$ decays exponentially as a function of the distance as in eqn (1.7) if the temperature is not at the critical point. The correlation function describes the degree of similarity of the states of spins separated by a distance r. Thus, eqn (1.7) implies that the correlation between spin states is very small beyond the distance ξ, called the *correlation length*. This correlation length increases rapidly as the temperature approaches the critical point and eventually diverges. The rate of this divergence is described by the exponent ν defined in eqn (1.8). When the temperature is tuned to be exactly at the critical point, the system sustains fluctuations of all length scales and the correlation function decays slowly in a power law manner as in eqn (1.9). This power is characterized by the exponent η.

Critical exponents are very basic quantities to characterize critical phenomena, and an important goal of the theory of critical phenomena is to develop a systematic method to calculate the values of critical exponents. Most importantly, there are simple relations between exponents (*scaling law*), which allow one to determine an exponent given the values of other exponents (i.e. not all exponents are independent). For example, the *Rushbrooke scaling law* is $\alpha + 2\beta + \gamma = 2$.

1.4   Scale transformation and renormalization group

An essential feature of critical phenomena is that fluctuations of all length scales appear simultaneously, causing non-analytic behavior of physical quantities. The presence of singularities make standard theoretical perturbative approaches inappropriate. Figures 1.5, 1.8 and 1.9 show such emergence of fluctuations in the two-dimensional $(d = 2)$ ferromagnetic Ising model (see Section 1.5) simulated by the Monte Carlo method to be detailed in Chapter 11.

Let us first have a look at Fig. 1.5(a) that shows the state of the system at a temperature slightly below the critical point, $T = 0.995T_c$. The two-dimensional Ising model with a linear size 486 was simulated, and a typical spin configuration is shown here with up spins in black and down spins in white. No external field h is applied. Most spins are in the up state and the system is magnetically ordered. Up spins are connected from one edge of the square to the opposite edge, whereas down spins exist only in isolated islands. Most of these islands have a length scale of a few millimeters in the

present printed scale (of the order of 10 spins in the number of sites). This length scale
is essentially the correlation length ξ.

Now, let us apply a *block-spin transformation* to the configuration of Fig. 1.5(a).
The result is depicted in Fig. 1.5(b). The block-spin transformation in this example
consists of replacing the neighboring $3 \times 3 = 9$ spins by a single spin according to the
majority rule illustrated in Fig. 1.6. Figure 1.5(b) can be regarded as a *coarse-grained*
version of 1.5(a) with the details of very short length scales washed out. The dominance
of black states is more prominent in (b) than in (a). The length scale is reduced to
1/3 of the original one and, correspondingly, the linear size of typical white regions
(correlation length) is reduced by the same factor. Figure 1.5(b) should therefore be
expanded (rescaled) by the factor 3 if we are to restore the original scale.

(a) (b)

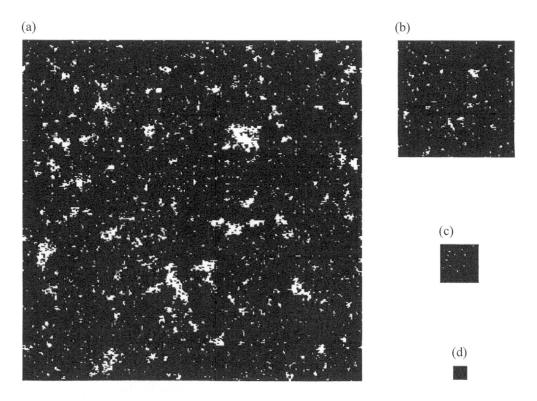

(c)

(d)

Fig. 1.5 (a) Typical spin configuration of the two-dimensional ferromagnetic Ising model at
$T = 0.995T_c$. (b), (c) and (d) are the results of successive block-spin transformations.

Fig. 1.6 Block-spin transformation in which we apply the majority rule to replace $3 \times 3 = 9$
Ising spins by a single spin.

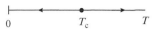

Fig. 1.7 The renormalization group transformation changes the effective temperature. If the initial temperature is below the critical point, the effective temperature decreases, whereas it increases when the initial value is above the critical temperature. The critical point corresponds to the (unstable) fixed point of the transformation.

Repeated applications of block-spin transformation to (b) yield (c) and then (d). These latter figures (c) and (d) are almost uniformly black. This is a consequence of consecutive eliminations of fluctuations of short length scales (i.e. white islands in the black sea) in order to focus our attention to the phenomenon of long length scales. This process allows us to extract the essential features of the macroscopic system near the critical point. This is the physical picture behind the *renormalization group*. A block-spin transformation is a step to realize the idea of a renormalization group transformation, which consists of elimination of short-length fluctuations (coarse graining) and *rescaling*.[2] The latter process of rescaling in the example of Fig. 1.5 is just to expand (b) by the factor 3 and reproduce the original scale of (a).

The all-black state corresponds to the maximum magnetization, which is physically realized at zero temperature. We may thus understand that the renormalization group gradually lowers the effective temperature to eventually reduce the system to the zero-temperature state of Fig. 1.5(d), see Fig. 1.7.

If we apply the same manipulations to the system exactly at the critical point $T = T_c$, we obtain Fig. 1.8. The external field is zero in this case too. The initial state (a) may give the reader an impression that the black state is slightly dominant over the white state (or vice versa). However, when we apply the renormalization steps as in (b), (c), (d), the system does not seem to approach an overwhelmingly black (or white) state. The physical reason behind this behavior is that there exist fluctuations (black and white islands) of all length scales at the critical point, from microscopic to macroscopic scales, which makes it impossible to eliminate short-length fluctuations to reduce the system to essentially the zero-temperature or infinite-temperature states. The system remains unchanged by the renormalization group. In fact, white islands of small to large length scales coexist in Fig. 1.8(a) and we cannot identify a typical length scale. For temperatures lower than the critical one, $T < T_c$, as in Fig. 1.5, fluctuations have a typical length scale ξ, and hence a few renormalization steps are sufficient to eliminate these short-length behaviors. A critical region is a fluctuation-dominated regime of the system. Exactly at the critical point, the system has an essentially different property, that of the absence of a typical length scale. This fact may be rephrased by stating that the effective temperature does not change by a renormalization group transformation if the system is at the critical point. The critical point corresponds to a *fixed point* of the renormalization group transformation.

[2] There indeed exists an additional important process of *renormalization* (amplitude change of microscopic degrees of freedom), as will be discussed in detail in Chapter 3.

(a)   (b)

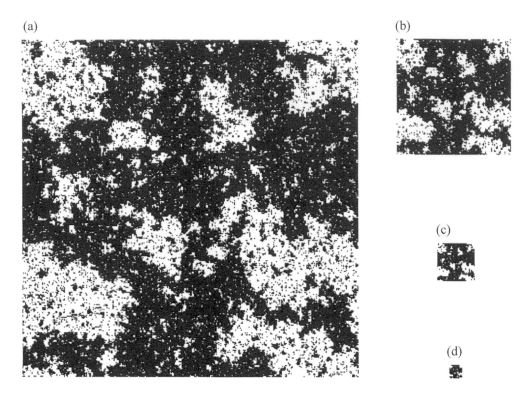

(c)

(d)

Fig. 1.8 Change of spin configurations by renormalization group transformations for a system at temperature $T = T_c$.

The final example of block-spin transformation is for a temperature higher than the critical point $T = 1.05T_c$. As shown in Fig. 1.9, a few transformations change the system state to a completely random one. The situation in (c) and (d) resembles the high-temperature limit of the system where spins rapidly change their states due to strong thermal agitation. In this sense, the renormalization-group transformation reveals that the system above the critical point has essentially the same properties as the system in the high-temperature limit. The correlation length is of the order of a few millimeters to a centimeter in (a), which is reduced to the very short distance between neighboring (block) spins in (c) and (d).

A general strategy to study critical phenomena is, as suggested by the examples mentioned above, to write a set of *renormalization group equations*, which describe how parameters such as the temperature change as the degree of coarse graining and rescaling is increased, and to analyze its solution around the (critical) fixed point. A fixed point is a point in the parameter space, to which corresponds a state (or fixed-point Hamiltonian) invariant under a renormalization group transformation. It has an associated correlation length ξ that is either infinity (at a critical fixed point) or zero (at a trivial fixed point in the high- or low-temperature limit, for example). These ideas are formulated in detail in later chapters.

(a)

(b)

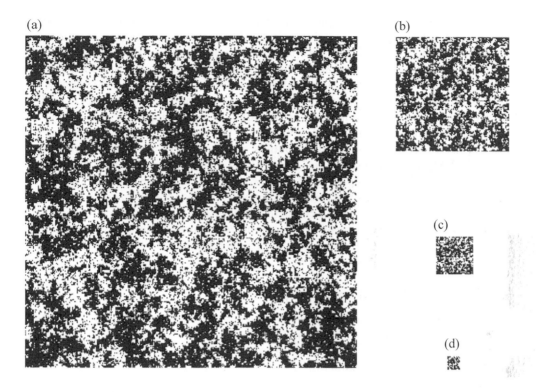

(c)

(d)

Fig. 1.9 Spin configuration at $T = 1.05T_c$ and its block-spin transformations.

The important concept of *universality* emerges from the idea of renormalization group that eliminates inessential short-range details and emphasizes increasingly macroscopic viewpoints. More precisely, an important consequence of universality is that quantities that describe the essential features of critical phenomena, typically the critical exponents, do not depend on the system details but are specified only by a few basic factors such as the symmetry of the system, range of the interactions (i.e. short or long range), and its spatial dimensionality (more precisely, the connectivity of its elementary degrees of freedom, e.g. spins in the Ising model, when the system is defined on a lattice). For example, two apparently different critical phenomena share the same critical exponents, one in the Ising model and the other in a simple liquid, as long as both are in three dimensions. These two distinct physical systems are said to belong to the same *universality class*. It is surprising that a model of magnetism shows essentially the same critical behavior as one for the liquid. The physical reason behind this behavior is that many characteristics of the system[3] gradually recede as the renormalization-group transformation proceeds and eventually only the essential factors, the spatial dimensionality and the symmetry of the system, survive.

---

[3] For example, whether or not the microscopic elements are located on discrete lattice sites (Ising model) or distributed continuously spatially (liquid).

The symmetry in the example of the Ising model in the absence of an external field is that all the microscopic ingredients (Ising spins) can be transformed into minus its values, $S_i \rightarrow -S_i$ ($\forall i$), with the Hamiltonian kept invariant. This is a *global* transformation that is mathematically represented by the group $\mathbb{Z}_2$, which consists of two elements $\{1, -1\}$ with the usual rule of multiplication. Correspondingly, the order parameter is a scalar quantity, the number of components being just one. These features are shared by the liquid–vapor transition in simple liquids.

Strong evidence to support such a viewpoint is provided by the experimental fact that the critical exponents of simple liquids agree with those for single-component magnets. Additional support for this view will be given from the explicit formulation of the renormalization group method in later chapters. Another quantity that shows universality, in addition to the critical exponents, is the ratio of *critical amplitudes*, the coefficient A in eqn (1.10), on the high-temperature ($t > 0$) and low-temperature ($t < 0$) sides of the phase transition. There are, of course, many properties that are non-universal, the critical temperature itself being a relevant example that depends on the microscopic details of the model.

1.5 Ising model and related systems

It is not necessary to construct and study a precise model Hamiltonian, which reflects the details of a real material of interest, in order to understand the critical phenomena shown by the material. Universality allows us to simplify the model to a very basic one, eliminating all inessential characteristics, if we are interested in the values of universal quantities such as the critical exponents. This fact provides the useful strategy that consists of adopting the simplest possible model system for investigation. Critical exponents are very robust against changes in the parameter values that specify the system details.

A very popular basic model for phase transitions and critical phenomena is the *Ising model* defined by the following equation, or Hamiltonian, or energy function

$$H = -J \sum_{\langle ij \rangle} S_i S_j - h \sum_i S_i. \qquad (1.11)$$

Here, S_i is the Ising spin ($S_i = \pm 1$) at *site* (lattice site) i, and $\langle ij \rangle$ denotes an interacting spin pair. An example is given in Fig. 1.10, where a neighboring pair

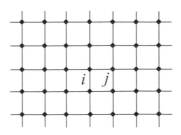

Fig. 1.10 Square lattice and a pair of nearest-neighbor sites $\langle ij \rangle$.

is shown on the *square lattice*. The coefficient J is the interaction constant (coupling constant) and h represents the external magnetic field expressed in units of energy. The Ising model is a simplified model of macroscopic systems with the number of elementary components equal to unity (that is, S_i is a scalar, not a vector quantity) and has been studied extensively as a model of magnetism. In this latter case J denotes the *exchange interaction*. For $J > 0$ the interaction is *ferromagnetic* and it is *antiferromagnetic* if $J < 0$.

The three-dimensional Ising model describes the critical behavior of simple liquids due to the effective symmetry shared by both systems (the Ising model and simple liquid), i.e. the fact that the number of components of their relevant variables is one. A more direct relation can also be established between the two systems using the idea of a *lattice gas*. Molecules of gas and liquid distribute continuously in space. Near the critical point, since fluctuations of long wavelength play a dominant role, it is allowed to discretize the space and ignore phenomena that occur only in short-range scales. Spatial discretization means to allow molecules to exist only on discrete lattice sites. More concretely, let the Ising spin describe whether or not a molecule exists at site i by the rule that $S_i = 1$ when there is one and $S_i = -1$, otherwise. Neighboring molecules are assumed to have an interaction $-\phi$ when $S_i = S_j = 1$. Then, the Hamiltonian reads

$$H = -\phi \sum_{\langle ij \rangle} \frac{1}{2}(1 + S_i) \cdot \frac{1}{2}(1 + S_j) - \mu \sum_i \frac{1}{2}(1 + S_i). \tag{1.12}$$

The chemical potential is denoted by μ. This equation is rewritten in the usual form of an Ising model using appropriate J and h as

$$H = -J \sum_{\langle ij \rangle} S_i S_j - h \sum_i S_i + \text{const.} \tag{1.13}$$

The density ρ is related to the average number of molecules per lattice site as

$$\rho v_0 = \frac{1}{2}(1 + \langle S_i \rangle), \tag{1.14}$$

where v_0 is the unit volume. Thus, the density of liquid is related to the magnetization of the Ising model $m = \langle S_i \rangle$. This is another way of justifying the fact that the magnetic and simple liquid–gas transitions belong to the same universality class.

The Ising model also represents a *binary alloy*, a mixture of two different species of atoms, A and B (Fig. 1.11). The state $S_i = 1$ denotes that site i is occupied by an atom A and $S_i = -1$ by an atom B. When the atoms interact with each other with the energy values J_{AA} (for atom pair AA) J_{BB} (for BB) and J_{AB} (for AB), the Hamiltonian reads

$$H = \frac{1}{4} \sum_{\langle ij \rangle} J_{AA}(1 + S_i)(1 + S_j) + \frac{1}{4} \sum_{\langle ij \rangle} J_{BB}(1 - S_i)(1 - S_j) \tag{1.15}$$

$$+ \frac{1}{4} \sum_{\langle ij \rangle} J_{AB} \left\{ (1 + S_i)(1 - S_j) + (1 - S_i)(1 + S_j) \right\}. \tag{1.16}$$

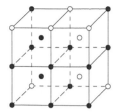 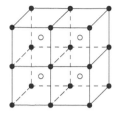

Fig. 1.11 Structure of a binary alloy. In this example, two types of atoms A and B, depicted in black and white, occupy the lattice sites of the body-centered cubic lattice. The left panel represents a disordered state and the right panel is for an ordered state.

This Hamiltonian may be rewritten as

$$H = -J \sum_{\langle ij \rangle} S_i S_j - h \sum_i S_i + \text{const.} \tag{1.17}$$

Here, $J = -(J_{AA} + J_{BB} - 2J_{AB})/4$ and $h \propto (J_{BB} - J_{AA})$. There is an extra constraint for the spin variables in the case of the binary alloy, which fixes the difference between the number of up spins ($S_i = 1$) and down spins ($S_i = -1$), corresponding to the fixed magnetization of magnetic materials: Since the total number of atoms of each type, A and B, is constant, the sum $\sum_i S_i$ is fixed to a given value. For instance, if the numbers of A and B atoms coincide, this constant is 0. According to experiments, critical exponents of binary alloys agree very well with those of the three-dimensional Ising model. An example is provided by the alloy beta-brass (a mixture of copper and zinc atoms), whose exponents are the same as those of the Ising antiferromagnet.

It will be useful to introduce a few other models in addition to the simplest Ising model. The Ising model with the binary value for spins, $S_i = \pm 1$, is sometimes called the spin-1/2 Ising model. The reason is that the z-component of the quantum spin operator takes only two values, $S_i^z = \pm 1/2$, if the magnitude of the spin operator is $S = 1/2$ in units of the Planck constant $\hbar$. A simple generalization of the spin-1/2 case is the spin-1 Ising model in which S_i takes three values $1, 0, -1$. The spin-1 Ising model often has an extra term in its Hamiltonian representing *anisotropy*,

$$H = -J \sum_{\langle ij \rangle} S_i S_j - D \sum_i S_i^2 - h \sum_i S_i. \tag{1.18}$$

The anisotropy term of strength D is constant in the spin-1/2 case, since $S_i^2 = 1$, but it is not when $S = 1$ in which case S_i^2 is 1 or 0. The consequences of the existence of this term will be discussed in the next chapter in relation to the tricritical point.

In the *q-state Potts model*, the spin variable S_i takes up to q values, where q is an integer equal to or greater than 2. If we write these values as $S_i = 1, 2, \cdots, q$, the interaction between two Potts spins is assumed to take two values depending on whether these two spins are in the same state or not,

$$H = -J \sum_{\langle ij \rangle} \delta_{S_i, S_j} - h \sum_i \delta_{S_i, 1}. \tag{1.19}$$

Here, δ_{S_i,S_j} is 1 when $S_i = S_j$ and is 0 otherwise, i.e. *Kronecker's symbol*. The external field is assumed to apply only if the spin is in state 1 in the above Hamiltonian. Other choices are possible, for example, that the field applies to state 2 $(\delta_{S_i,2})$ or to a few states $(\delta_{S_i,1} + \delta_{S_i,2})$. The $q = 2$ Potts model reduces to the Ising model by appropriate changes in the definition of coefficients (Exercise 1.1). The Potts model shows a rich variety of phase transitions and critical phenomena according to the value of q and the spatial dimension and has been investigated extensively. The $q = 1$ limit of the Potts model is closely related to the problem of *percolation* as will be explained in Section 8.3.

The symmetry of the Hamiltonian may certainly change with q in the Potts model, but it can also change when the number of components of the spin is greater than 1, in which case the spin is a vector $\boldsymbol{S}_i$,

$$H = -J \sum_{\langle ij \rangle} \boldsymbol{S}_i \cdot \boldsymbol{S}_j - \sum_i \boldsymbol{h} \cdot \boldsymbol{S}_i. \tag{1.20}$$

The system with two components $\boldsymbol{S}_i = (S_i^x, S_i^y)$ is the *XY model*, and the three-component system is called the *Heisenberg model*. The magnitude of the spin variable $\boldsymbol{S}_i^2$ is usually fixed to 1. Consequently, the Hamiltonian of the XY model is often written as

$$H = -J \sum_{\langle ij \rangle} \cos(\phi_i - \phi_j) - h \sum_i \cos \phi_i, \tag{1.21}$$

where $\boldsymbol{S}_i = (\cos \phi_i, \sin \phi_i)$ and $\boldsymbol{h}$ has been chosen as $\boldsymbol{h} = (h, 0)$. These vector-spin models have different critical exponents depending on the number of components of the spin $\boldsymbol{S}_i$. In general, when $\boldsymbol{S}_i$ has n components, the system is called the *n-vector model*.

Rigorously speaking, spins are quantum-mechanical operators and we have to justify the classical treatment mentioned so far. An intuitive reason is that critical phenomena are caused by cooperation of a very large number of microscopic degrees of freedom. Though each of these degrees of freedom may be quantum mechanical, the net behavior of the system is essentially macroscopic, and quantum-mechanical effects usually do not show up explicitly. This picture is confirmed by rich data of agreement of critical exponents between classical theories and experiments. It is, nevertheless, necessary to seriously consider quantum effects when critical phenomena are observed at extremely low temperatures. We then should consider *quantum spin systems* by regarding $\boldsymbol{S}_i$ in eqn (1.20) as quantum-mechanical operators. The important subject of *quantum phase transitions* is outside the scope of the present volume except for a few simple examples.

EXERCISE 1.1 Show that the two-state Potts model is equivalent to the Ising model. Since S_i in the two-state Potts model takes two values, 1 and 2, it will be useful to rewrite these two values in terms of Ising variables (to be denoted as $\sigma_i = -1, 1$ to distinguish it from the Potts variable S_i) and to express Kronecker's delta δ_{S_i,S_j} by the product $\sigma_i \sigma_j$. A similar change will be necessary for the external field term.

2
Mean-field theories

The basic strategy of theoretical studies of phase transitions and critical phenomena is to solve models, such as the ones described in the last section of Chapter 1, according to the prescription of statistical mechanics. This program is actually quite hard to follow closely. As an example, the total number of states for the N-spin Ising model is 2^N, since each spin may take one of the two values, 1 and -1. This exponential number increases very rapidly with N, and it soon becomes impossible to calculate the partition function exactly in a straightforward manner as N reaches a moderate value. For instance, $N = 10$ yields $2^N = 1024$, and as N increases to $N = 100, 1000, 10\,000$, 2^N explodes from $1.27 \times 10^{30}, 1.07 \times 10^{301}$ to 2.00×10^{3010}. In some limited cases we may derive exact solutions by ingenious methods, as described in Chapter 9. However, in general, we have to resort to approximate methods to understand the essential features of the physical phenomena under consideration. One of the most common and important approximations is called mean-field. In the present chapter we explain the mean-field approximation, the Landau theory, the infinite-range model, and the Bethe approximation, and show that all these (mean-field) theories are essentially equivalent to each other. Also described are the Landau theory of tricritical behavior, correlation functions, the limit of applicability of the mean-field theory, and dynamic critical phenomena. These mean-field solutions provide a reasonable starting point for more advanced methods including the renormalization group.

2.1 Mean-field approximation

Let us first explain the *mean-field approximation* for the Ising model. The basic strategy is to focus our attention on a single spin and replace the neighboring spins by their averages, as illustrated in Fig. 2.1. Then, the problem reduces to a single-variable case, and the number of degrees of freedom appearing in the computation of the partition function is drastically reduced from 2^N to 2. The problem of many interacting particles is replaced by a non-interacting one, which greatly facilitates the theoretical treatment. As will be shown below, this procedure is equivalent to the approximation where one ignores the deviations (fluctuations) from the average value of the spin variables. We explain this latter method since it provides a transparent point of view on the essence of the mean-field approximation.

Let us separate the Ising spin variable S_i into its thermal average $m = \langle S_i \rangle$ and the deviation (fluctuation[1]) from the average $\delta S_i = S_i - \langle S_i \rangle$ in the Ising model

---

[1] The term fluctuation is used in a few slightly different ways. In the previous chapter a fluctuation meant that, within a spatial region, the degree of freedom takes a different value from that in the

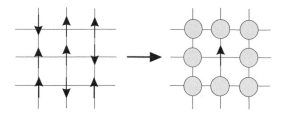

Fig. 2.1 In the mean-field approximation spin variables surrounding a given spin are replaced by their average values (shown in gray circles).

Hamiltonian

$$H = -J \sum_{\langle ij \rangle} S_i S_j - h \sum_i S_i. \tag{2.1}$$

We have $S_i = m + \delta S_i$ and ignore the second-order terms in δS_i assuming that fluctuations are not very significant. We expect that $\langle S_j \rangle$ does not depend on the index j due to the spatial uniformity of the system (space translation symmetry). This may seem a very crude approximation, but it turns out that qualitatively reliable results can be derived on critical exponents as long as the criterion of validity (*Ginzburg criterion*) described in Section 2.10 is satisfied. Equation (2.1) now reads

$$H = -J \sum_{\langle ij \rangle} (m + \delta S_i)(m + \delta S_j) - h \sum_i S_i$$

$$\approx -Jm^2 N_B - Jm \sum_{\langle ij \rangle} (\delta S_i + \delta S_j) - h \sum_i S_i. \tag{2.2}$$

Here, N_B is the total number of interacting spin pairs, $\sum_{\langle ij \rangle} 1 = N_B$. We mainly consider the cases in which only *nearest-neighbor* pairs of spins interact with each other as depicted in Fig. 2.2. Then, N_B is the number of *bonds* or the number of interacting pairs.

We focus our attention on site i in the interacting pair in the second line of the above equation. We find that δS_i appears four times as seen in Fig. 2.2: Bonds to the up, down, right, and left neighbors. If we write z for the *coordination number* (the number of bonds emanating from a site), we have

$$H = -Jm^2 N_B - Jmz \sum_i \delta S_i - h \sum_i S_i. \tag{2.3}$$

It is convenient to further rewrite δS_i in terms of S_i using the definition $\delta S_i = S_i - m$,

$$H = -Jm^2 N_B - Jmz \sum_i (S_i - m) - h \sum_i S_i$$

$$= N_B Jm^2 - (Jmz + h) \sum_i S_i, \tag{2.4}$$

surrounding regions. In the present section, the deviation of the value of a very local degree of freedom from the average is called a fluctuation.

Fig. 2.2 Site i and its nearest neighbors on the square lattice. The number of nearest neighbors z is 4 in this case.

where the relation $N_{\mathrm{B}} = zN/2$ has been used. For example, the square lattice ($z = 4$) has $N_{\mathrm{B}} = 2N$ (assuming periodic boundary conditions).

Equation (2.4) coincides with the following Hamiltonian, which is obtained from the original Hamiltonian (2.1) after replacement of S_j (those interacting with S_i) by the average m,

$$H \approx \sum_i H_i = -(Jmz + h) \sum_i S_i, \tag{2.5}$$

the difference between this (2.5) and eqn (2.4) being only in the additive constant. The physical picture of Fig. 2.1 is realized in the replacement of S_j by m.

In eqn (2.4) the effects of interactions of spin S_i with the neighboring spins are expressed in the same form as those of an external field of strength Jmz. We thus call such a term an *effective field* or a *molecular field*. Correspondingly, the mean-field approximation is also called the *molecular-field theory*.

The problem has now been simplified to a single-site case as seen in eqn (2.5), where no interactions between spins exist explicitly. All S_i can be treated independently and one may think that the problem has been solved. It is not the case yet, however. The quantity m has been introduced as the average magnetization and we should specify its value. To determine m, we note that m in eqn (2.5) is the average $\langle S_j \rangle$ of spins neighboring to S_i. Therefore, because of translational symmetry, we require that m be equal to $\langle S_i \rangle$, which is expressed as the *self-consistent equation*

$$m = \frac{\displaystyle\sum_{S_i = \pm 1} S_i e^{\beta(Jmz + h)S_i}}{\displaystyle\sum_{S_i = \pm 1} e^{\beta(Jmz + h)S_i}} = \tanh \beta(Jmz + h). \tag{2.6}$$

The symbol β is for the *inverse temperature*, $\beta = 1/k_{\mathrm{B}}T$.[2] For simplicity we will adopt the unit that reduces the Boltzmann constant k_{B} to 1 throughout this book. It is not difficult, if necessary, to recover the formulas explicitly including k_{B} by dimensional analysis.

Equation (2.6) determines m as a function of the external parameters h and β and is also called the (mean-field) *equation of state*. It is not possible to explicitly

[2] Not to be confused with the critical exponent β.

(a)

(b)

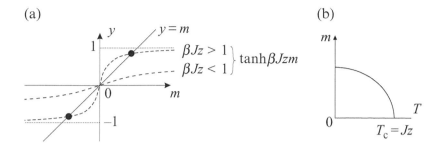

Fig. 2.3 (a) Graphical solution of the equation of state for the mean-field approximation of the Ising model. When $\beta Jz > 1$, the line $y = m$ and the curve $y = \tanh \beta Jzm$ cross at two points with $m \neq 0$ as indicated by black dots. (b) The solution $m(T)$ marked by the black dot in (a) is depicted as a function of T. Only the solution with $m > 0$ is shown here.

solve eqn (2.6) for m, but we may develop the following argument for the behavior of the solution by using a graphical representation of the left- and right-hand sides. Let us assume $h = 0$ for simplicity. As shown in Fig. 2.3(a), the slope of the right-hand side $\tanh(\beta Jmz)$ at the origin is larger than 1 when $\beta Jz > 1$ and the equation of state has non-vanishing solutions. This means that the system has a finite *spontaneous magnetization*, the non-vanishing magnetization that exists in the limit of a vanishing external field $h \to 0$, below the critical point $T_c = Jz$. Figure 2.3(b) represents this situation. The sign of m is determined by whether h approaches 0 from above ($h = 0+$) or below ($h = 0-$). Only the solution $m > 0$ is shown in Fig. 2.3(b), but there exists a corresponding negative solution with the same absolute value. The absolute value of the solution $m \neq 0$ increases as the temperature decreases. In this way, we have reached a qualitative understanding of Fig. 1.3(c) from a mean-field perspective.

The mean-field approximation presented in this section represents one possible way to realize a mean-field theory. A mean-field theory constitutes a general strategy to reduce the original problem of exponential complexity (there are $2^N = \exp(N \log 2)$ possible configurations) into one of polynomial complexity, e.g. N^α with $\alpha \geq 0$. In our particular example of the Ising model, we reduced 2^N to 2. However, we could have applied the same type of approximation to a cluster of N_c spins instead of a single spin, with a reduction from 2^N to 2^{N_c}. An example of this latter case is the Bethe (or Bethe-Peierls) approximation, to be discussed in Section 2.8.

> **EXERCISE 2.1** Consider the model Hamiltonian of eqn (2.1) for a collection of spins of magnitude S with $S_i = -S, -S+1, \cdots, S-1, S$. Determine the critical temperature of the system by using the mean-field approximation.

2.2 Critical exponents of the mean-field theory

We next study critical exponents of the mean-field approximation. As one sees in eqn (1.5), the critical exponent β describes how the magnetization m vanishes as the temperature increases toward the critical point in the absence of an external field h. Since we are interested in the region where m is very small, it is legitimate to expand the right-hand side of eqn (2.6) around $m = 0$. For $h = 0$ we then obtain

$$m = \beta J z m - \frac{1}{3}(\beta J z)^3 m^3 + \cdots . \tag{2.7}$$

Since we are interested in solutions with $m \neq 0$, we divide both sides of this equation by m and solve the result for m to find

$$m = \pm \left(\frac{3(Jz - T)}{(\beta J z)^2 J z} \right)^{1/2} \approx \pm \left(\frac{3(T_c - T)}{T_c} \right)^{1/2}, \tag{2.8}$$

where we used $\beta J z \approx 1$ near the critical point. The critical exponent β is therefore $1/2$.

The value of the critical exponent γ that describes the divergence of the magnetic susceptibility χ, is 1. To confirm this fact, it is useful to notice that m and h both approach 0 with the same order of magnitude as $h \to 0$ when the temperature is above the critical point $T > T_c$ as in eqn (1.2). We therefore expand the equation of state (2.6) assuming m and h are small quantities of the same order of magnitude,

$$m = \beta J z m + \beta h + \cdots . \tag{2.9}$$

By rewriting this relation to obtain χ defined by $m = \chi h$, we find

$$\chi = \frac{1}{T - T_c}. \tag{2.10}$$

Since the susceptibility diverges inversely proportionally to the temperature difference $T - T_c$, the critical exponent is $\gamma = 1$ for $T > T_c$. As seen in Exercise 2.2, the critical exponent assumes the same value for $T < T_c$. We therefore have $\gamma = \gamma'$ in eqn (1.3).

> **EXERCISE 2.2** Show that the mean-field value of the critical exponent γ' for the magnetic susceptibility below the critical point $T < T_c$ has the same value as the high-temperature counterpart $\gamma = \gamma' = 1$. It will be useful to first differentiate both sides of the equation of state (2.6) with respect to h and then use the facts that the susceptibility is given by $\chi = \partial m / \partial h$ ($h \to 0$) and that the magnetization m is almost zero near the critical point.

To study the critical exponent α that characterizes the rate of divergence of the specific heat, it is useful to notice that the specific heat C is calculated from the temperature dependence of m derived above: When $h = 0$, the mean-field Hamiltonian (2.4)

$$H = N_B J m^2 - J m z \sum_i S_i \tag{2.11}$$

is reduced to $H = 0$, if $T > T_c$, since $m = 0$. The specific heat, the temperature derivative of the energy, is therefore 0. In the low-temperature region $T < T_c$, on the other hand, the Hamiltonian does not vanish as m is finite. The specific heat consequently has a positive value. These considerations lead us to the temperature

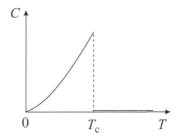

Fig. 2.4 The specific heat does not diverge but has a jump according to the mean-field approximation.

dependence of the specific heat as depicted in Fig. 2.4 with a jump at T_c. This jump is described by the exponent $\alpha = 0$ because it implies $C \propto |T - T_c|^0$, that is, the specific heat approaches a finite constant as the temperature reaches the critical value, which is indeed the case both above and below T_c as shown above. This also confirms that $\alpha = \alpha'(= 0)$.

EXERCISE 2.3 Determine the average energy per spin and specific heat $(h = 0)$ of the Ising model in the mean-field approximation.

To find the value of the critical exponent δ that describes the magnetic-field dependence of the magnetization exactly at $T = T_c$, it is useful to expand the equation of state under the assumption that both m and h are small (but not necessarily of the same order), thus providing the h dependence of m. The third-order expansion of the right-hand side of the equation of state (2.6) with h kept non-vanishing yields

$$m \approx \beta_c(Jzm + h) - \frac{\beta_c^3}{3}(Jzm + h)^3, \tag{2.12}$$

where β_c stands for $1/T_c = 1/Jz$. This equation can be rewritten as

$$\beta_c h \approx \frac{1}{3}\left\{ (\beta_c h)^3 + 3(\beta_c h)^2 m + 3(\beta_c h)m^2 + m^3 \right\}. \tag{2.13}$$

Using the definition $h \approx m^\delta$, we write the above equation just in terms of the order of magnitude of m, $[m]$, dropping the coefficients,

$$[m^\delta] \approx [m^{3\delta}] + [m^{2\delta+1}] + [m^{\delta+2}] + [m^3]. \tag{2.14}$$

We have to choose $\delta = 3$ so that the lowest-order terms of both sides are consistent with each other.

The results for the mean-field critical exponents are summarized in the following table:

Exponent	Mean-field value
α	0
β	$\frac{1}{2}$
γ	1
δ	3

A closer look at the above-mentioned derivation of critical exponents reveals that universality is realized even within the mean-field approximation since the values of the critical exponents depend just on some symmetry properties and not on other details of the system. For example, the result $\beta = 1/2$ emerges from the fact that the right-hand side of eqn (2.7) is composed of the first- and third-order terms of m: By dividing both sides of this equation by m, we reduce m^3 in the second term of the right-hand side to m^2 and the left-hand side to a constant. Then, the magnetization behaves as the square root of the temperature difference. This means that the coefficient $1/3$ of the second term of the right-hand side of eqn (2.7) has no influence on the critical exponent. Only the negative sign matters. We further find that the explicit form as a hyperbolic tangent of the right-hand side of the equation of state (2.6) does not come into play, but the only ingredient that affects the result is that $\tanh(\beta Jmz)$ is an odd function of m with a negative coefficient in the third order. A similar comment applies to the critical exponent γ. The only requirement to reproduce $\gamma = 1$ is that the right-hand side of the equation of state is an odd function with a positive coefficient in the first order of the expansion. A different value of this positive coefficient leads to a different value of the critical point with the same critical exponent. We notice here the non-universality of the value of the critical point.

EXERCISE 2.4 Let us study the Heisenberg model by the mean-field approximation. The Hamiltonian is

$$H = -J \sum_{\langle ij \rangle} \boldsymbol{S}_i \cdot \boldsymbol{S}_j - h \sum_i S_i^z. \tag{2.15}$$

Here, $\boldsymbol{S}_i$ is a classical vector of unit length having three components S_i^x, S_i^y, S_i^z. Since the external field is applied along the z-axis, the magnetization $\boldsymbol{m} = \langle \boldsymbol{S}_i \rangle$ is also parallel to the z-axis. Derive the mean-field Hamiltonian similar to eqn (2.4) by the same argument as in the Ising model. Also, derive the self-consistent equation corresponding to eqn (2.6) using the log-derivative of the partition function. Calculate the critical point and exponent β and confirm that the latter agrees with the Ising case.

2.3 Landau theory

The *Landau theory* is a variant of the mean-field theory, which does not include the elementary degrees of freedom of the statistical model. It is a phenomenological theory in that no microscopic variables, such as the Ising spins, are used and the free energy is written as a function of the magnetization (i.e. the order parameter) from

(a)   (b)   (c)

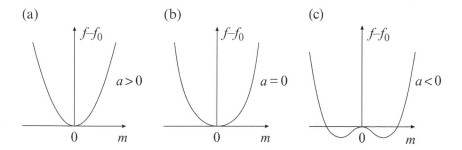

Fig. 2.5 The m dependence of the Landau free energy. The locations of minima change according to whether the temperature is above the critical point (a), at the critical point (b) or below (c).

symmetry considerations alone. The condition of thermal equilibrium is realized as a minimization of the free energy.

Let us first discuss the simple case of $h = 0$, i.e. no external field. The free energy per microscopic degree of freedom or per unit volume will be written as f and is regarded as a function of magnetization. The magnetization per spin $m = \langle S_i \rangle$ changes its sign, $m \to -m$, if we change the signs of all the spins, $S_i \to -S_i$, $\forall i$. In the absence of external field h, the Hamiltonian (2.1) is a bilinear form of the spin variables, and remains invariant under the overall inversion of the sign of the spins. This $\mathbb{Z}_2$ transformation represents a *global symmetry*. Consequently, the free energy remains invariant under the same operation. Thus, the free energy $f(m)$ is an even function of the magnetization m.

Since we are interested in critical phenomena, the temperature is close to the critical point and the magnetization m assumes a very small value. This would allow us to expand the free energy in even powers of m and retain only the lowest-order terms. This *Landau free-energy expansion*[3] (analytic expansion in terms of the order parameter) for the Ising universality class reads

$$f(m) = f_0 + am^2 + bm^4, \tag{2.16}$$

where f_0, a and b are constants as functions of m but have temperature dependence. Generally, the Landau free energy is determined by writing all possible scalar invariants in terms of powers and products of the order parameter components. Thus, relevant symmetries of the original microscopic model are preserved at a coarse-grained level of description. See Section 5.5.

Thermal equilibrium is realized by minimization of $f(m)$ for a given h (which is 0 for the moment). It is convenient to graphically show the functional form of

[3] Note that the Landau free energy $f(m)$ in eqn (2.16), with m a given magnetization, is not the true thermodynamic free energy as a function of T and h, since not all microscopic configurations (giving other values of m) are included. As will be explained in Chapter 5, the Landau theory can be obtained as a saddle-point approximation of a certain field theory. The value of the approximate equilibrium free energy $f(m_0)$ is obtained by minimization of $f(m)$ with respect to m. Then, thermodynamic properties such as the specific heat are determined by differentiation of $f(m_0)$ with respect to the corresponding parameters.

eqn (2.16) to identify the locations of minima. We first notice that b should be positive. Otherwise, the free energy $f(m)$ decreases indefinitely as $|m|$ increases, which implies an instability. The m dependence of $f(m)$ is illustrated in Fig. 2.5 for three possible values of the coefficient a. For $a > 0$, the minimum is at $m = 0$ and thus there is no spontaneous magnetization (Fig. 2.5(a)). When a is exactly 0, the Landau expansion (2.16) starts from the fourth order and $f(m)$ is very flat at the origin, Fig. 2.5(b), but still the equilibrium magnetization remains 0. As soon as a becomes negative, two minima emerge away from $m = 0$ and the absolute value $|m_0|$ at these points grows with decreasing a, Fig. 2.5(c). The original Hamiltonian and free energy are symmetric (invariant) under a change of sign of m, but the realized state for $a < 0$ does not have such a symmetry since only one of the two minima is actually realized in a physical system. In other words, the thermodynamic free energy is invariant under the overall change of the sign of spins ($S_i \to -S_i$, $\forall i$) but the realized equilibrium state has no such symmetry. This phenomenon is called *spontaneous symmetry breaking*. A small external field or the initial condition of time evolution of the system determines which of the two states is actually realized, which is also called *ergodicity breaking* because only a part of the phase space is reached by the system. This situation is common with the mean-field approximation explained in Section 2.1.

Since the equilibrium position (minimum) of $f(m)$ changes at $a = 0$, we may identify $a = 0$ with the critical point $T = T_c$. This observation would allow us to choose an odd power of kt as a, where k is a positive constant and $t = (T - T_c)/T_c$ is the deviation of the temperature from the critical point normalized by T_c, also known as the *reduced temperature*. The simplest choice is $a = kt$, for which clearly $a > 0$ above the critical point and $a < 0$ below. The temperature dependence of b does not affect the qualitative behavior of the free energy around the critical point or the critical exponents, and we therefore take b as a constant, independent of temperature.

Let us evaluate the resulting critical exponents of the Landau theory. The exponent β is determined by the temperature dependence of m that minimizes the free energy at the low-temperature side $a < 0$, i.e. $T < T_c$. Differentiation of the free energy gives the minimization condition

$$\frac{\mathrm{d}f}{\mathrm{d}m} = 2am + 4bm^3 = 0. \tag{2.17}$$

Thus, the equilibrium value of magnetization, m_0, is

$$m_0 = \sqrt{-\frac{a}{2b}} = \sqrt{\frac{k(T_c - T)}{2bT_c}}, \tag{2.18}$$

from which we conclude $\beta = 1/2$.

To find the value of the critical exponent α, we differentiate the minimum (equilibrium) value of the free energy with respect to temperature and study how the specific heat depends on temperature. Below the critical point, we find

$$f = f_0 + am_0^2 + bm_0^4 = f_0 - \frac{k^2(T - T_c)^2}{4bT_c^2}. \tag{2.19}$$

The specific heat is therefore finite at the critical point. Above the critical temperature, f is a constant f_0 since $m_0 = 0$, and the specific heat vanishes. We conclude $\alpha = 0$.

As for the critical exponent δ that describes the field dependence of the magnetization exactly at the critical point, we add the external field term $-hm$ to the free energy, and differentiate the latter with respect to m to determine its equilibrium value,

$$\frac{\mathrm{d}f}{\mathrm{d}m} = 2am + 4bm^3 - h = 0. \tag{2.20}$$

Since $a = 0$ at $T = T_c$, we find $\delta = 3$ from $m^3 \propto h$.[4]

The susceptibility $\chi = \partial m / \partial h$ is evaluated from eqn (2.20) as

$$\chi = \frac{1}{2a + 12bm^2}. \tag{2.21}$$

For $T > T_c$, we thus have

$$\chi = \frac{1}{2a} = \frac{T_c}{2k(T - T_c)}, \tag{2.22}$$

from which $\gamma = 1$ is concluded. If $T < T_c$,

$$\chi = \frac{1}{2a + 12b(-a/2b)} = \frac{T_c}{4k(T_c - T)}, \tag{2.23}$$

and hence we find $\gamma' = 1$. The critical exponent γ takes the same value above and below the critical point. The *critical amplitudes*, the coefficient of $|T - T_c|$, are different by a factor of 2; compare eqns (2.22) and (2.23). Universality, nevertheless, manifests itself in the ratio 2 between these critical amplitudes; the ratio does not depend on the coefficient k or the critical temperature T_c.

> **EXERCISE 2.5** Confirm that the ratio of critical amplitudes for the magnetic susceptibility has the universal value of 2 according to the mean-field approximation of Sections 2.1 and 2.2. It will be useful to make use of the computations appearing in Exercise 2.2.

The Landau theory uses only the symmetry properties of the free energy $f(m)$, and the resulting values for the critical exponents do not reflect the details of model systems. Consequently, the critical exponents do not depend on the spatial dimensionality and/or the number of spin components, e.g. the Ising model has a single component and the XY model has two, and the Heisenberg model has three components in its microscopic spin variable. This latter independence of dimensionality and number of components is characteristic of the Landau theory and mean-field theory, and is in general incorrect.

The Landau theory shares its values of critical exponents with the mean-field approximation of Section 2.1. In this sense, these two theories are equivalent to each other. It is further possible to derive the Landau expansion (2.16) from the mean-field

---

[4] To simplify notation, we use the same symbol m, instead of m_0, for the equilibrium value of magnetization.

approximation of Section 2.1 as follows. The free energy as a function of m for the mean-field Hamiltonian (2.4) is, if $h = 0$,

$$f = \frac{1}{N}\left(N_{\mathrm{B}}Jm^2 - T\log\sum_{\{S_i = \pm 1\}} e^{\beta Jmz\sum_i S_i}\right)$$

$$= \frac{zJm^2}{2} - T\log 2\cosh(\beta Jmz). \tag{2.24}$$

This expression is expanded to fourth order as

$$f = -T\log 2 - \frac{Jz(Jz\beta - 1)}{2}m^2 + \frac{1}{12}(Jz)^4\beta^3 m^4. \tag{2.25}$$

This is of the same form as eqn (2.16). In particular, we confirm the important features that the coefficient of the second-order term vanishes at the critical point, and that the fourth-order term has a positive coefficient.

EXERCISE 2.6 The van der Waals equation of state characterizing a gas–liquid transition

$$\left(P + \frac{N^2 a}{V^2}\right)\left(\frac{V}{N} - b\right) = T \tag{2.26}$$

can be regarded as a mean-field theory, where $P = -\partial F/\partial V$ is pressure, V volume, N number of atoms, $a > 0$ is a measure of the attraction between atoms, and $b > 0$ the excluded volume due to the finite atomic size. Define the volume per atom $v = V/N$, then determine the critical values v_c, P_c, T_c, and the ratio $P_c v_c/T_c$. Calculate the critical exponents δ and γ of the van der Waals fluid, and compare them with those determined in the mean-field theory of a ferromagnet. Noticing the correspondence to the magnetic quantities $h \to p$, $m \to \mathcal{V}$, where $p = (P - P_c)/P_c$ and $\mathcal{V} = (v - v_c)/v_c$ are the reduced pressure and volume, respectively, we define the critical exponent δ as

$$p \propto \mathcal{V}^\delta, \tag{2.27}$$

and the exponent γ as the rate of divergence of the isothermal compressibility $(v \to v_c)$

$$\kappa_T = -\frac{1}{v}\left.\frac{\partial v}{\partial P}\right|_T \propto (T - T_c)^{-\gamma}. \tag{2.28}$$

EXERCISE 2.7 Strictly speaking, the Landau free-energy expansion is valid under the assumption that the order parameter m vanishes as $T \to T_c$. Consider the situation where cubic terms are allowed

$$f = f_0 + am^2 + bm^4 + cm^3 \quad (a, b > 0), \tag{2.29}$$

and determine the equilibrium value of m as a function of temperature. Show that this situation represents a first-order transition. Hint: f should be minimized for equilibrium.

2.4 Landau theory of the tricritical point

The Landau theory assumes a positive coefficient b for the fourth-order (quartic) term. Under certain circumstances, however, a negative b describes the system behavior more appropriately. A typical example is the critical behavior around a *tricritical point*, where three lines of critical points meet. Let us see how the Landau theory is modified when b is negative.

As explained already, the Landau expansion to fourth order (2.16) with negative b leads to a thermodynamic instability because the equilibrium value of m is indefinitely large, i.e. unbounded. This suggests that one needs to include a sixth-order term,

$$f = \frac{1}{2}am^2 + \frac{1}{4}bm^4 + \frac{1}{6}cm^6 - hm. \tag{2.30}$$

The coefficients in this equation involve rational numbers so that equations appearing later look simpler after differentiation. In the following, we will show that the Landau free energy f displays a phase transition whose order depends upon the sign of b. Thermodynamic stability requires $c > 0$, and the signs of a and b are now arbitrary. For $b > 0$, the situation is the same as in the previous section: The sign of $a = kt$ determines the equilibrium value of m. Similar is the case for $b = 0$ with some modifications in the critical exponents because the starting order of m is six when $a = 0$. We will come back to this point later.

The new case $b < 0$ needs a careful analysis. We assume $h = 0$ for simplicity and study the a dependence of $f(m)$ with b and c fixed to negative and positive constants, respectively. First, at high temperature, a is positive and large, and $f(m)$ has a simple structure with the minimum at $m = 0$. As the temperature decreases, a becomes smaller. In the range $a < b^2/4c (\equiv a_0)$, the negative coefficient of the fourth-order term causes local minima at non-vanishing m, see Fig. 2.6(a).[5] The local minimum value of the free energy at the solution $m \neq 0$ is higher than the global minimum at $m = 0$ when a is slightly smaller than a_0 as depicted in Fig. 2.6(a). Then, the state with $m = 0$ remains globally stable. The states with $m \neq 0$ have local stability around each local minimum and are called *metastable states*.

A further decrease of temperature causes a to be smaller than $a_1 = 3b^2/16c$, in which case the global minimum shifts from $m = 0$ to the two symmetrically located states with $m \neq 0$. Consequently, the magnetization of the equilibrium state jumps from $m = 0$ to $m \neq 0$, see Fig. 2.6(b). This is a first-order phase transition. For $a < 0$ the local stability of the solution $m = 0$ is lost, Fig. 2.6(d).

---

[5] The threshold of a for the minima to appear, $a_0 = b^2/4c$, can be derived from the condition that the minimization equation $\partial f/\partial m = m(a + bm^2 + cm^4) = 0$ starts to have non-vanishing real-valued solutions.

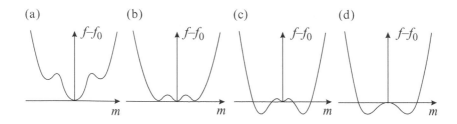

Fig. 2.6 For a negative coefficient of the fourth-order term, b, the Landau free energy describes a first-order phase transition. (a) $3b^2/16c = a_1 < a < a_0 = b^2/4c$, (b) $a = a_1$, (c) $0 < a < a_1$, (d) $a < 0$.

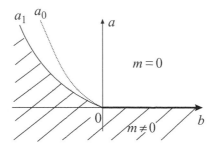

Fig. 2.7 A phase diagram that contains a tricritical point at the origin. The hatched region is the ferromagnetic phase with $m \neq 0$. Metastable states appear below the curve a_0. The transition is second order to the right of the origin and it is first order to the left.

EXERCISE 2.8 Derive the transition point $a_1 = 3b^2/16c$ of the first-order transition.

The result of these analyses is summarized in Fig. 2.7. When $b > 0$, the critical point is at $a = 0$ and a spontaneous magnetization exists when $a < 0$. If $b < 0$, metastable states appear at $a = a_0$, and states with spontaneous magnetization become globally stable at $a = a_1$ through a first-order transition. The point $b = 0$ is at the border of these two distinct situations, and the origin $a = b = 0$ is a special point, called a tricritical point.

The special point $a = b = 0$ is called a tricritical point for the following reason. A similar analysis in the presence of an external field reveals the structure of the phase diagram in a three-dimensional space with an additional axis for h. If we choose the h-axis perpendicular to the plane of Fig. 2.7, the line of first-order transitions for $b < 0$, $a = a_1$, extends as a plane through the region with $h \neq 0$. The first-order transition line $a = a_1$ drawn in Fig. 2.7 for $h = 0$ is a cross-section of this plane. For values of $|h|$ larger than a certain value, this plane of first-order transitions terminates with a line of second-order transitions along the boundary. No transitions exist beyond this line. This line of second-order transitions starts at the origin $a = b = h = 0$ and extends to both regions of $h > 0$ and $h < 0$. Hence, these two lines for $b < 0$ and another line of second-order transitions for $b > 0$, i.e. $a = h = 0$, merge at the origin. This observation justifies calling the origin $a = b = h = 0$ a tricritical point.

Critical exponents for the tricritical point are evaluated as follows. If we take the derivative of the free energy with respect to m at $b = h = 0$, we obtain

$$m^4 = -\frac{a}{c} = \frac{k(T_c - T)}{cT_c}. \tag{2.31}$$

We therefore conclude $\beta = 1/4$. As for α, the minimum value of $f(m)$ at $m^4 = -a/c$ (valid below the critical point) is found to be $f \propto |t|^{3/2}$. The second-order derivative of this free energy with respect to temperature reveals $\alpha = 1/2$. To estimate δ, we study the minimization condition of the Landau free energy in the presence of an external field, as was the case for the ordinary critical point,

$$am + cm^5 - h = 0. \tag{2.32}$$

Since $a = 0$ at the critical point, we find $m^5 = h/c$, and thus $\delta = 5$. The exponent γ is evaluated by using the expression for the susceptibility

$$\chi = \frac{1}{a + 5cm^4}, \tag{2.33}$$

which was obtained by differentiation of both sides of eqn (2.32). It is easy to see from this equation that $\chi = 1/a$ and $\chi = -1/4a$ from above and below the critical point, respectively. We therefore have $\gamma = \gamma' = 1$. The following table summarizes the mean-field critical exponents at the tricritical point:

Exponent	Mean-field value at the tricritical point
α	$\frac{1}{2}$
β	$\frac{1}{4}$
γ	1
δ	5

An example of a system that displays tricriticality is the spin-1 Ising model of eqn (1.18) with $S_i = 1, 0, -1$, also known as the *Blume–Capel model*. It may be viewed as a classical spin-1 magnet or as a lattice gas with $S_i = 0$ representing vacancies or impurities in an otherwise spin-1/2 ($S_i = \pm 1$) Ising model. This model may describe liquid ^{4}He–^{3}He mixtures, where $S_i = 0$ is identified with the presence of a ^{3}He atom, while $S_i = \pm 1$ with a ^{4}He atom on site i. One treats the interaction part of the Hamiltonian (1.18) by the mean-field approximation developed in Section 2.1 and expands the resulting free energy to sixth order in m, as mentioned in the last part of Section 2.3. One then finds that the coefficient of the fourth-order term changes sign depending on the values of Jz and D.

EXERCISE 2.9 Apply the mean-field approximation to the Hamiltonian (1.18) with $h = 0$ and series-expand the free energy to sixth order in m. Show that the coefficient of the fourth-order term can change sign when D is negative. Also confirm that the coefficient of the sixth-order term is positive when the sign of the coefficient of the fourth-order term changes. It will be useful to carry out the expansion using symbol-manipulation software on a computer as the algebra is lengthy.

One may now wonder what happens if the coefficients of the fourth- and sixth-order terms are both zero. It is easy to imagine that new types of critical behavior may emerge, and a different set of critical exponents would result, from the effects of the eighth-order term. It is indeed possible to derive a series of mean-field critical exponents in this manner. However, the experimental realization of these situations is actually difficult because the system is assumed to have very special values of adjustable parameters, as implied by the vanishing of many coefficients. For example, the ordinary critical point is realized by tuning h and T to their critical values, $h = 0$ and $T = T_c$. The tricritical point of the spin-1 Ising model needs an additional adjustment of the anisotropy parameter D to its critical value. Higher-order critical points need more and more tuning of the parameters, making these situations virtually impossible to realize experimentally.

2.5   Infinite-range model

The theory described in Section 2.1 is an approximation to analyze model systems. The *infinite-range model* is a very interesting system because it can be solved exactly in the thermodynamic limit and the result coincides with the mean-field solution. Let us first derive the solution and next show how the mean-field approximation gives the exact result.

The infinite-range model with Ising spins is defined by the following Hamiltonian,

$$H = -\frac{J}{2N} \sum_{i \neq j} S_i S_j = -\frac{J}{2N} \left(\left(\sum_i S_i \right)^2 - \sum_i S_i^2 \right). \tag{2.34}$$

The summation runs over all distinct pairs of i and j, i.e. $i = 1, 2, \cdots N$ and $j(\neq i) = 1, 2, \cdots N$. The factor $1/2$ is necessary because a single pair appears twice; for instance the pair $(1, 2)$ is counted twice as $i = 1, j = 2$ and as $i = 2, j = 1$. All spin pairs have the same interaction J/N, which may be interpreted as if the range of the interactions extends to the infinitely distant sites. This is the origin of the name infinite-range model. Although it may look very artificial, this model is important because the mean-field method gives the exact solution, which makes it possible to construct a mean-field-type theory for problems in which it is not easy to do so in a conventional manner. Indeed, the mean-field theory of spin glasses is an example of such an approach, as will be detailed in Chapter 8.

Our analysis of the infinite-range model starts from the definition of the partition function. Using the standard notation $K = \beta J = J/T$, we have

$$Z = \sum_{\{S_i = \pm 1\}} \exp \left\{ \frac{K}{2N} \left(\sum_i S_i \right)^2 \right\}, \tag{2.35}$$

where we excluded the second term in eqn (2.34). This latter term is small, of relative order $\mathcal{O}(N^{-1})$ compared to the leading first term, and therefore has no influence on the solution in the thermodynamic limit, as described below. It will be useful to remove the square in the exponent using the Gaussian integral

$$e^{ax^2/2} = \sqrt{\frac{aN}{2\pi}} \int_{-\infty}^{\infty} dm\, e^{-Nam^2/2+\sqrt{N}amx}. \tag{2.36}$$

An application of this formula rewrites the partition function (2.35) as

$$Z = \sum_{\{S_i=\pm 1\}} \sqrt{\frac{KN}{2\pi}} \int_{-\infty}^{\infty} dm\, e^{-NKm^2/2+Km\sum_i S_i}. \tag{2.37}$$

The spin variables S_i are now independent of each other, and the summation is evaluated to give the result

$$Z = \sqrt{\frac{KN}{2\pi}} \int_{-\infty}^{\infty} dm\, e^{-NKm^2/2+N\log(2\cosh Km)}. \tag{2.38}$$

The problem has now been reduced to a single integral. It is impossible to evaluate this integral in closed form in general. However, since the exponent of the integrand is proportional to N, we can use the *saddle-point method* (the method of *steepest descents*) as long as we are interested in the asymptotic behavior in the limit of large N. A brief account on the saddle-point method is given in Appendix A.1. In short, this method is a prescription to evaluate an integral whose integrand has a very sharp peak; the value of the integral is asymptotically equal to the value of the integrand at the peak. In the case of eqn (2.38), the exponent of the integrand is N times the function $g(m) = -Km^2/2 + \log(2\cosh Km)$. In the large-$N$ limit $e^{Ng(m)}$ has a very sharp peak at the point where $g(m)$ reaches its maximum, which allows us to apply the saddle-point method.

The partition function is thus evaluated as

$$Z \approx \sqrt{\frac{KN}{2\pi}} e^{-NKm^2/2+N\log(2\cosh Km)}. \tag{2.39}$$

It should be remembered that m in this expression has a specific value that maximizes the exponent $g(m)$. The corresponding free energy reads

$$\beta f(m) = -g(m) = \frac{Km^2}{2} - \log(2\cosh Km). \tag{2.40}$$

The factor $\sqrt{KN/2\pi}$ on the right-hand side of eqn (2.39) can be ignored in the limit $N \to \infty$ since it is smaller than the other terms, as one sees by taking the logarithm of the right-hand side. This result (2.40) agrees with the mean-field free energy (2.24) after replacements of J with J/N and z with N.[6]

To maximize the integrand we take the derivative of $f(m)$ and set it to zero (the saddle-point condition). The result is $m = \tanh Km$. This is the same equation as the equation of state of the mean-field approximation (2.6) with $h = 0$ and Jz replaced by $(J/N) \cdot N$. Thus, the variable m, artificially introduced for the Gaussian integral,

---

[6] The coordination number z in the infinite-range model is the number of all spins other than itself, $N - 1 \approx N$.

actually represents the magnetization. It is indeed verified that the saddle-point (extremum) condition for the exponent of the integrand in eqn (2.37) is found to be

$$m = \frac{1}{N} \sum_{i=1}^{N} S_i, \tag{2.41}$$

which explicitly shows that m is for magnetization.[7]

It is not difficult to understand the physical reason why the mean-field approximation provides the exact solution to the infinite-range model. The Hamiltonian of the infinite-range model is written as, if we ignore the contribution of the $i = j$ terms (which is negligible in the limit of large N),

$$H = -\frac{J}{2} \sum_{i=1}^{N} S_i \left(\frac{1}{N} \sum_{j=1}^{N} S_j \right). \tag{2.42}$$

The quantity in parentheses on the right-hand side is identified with the magnetization m in the limit of large N as one sees in eqn (2.41). We are thus allowed to replace the expression in parentheses with its average value m, and the problem reduces to a single-body (non-interacting) case.

In the infinite-range model each spin interacts with $N - 1$ other spins (which is an infinite number in the thermodynamic limit $N \to \infty$). It is useful to note here that the coordination number is $z = 2d$ in the d-dimensional *hypercubic lattice*, an extension of the three-dimensional cubic lattice to an arbitrary space dimension d. Consequently, the infinite-range model may be closely related with a system in a very high-dimensional space. It would then be expected that critical exponents of the infinite-range model, the mean-field values, accurately describe the behavior of sufficiently high-dimensional systems. It will indeed be shown later that the mean-field theories give the exact critical exponents for dimensions larger than the *upper critical dimension* d_{uc} (which is usually $d_{uc} = 4$). The physical picture is that the number of interacting partners of a spin is large for high-dimensional cases, which would yield a large 'force' to fix the spin under consideration into a specific direction. Such a large force is expected to reduce fluctuations, resulting in reliability of the mean-field theories in which fluctuations are ignored.

2.6 Variational method

The following variational approach provides another viewpoint to understand the physics behind the mean-field theories. The source of difficulty to perform exact calculations of physical quantities lies in the non-trivial structure of the probability distribution function $P(S_1, \cdots, S_N) = e^{-\beta H}/Z$ with, for example, the Hamiltonian (2.1), where the degrees of freedom $S_1, S_2, \cdots, S_N$ are coupled with each other. It may

---

[7] Rigorously speaking, the right-hand side of eqn (2.41) is a stochastic variable, whereas the left-hand side is not, the latter being the expectation value of each term on the right-hand side. This implies that the left-hand side may take a different value from the right-hand side. However, in the limit of large N, the probability for a finite difference between both sides vanishes, as will be explained in some detail in Chapter 8 under the name of a self-averaging property.

thus be useful to employ an approximation that decouples the distribution function into a product of simpler functions, following the spirit of the mean-field approximation typically represented by eqn (2.5). The key element here is the decoupling of the distribution function, not the Hamiltonian. We therefore approximate the full distribution by the product of single-site functions,

$$P(S_1, S_2, \cdots, S_N) \approx \prod_i P_i(S_i), \tag{2.43}$$

and determine $P_i(S_i)$ by the general variational principle of statistical mechanics that minimizes the free energy $F = E - TS$. Here, the internal energy E is the expectation value of the Hamiltonian and S is the entropy (not to be confused with spin). Under the above approximation, we find, noting that the entropy is the average of the logarithm of the inverse probability, $\log(1/P) = -\log P$,

$$
\begin{aligned}
F &= \sum_{\{S_i\}} \left\{ H \prod_i P_i(S_i) \right\} - T \sum_{\{S_i\}} \left\{ \prod_i P_i(S_i) \left(-\sum_i \log P_i(S_i) \right) \right\} \\
&= -J \sum_{\langle ij \rangle} \sum_{S_i, S_j} S_i S_j P_i(S_i) P_j(S_j) - h \sum_i \sum_{S_i} S_i P_i(S_i) \\
&\quad + T \sum_i \sum_{S_i} P_i(S_i) \log P_i(S_i),
\end{aligned}
\tag{2.44}
$$

where we have used the normalization $\sum_{S_k} P_k(S_k) = 1$ for k other than i and j. The variational principle consists of changing the function P_i slightly to $P_i + \delta P_i$, and demand that the resulting change δF of the free energy be vanishing. This amounts to formally differentiating the free energy with respect to P_i and setting the result to zero. If we incorporate the normalization condition by an additional term in the free energy using a Lagrange multiplier, $\lambda(\sum_{S_i} P_i(S_i) - 1)$, we find

$$\frac{\delta F}{\delta P_i} = -J \sum_j S_i m_j - h S_i + T \log P_i(S_i) + T + \lambda = 0, \tag{2.45}$$

where we have written m_j for $\sum_{S_j} S_j P_j(S_j)$. The summation in the first term is restricted to neighboring sites of i. The condition (2.45) is solved for the distribution function as

$$P_i(S_i) = \frac{\exp\left(\beta J \sum_j S_i m_j + \beta h S_i \right)}{Z_{\mathrm{MF}}}, \tag{2.46}$$

where Z_{MF} is the normalization factor. In the case of uniform magnetization $m_j (= m)$, the result (2.46) together with the decoupling (2.43) leads to the distribution $P(S_1, \cdots, S_N) \propto e^{-\beta H}$, with H being identical to the mean-field Hamiltonian (2.5).

The analysis so far has been general in that it did not use the values of the Ising spins $S_i = \pm 1$, and thus applies to many other cases. It is instructive to use the values of the Ising spins now explicitly and see its consequences. Since S_i takes only two values ± 1, we can always write any function of S_i as a sum of a constant and a term proportional to S_i because all higher-order terms reduce to one of these two since $S_i^2 = S_i^4 = \cdots = 1$, and $S_i^3 = S_i^5 = \cdots = S_i$. Thus, we may write

$$P_i(S_i) = \frac{1 + m_i S_i}{2}, \tag{2.47}$$

which is compatible with the previous notation $m_i = \sum_{S_i} S_i P_i(S_i)$ and the normalization $\sum_{S_i} P_i(S_i) = 1$. Substitution of eqn (2.47) into eqn (2.44) yields

$$F = -J \sum_{\langle ij \rangle} m_i m_j - h \sum_i m_i$$
$$+ T \sum_i \left(\frac{1 + m_i}{2} \log \frac{1 + m_i}{2} + \frac{1 - m_i}{2} \log \frac{1 - m_i}{2} \right). \tag{2.48}$$

Variation of this expression with respect to m_i, which is effectively a differentiation of F with respect to m_i, leads to

$$m_i = \tanh \beta \left(J \sum_j m_j + h \right). \tag{2.49}$$

This is identical to eqn (2.6) if the magnetization is uniform ($m_i = m$, $\forall i$).

2.7   Antiferromagnetic Ising model

The mean-field theories described in most parts of the present chapter deal with the ferromagnetic Ising model, in which neighboring spins tend to align parallel to each other. In many materials, however, antiferromagnetic interactions exist,

$$H = J \sum_{\langle ij \rangle} S_i S_j - h \sum_i S_i \quad (J > 0), \tag{2.50}$$

where the stable configuration of a pair of neighboring spins is antiparallel, $S_i S_j = -1$. For simplicity let us consider the case of a *two-sublattice system*, in which each site belongs to one of the two sublattices (A or B), and the neighboring site always belongs to the other sublattice (B or A) as depicted in Fig. 2.8. Many lattices including the square lattice in two dimensions and the simple cubic lattice in three dimensions have this structure. A typical exception is the triangular lattice, in which sites are classified into three sublattices and the analysis becomes much more complicated than in the two-sublattice case explained below.

The variational approach in the previous section provides a good starting point because the results for the free energy (2.48) and the self-consistent equation (2.49) remain valid just by a sign change of J since we did not use the sign of J in their derivations. We therefore use eqn (2.49) and write $m_i = m_A$ for i on sublattice A and $m_i = m_B$ for sublattice B. Since all neighboring sites of $i \in A$ belong to sublattice B and vice versa, eqn (2.49) with $J \to -J$ is expressed as

$$m_{\mathrm{A}} = \tanh\beta\big(-Jzm_{\mathrm{B}} + h\big) \qquad (2.51)$$

$$m_{\mathrm{B}} = \tanh\beta\big(-Jzm_{\mathrm{A}} + h\big) \qquad (2.52)$$

for $i \in \mathrm{A}$ and $i \in \mathrm{B}$, respectively. We next analyze the properties of the solution to these equations.

A first observation is that the same equation as in the ferromagnetic case, $m = \tanh(\beta Jzm)$, results from eqns (2.51) and (2.52) when $h = 0$, if we choose $m_{\mathrm{A}} = -m_{\mathrm{B}} = m$. This means that the spins on sublattice B have exactly the same properties, except for the opposite orientation, as those on sublattice A, as naturally expected from the antiferromagnetic interactions. Thus, the system develops a spontaneous *staggered magnetization*, an alternating configuration of up and down spins, below the critical point $T_{\mathrm{c}} = T_{\mathrm{N}} = Jz$. The critical temperature for an antiferromagnet is often termed the *Néel temperature*, from which the symbol T_{N} comes.

The magnetic susceptibility χ_{AF}, however, does not diverge at $T = T_{\mathrm{N}}$, in contrast to a ferromagnet because the spins do not align along the same orientation below T_{N}, and therefore a uniform field is not effective to cause a macroscopic response around the critical temperature. To see how χ_{AF} behaves, we differentiate both sides of eqns (2.51) and (2.52) with respect to h and then take the zero-field limit $h \to 0$ to obtain a set of equations satisfied by the sublattice susceptibilities $\chi_{\mathrm{A}} = \partial m_{\mathrm{A}}/\partial h|_{h\to 0}$ and $\chi_{\mathrm{B}} = \partial m_{\mathrm{B}}/\partial h|_{h\to 0}$ as

$$\chi_{\mathrm{A}} = \beta(-Jz\chi_{\mathrm{B}} + 1)\operatorname{sech}^2(\beta Jzm) \qquad (2.53)$$

$$\chi_{\mathrm{B}} = \beta(-Jz\chi_{\mathrm{A}} + 1)\operatorname{sech}^2(\beta Jzm), \qquad (2.54)$$

where we have used $m_{\mathrm{A}} = -m_{\mathrm{B}} = m$ for $h \to 0$. The solution $\chi_{\mathrm{A}} = \chi_{\mathrm{B}}$ can be identified with the total susceptibility per spin χ_{AF},

$$\chi_{\mathrm{AF}} = \beta(-Jz\chi_{\mathrm{AF}} + 1)\operatorname{sech}^2\beta Jzm = \beta(-Jz\chi_{\mathrm{AF}} + 1)\left(1 - m^2\right), \qquad (2.55)$$

where we have used the relation $1/\cosh^2(\beta Jzm) = 1 - \tanh^2(\beta Jzm) = 1 - m^2$. For $T \geq T_{\mathrm{N}}$, there is no spontaneous staggered magnetization, $m = 0$, and the susceptibility has a simple form according to eqn (2.55),

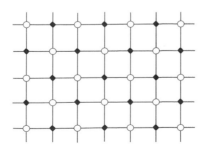

Fig. 2.8 Black dots denote sites on sublattice A and white dots are sites for sublattice B. In antiferromagnets, at low temperatures, Ising spins on sublattice A point up, whereas those on B point down (or vice versa).

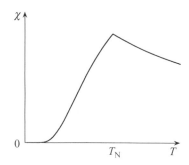

Fig. 2.9 The magnetic susceptibility of an antiferromagnetic Ising model according to the mean-field approximation. There is no divergence, but a cusp develops at the Néel temperature.

$$\chi_{\mathrm{AF}} = \frac{1}{T + Jz} = \frac{1}{T + T_{\mathrm{N}}}. \tag{2.56}$$

Thus, the susceptibility does not diverge at the Néel temperature but has a finite value. On the low-temperature side, $T < T_{\mathrm{N}}$, χ_{AF} behaves as

$$\chi_{\mathrm{AF}} = \frac{1 - m^2}{T + Jz(1 - m^2)} = \frac{1 - m^2}{T + T_{\mathrm{N}}(1 - m^2)}. \tag{2.57}$$

These results are depicted in Fig. 2.9.

The susceptibility has a cusp at the Néel temperature and decreases quickly towards 0 as $T \to 0$.

One can mathematically introduce an external *staggered magnetic field* with $+h$ on sites A and $-h$ on sites B, instead of the uniform h, and compute the resulting *staggered magnetic susceptibility* $\partial m_{\mathrm{A}}/\partial h|_{h \to 0}$. Then, one finds a divergent staggered susceptibility at the critical point. Notice that such staggered ordering magnetic field is in general impossible to realize by direct experimental means. In this case one says that there is no physical external probe that couples to the order parameter.

2.8   Bethe approximation

Let us return to ferromagnetic systems and study the *Bethe approximation* (also called *Bethe-Peierls approximation*), a straightforward and useful approach to improve over the mean-field approximation of Section 2.1. The latter approximation treats exactly the degree of freedom of a single spin and replaces all the other variables by their mean values. In the Bethe approximation, nearest neighbors are treated without approximation and the spins beyond those neighbors are approximated by their average, as illustrated in Fig. 2.10. For the Ising model of eqn (2.1), the self-consistent Hamiltonian of the central cluster is

$$H = -J \sum_{i=1}^{z} S_i S_0 - h S_0 - h \sum_{i=1}^{z} S_i - h_1 \sum_{i=1}^{z} S_i, \tag{2.58}$$

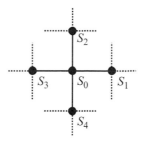

Fig. 2.10 The Bethe approximation treats neighboring spins exactly, and those beyond nearest neighbors are replaced by their average.

where S_0 is the central spin we focus our attention on, $S_1, S_2, \cdots, S_z$ are the neighboring spins, h is the uniform external field, and h_1 is the effective field that expresses the influence of spins beyond nearest neighbors.

The problem is solved within the Bethe approximation if we find the value of the unknown h_1 in eqn (2.58). Similarly in spirit to the self-consistent mean-field equation (2.6), we require that the average of the central spin $\langle S_0 \rangle = m_0$ be equal to that of the neighboring ones $\langle S_i \rangle = m_1$ ($i = 1, \cdots, z$), i.e. $m_0 = m_1 = m$.

$$\langle S_0 \rangle = \langle S_i \rangle. \tag{2.59}$$

In order to calculate these averages, we first write the partition function for the Hamiltonian (2.58) with h_1 kept unknown,

$$Z = \sum_{S_0, \cdots, S_z} \exp\left(K \sum_i S_i S_0 + \beta h S_0 + \beta h \sum_i S_i + \beta h_1 \sum_i S_i \right), \tag{2.60}$$

where $K = \beta J$. It is useful to separate the cases of S_0 being fixed to 1 and of $S_0 = -1$, since the other spins from S_1 to S_z are then independent variables and we can easily perform the summation to find

$$Z_\pm = e^{\pm \beta h} \left(2 \cosh(\pm K + \beta h + \beta h_1) \right)^z, \tag{2.61}$$

where Z_+ is for $S_0 = 1$ and Z_- for $S_0 = -1$. Then, the total partition function is the sum of these terms

$$Z = Z_+ + Z_- = e^{\beta h} \left(2 \cosh(K + \beta h + \beta h_1) \right)^z + e^{-\beta h} \left(2 \cosh(-K + \beta h + \beta h_1) \right)^z. \tag{2.62}$$

The magnetization variables m_0 and m_1 can be expressed in terms of $Z_\pm$. The probability to get $S_0 = 1$ is Z_+/Z and that of $S_0 = -1$ is Z_-/Z, from which we have

$$m_0 = \frac{Z_+ - Z_-}{Z}. \tag{2.63}$$

As for m_1, the logarithmic derivative of the partition function Z with respect to βh_1 gives the sum of $\langle S_i \rangle$ over $i = 1, \cdots z$, i.e. $\sum_i \langle S_i \rangle = z m_1$. We thus find, using eqn (2.62),

$$m_1 = \frac{\partial \log Z}{z \, \partial(\beta h_1)} = \frac{Z_+ \tanh(K + \beta h + \beta h_1) + Z_- \tanh(-K + \beta h + \beta h_1)}{Z}. \quad (2.64)$$

From the condition $m_0 = m_1$, we find

$$e^{2\beta h_1} = \left(\frac{\cosh(K + \beta h + \beta h_1)}{\cosh(-K + \beta h + \beta h_1)} \right)^{z-1}. \quad (2.65)$$

This is the self-consistent equation for the effective field h_1, which can in principle be graphically solved by the method used in Section 2.1.

EXERCISE 2.10 Derive eqn (2.65).

In the Bethe approximation the critical point can be obtained from eqn (2.65). In this equation we set $h = 0$, take the logarithm, and expand the right-hand side to third order in βh_1,

$$\frac{2\beta h_1}{z-1} = 2 \tanh K \cdot \beta h_1 - \frac{2 \sinh K}{3 \cosh^3 K} (\beta h_1)^3 + \cdots. \quad (2.66)$$

A phase transition occurs when the coefficients of the linear terms of both sides coincide, as in the mean-field approximation,

$$\frac{1}{z-1} = \tanh K_c \quad \Longleftrightarrow \quad T_c = \frac{2J}{\log\left(\dfrac{z}{z-2}\right)}, \quad (2.67)$$

where K_c stands for J/T_c. This result coincides with the mean-field value $T_c = zJ$ in the limit of large z. For finite z, eqn (2.67) represents an improvement over the mean-field approximation. For instance, in the case of the two-dimensional square lattice with $z = 4$, the critical points are $T_c/J = 4, 2.8854$ and 2.2692 for the mean-field approximation, the Bethe approximation, and the exact solution, respectively. In one dimension, with $z = 2$, the mean-field approximation predicts $T_c/J = 2$, whereas the Bethe approximation gives the exact result $T_c = 0$. The effects of fluctuations are better taken into account in the Bethe approximation than in the mean-field approximation of Section 2.1.

Critical exponents remain unchanged from the mean-field values. To estimate the exponent β, let us expand the right-hand side of eqn (2.63) in powers of βh_1 in the absence of external field h. We find a linear term, and therefore the spontaneous magnetization is proportional to the effective field h_1. This motivates us to study the temperature dependence of the effective field. Equation (2.66) is suitable for this purpose: The role of m in the mean-field relation (2.7) is replaced here by βh_1. Accordingly, using the same argument as in the mean-field case, we find that βh_1 is proportional to $(T_c - T)^{1/2}$, and thus the critical exponent β is equal to $1/2$.

To investigate the critical exponent α, we notice that the internal energy is finite in the high-temperature (disordered) region $T > T_c$, in contrast to the mean-field approximation, since nearest-neighbor interactions are taken into account explicitly. The corresponding specific heat is finite in qualitative agreement with experiments. In the low-temperature region the specific heat is also finite, as in the mean-field case.

Hence, the value of the specific heat is improved quantitatively over the mean-field approximation but still remains finite. This implies that the exponent α is unchanged, i.e. $\alpha = 0$. Similar conclusions are drawn for the other critical exponents γ and δ.

EXERCISE 2.11 Calculate γ and δ in the Bethe approximation. It will be sufficient to evaluate γ for $T > T_c$, as the other case $T < T_c$ is a little complicated. Write explicitly the expansion (2.66) of the right-hand side of the self-consistent equation to first order with an external field h included. The third-order term will be unnecessary as it will not influence γ in the high-temperature region. The result will have the same form as the mean-field equation (2.9), which allows us to apply the same argument as in the mean-field case to derive γ. The other exponent δ can be evaluated by a modification of the mean-field case.

Another interesting way to understand the Bethe approximation is through the concept of the *cavity method*. Assume $h = 0$ for simplicity. The effective field h_1 acting on S_1 in eqn (2.58) may be regarded as the accumulated effect of many spins beyond S_1, represented as dotted lines connected to S_1 in Fig. 2.11, in the absence of interaction between S_0 and S_1 since this last interaction is separately taken into account. Then, the effect of the interaction between S_0 and S_1 onto S_0 is calculated by taking the trace over S_1 as,

$$\sum_{S_1 = \pm 1} e^{K S_0 S_1 + \beta h_1 S_1} \equiv A e^{\beta \hat{h}_1 S_0}, \tag{2.68}$$

where A is a constant, and $\hat{h}_1$, called the *cavity bias*, satisfies the following relation

$$\beta \hat{h}_1 = \frac{1}{2} \log \frac{1 + \tanh K \tanh \beta h_1}{1 - \tanh K \tanh \beta h_1} = \tanh^{-1}(\tanh K \tanh \beta h_1), \tag{2.69}$$

since $\tanh^{-1} x = \frac{1}{2} \log(1 + x)/(1 - x)$. This expression can be verified by equating both sides of eqn (2.68) for the cases of $S_0 = 1$ and $S_0 = -1$ and then by eliminating A. The effect of the other spins $S_2, S_3, \cdots, S_{z-1}$ onto S_0 are simply taken into account as the sum of cavity biases because these effects are considered independent of each other in the Bethe approximation,[8]

$$\exp(\beta h_0) \equiv \exp\left(\beta \sum_{j=1}^{z-1} \hat{h}_j \right). \tag{2.70}$$

This h_0 is the *cavity field* at site 0, i.e. the effective field in the absence of the remaining interaction between S_0 and S_z, drawn as a thin line in Fig. 2.11. From eqns (2.69) and (2.70), it follows that

$$\beta h_0 = \sum_{j=1}^{z-1} \tanh^{-1}(\tanh K \tanh \beta h_j). \tag{2.71}$$

---

[8] Here, it is assumed that there are no direct or indirect interactions among $S_1, \cdots, S_z$ except for the indirect interaction via S_0. This assumption is the basis of the Bethe approximation.

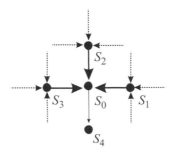

Fig. 2.11 Cavity fields propagate from sites $1, 2, \cdots, z - 1 (= 3$ in this example) to site 0.

Since the cavity field is expected to be uniform everywhere due to the equivalence of all sites, we write h_1 for all hs, $h_0 = h_1 = \cdots = h_{z-1}$, to adjust the notation to eqn (2.58),

$$\beta h_1 = (z - 1) \tanh^{-1}(\tanh K \tanh \beta h_1). \tag{2.72}$$

This equation is equivalent to the logarithm of eqn (2.65) for $h = 0$, as can be easily verified. One can get better results by considering larger clusters.

2.9 Correlation function

It is impossible to analyze the properties of correlation functions by the simple mean-field theories developed so far. The reason is that the spatial dependence of the spin variables and their interactions are not taken into account. To consider this situation, we discuss a generalization of the Landau theory in the present section. For simplicity, we deal only with the high-temperature region (disordered phase) above the critical point.

Suppose that the magnetization has some spatial dependence, which we write $\phi(\boldsymbol{r})$. This quantity may be regarded as the local average of the spin variables in the vicinity of point $\boldsymbol{r}$. The corresponding *two-point correlation function* is written as

$$G(\boldsymbol{r}) = \langle \phi(\boldsymbol{r})\phi(0) \rangle. \tag{2.73}$$

Now, let us generalize the Landau free energy (2.16) without the quartic term to the following form,

$$F = \int \left(a\,\phi(\boldsymbol{r})^2 + b\big(\nabla\phi(\boldsymbol{r})\big)^2 \right)\,\mathrm{d}\boldsymbol{r}. \tag{2.74}$$

Here, a is proportional to the temperature difference from the critical point, $a = kt$, as before. The second term $(\nabla\phi(\boldsymbol{r}))^2$ represents a ferromagnetic interaction that suppresses large absolute values of the derivative $\nabla\phi(\boldsymbol{r})$ and favors a uniformly magnetized state $\nabla\phi(\boldsymbol{r}) = 0$, in which case $\phi(\boldsymbol{r})$ reduces to the uniform magnetization m. The coefficient b of the second term is a positive constant known as *stiffness*.

It turns out that b contains information about the range of the interactions of the original microscopic system approximated by F. By dimensional analysis it goes as

$b \sim R^2$ in terms of the range of the exchange interaction R because the second term on the right-hand side of eqn (2.74) involves the spatial derivative squared, which should scale as R^{-2} if R can be regarded as a typical microscopic length scale of the system.

The capital F on the left-hand side is meant to stand for the total Landau free energy, the integral of the local free energy, written as $f(m)$ in eqn (2.16), over the whole volume of the system. We dropped the quartic term as the critical exponents above the critical point $T > T_c$ can be evaluated only from the quadratic expression (2.74) as one may remember from the case of γ in eqn (2.22). Equation (2.74) is called the *Gaussian model* due to its quadratic form. More systematic discussions on this type of field-theoretical descriptions of statistical systems will be developed in Chapter 5.

Computation of the correlation function from the free energy (2.74) is facilitated by Fourier transformation of basic variables,

$$\phi(\boldsymbol{r}) = \frac{1}{(2\pi)^d} \int \mathrm{d}\boldsymbol{q} \, \mathrm{e}^{\mathrm{i}\boldsymbol{q}\cdot\boldsymbol{r}} \tilde{\phi}(\boldsymbol{q}), \quad \tilde{\phi}(\boldsymbol{q}) = \int \mathrm{d}\boldsymbol{r} \, \mathrm{e}^{-\mathrm{i}\boldsymbol{q}\cdot\boldsymbol{r}} \phi(\boldsymbol{r}). \tag{2.75}$$

Likewise, for a finite system of volume Ω the Fourier transform is defined as

$$\phi(\boldsymbol{r}) = \frac{1}{\Omega} \sum_q \mathrm{e}^{\mathrm{i}\boldsymbol{q}\cdot\boldsymbol{r}} \tilde{\phi}(\boldsymbol{q}), \quad \tilde{\phi}(\boldsymbol{q}) = \int_\Omega \mathrm{d}\boldsymbol{r} \, \mathrm{e}^{-\mathrm{i}\boldsymbol{q}\cdot\boldsymbol{r}} \phi(\boldsymbol{r}), \tag{2.76}$$

and the Kronecker delta satisfies

$$\int_\Omega \mathrm{d}\boldsymbol{r} \, \mathrm{e}^{\mathrm{i}(\boldsymbol{q}-\boldsymbol{q}')\cdot\boldsymbol{r}} = \Omega \, \delta_{q,q'}. \tag{2.77}$$

This suggests the following relation between sums and integrals, and Dirac delta and Kronecker delta in the infinite volume limit $\Omega \to \infty$

$$\frac{1}{\Omega} \sum_q \to \frac{1}{(2\pi)^d} \int \mathrm{d}\boldsymbol{q}, \quad \text{and} \quad \Omega \, \delta_{q,q'} \to (2\pi)^d \delta(\boldsymbol{q} - \boldsymbol{q}'). \tag{2.78}$$

With the aid of the Fourier expression of the Dirac delta function

$$\frac{1}{(2\pi)^d} \int \mathrm{d}\boldsymbol{r} \, \mathrm{e}^{\mathrm{i}\boldsymbol{q}\cdot\boldsymbol{r}} = \delta(\boldsymbol{q}), \tag{2.79}$$

the free energy (2.74) is found to be rewritten as

$$F = \int \frac{\mathrm{d}\boldsymbol{q}}{(2\pi)^d} \left(kt + bq^2 \right) \tilde{\phi}(\boldsymbol{q}) \tilde{\phi}(-\boldsymbol{q}). \tag{2.80}$$

It is seen in eqn (2.80) that the degrees of freedom with different wave numbers $\boldsymbol{q}$ are summed up independently. This fact enables us to compute various physical quantities straightforwardly. In particular, the partition function of the model in the presence of an external inhomogeneous field $h(\boldsymbol{r})$ is

$$Z_G = \int \left(\prod_{q'} \mathrm{d}\tilde{\phi}(\boldsymbol{q}') \right) \exp \left(-\beta F + \frac{\beta}{(2\pi)^d} \int \mathrm{d}\boldsymbol{q} \, \tilde{h}(\boldsymbol{q}) \tilde{\phi}(-\boldsymbol{q}) \right), \tag{2.81}$$

which is taken as a *functional integral* over the configuration space of $\{\tilde{\phi}(\boldsymbol{q})\}$. We will come back to field-theoretical representations of partition functions in Chapter 5.

The expression of the correlation function in terms of Fourier components is

$$G(\boldsymbol{r}) = \langle \phi(\boldsymbol{r})\phi(0)\rangle = \frac{1}{(2\pi)^d}\int \mathrm{d}\boldsymbol{q}\,\langle \tilde{\phi}(\boldsymbol{q})\tilde{\phi}(-\boldsymbol{q})\rangle \mathrm{e}^{\mathrm{i}\boldsymbol{q}\cdot\boldsymbol{r}}. \tag{2.82}$$

We notice here that $\tilde{\phi}(-\boldsymbol{q}) = \tilde{\phi}(\boldsymbol{q})^*$ holds because $\phi(\boldsymbol{r})$ is real, where $*$ stands for complex conjugation. We thus evaluate $\langle \tilde{\phi}(\boldsymbol{q})\tilde{\phi}(-\boldsymbol{q})\rangle = \langle |\tilde{\phi}(\boldsymbol{q})|^2\rangle \equiv \tilde{G}(\boldsymbol{q})$. Since the free energy (2.80) is a quadratic form composed of independent $\boldsymbol{q}$-components, physical quantities can be calculated by Gaussian integrals, where the integration variables are $\{\tilde{\phi}(\boldsymbol{q})\}$ as in eqn (2.81). Since $\tilde{\phi}(\boldsymbol{q})$ is complex, we should integrate over its absolute value and phase. The phase actually does not appear in the free energy (2.80), and therefore its integration simply gives a constant. Only the integration over the absolute value $|\tilde{\phi}(\boldsymbol{q})| \equiv y_q$ should be performed. If we regard F as the effective Hamiltonian of a coarse-grained system, the correlation function reads

$$\tilde{G}(\boldsymbol{q}) = \frac{\int \left(\prod_{q'} \mathrm{d}\tilde{\phi}(\boldsymbol{q}')\right)|\tilde{\phi}(\boldsymbol{q})|^2 \mathrm{e}^{-\beta F}}{\int \left(\prod_{q'} \mathrm{d}\tilde{\phi}(\boldsymbol{q}')\right)\mathrm{e}^{-\beta F}} = \frac{\int \left(\prod_{q'} \mathrm{d}y_{q'}\right)y_q^2 \mathrm{e}^{-\beta F}}{\int \left(\prod_{q'} \mathrm{d}y_{q'}\right)\mathrm{e}^{-\beta F}}. \tag{2.83}$$

All wave numbers $\boldsymbol{q}'$ other than the one under consideration $\boldsymbol{q}$ give the same contribution in the numerator and denominator and hence cancel out. Only the integral over the specific $\boldsymbol{q}$ yields a non-trivial result. After the replacement $c_q = (kt + bq^2)/(2\pi)^d$, we have

$$\tilde{G}(\boldsymbol{q}) = \frac{\int \mathrm{d}y_q\, y_q^2 \exp(-\beta c_q y_q^2)}{\int \mathrm{d}y_q \exp(-\beta c_q y_q^2)} = \frac{1}{2\beta c_q} = \frac{(2\pi)^d T}{2(kt + bq^2)}. \tag{2.84}$$

The original correlation function is given by the Fourier transformation of eqn (2.84),

$$G(\boldsymbol{r}) = \frac{T}{2}\int \mathrm{d}\boldsymbol{q}\, \mathrm{e}^{\mathrm{i}\boldsymbol{q}\cdot\boldsymbol{r}}\frac{1}{kt + bq^2}. \tag{2.85}$$

As shown in Exercise 2.12, the asymptotic form of $G(\boldsymbol{r})$ in the limit of large $r \gg \xi = \sqrt{b/kt}$ for positive t ($T > T_c$) is the *Ornstein–Zernike formula*,

$$G(\boldsymbol{r}) \propto r^{-(d-1)/2}\mathrm{e}^{-r/\xi}. \tag{2.86}$$

This agrees with the expression in eqn (1.7) with $\tau = (d-1)/2$. From $\xi = \sqrt{b/kt}$ we find $\nu = 1/2$. When the temperature is exactly at the critical point $t = 0$, eqn (2.86) does not apply and we should refer back to the integral of eqn (2.85). Let us multiply $\boldsymbol{q}$ in eqn (2.85) by $1/r$ to extract all the r dependence out of the integral to find

$$G(\boldsymbol{r}) \propto r^{-d+2}. \tag{2.87}$$

The critical exponent η is found to be $\eta = 0$ according to the definition of eqn (1.9).[9]

The following table summarizes the mean-field values of critical exponents related to correlation functions:

Exponent	Mean-field value
ν	$\frac{1}{2}$
η	0

EXERCISE 2.12 Let us perform the integral (2.85). We write the target integral in a more general form as

$$g(\boldsymbol{r}) = \int_{-\infty}^{\infty} dq_1 \cdots dq_d \, \frac{e^{i(q_1 r_1 + \cdots + q_d r_d)}}{q_1^2 + \cdots + q_d^2 + a^2}. \tag{2.88}$$

(a) First, use

$$\frac{1}{b} = \int_0^{\infty} du \, e^{-bu} \quad (b > 0) \tag{2.89}$$

and raise the denominator of the integrand in eqn (2.88) to the exponent to separate the integral for each q_i $(i = 1, \cdots, d)$. (b) Next, carry out the integral for each q_i and derive the formula

$$g(\boldsymbol{r}) = \pi^{d/2} \int_0^{\infty} du \, u^{-d/2} \exp\left(-a^2 u - \frac{r^2}{4u}\right). \tag{2.90}$$

(c) The above integral is expressed in terms of the modified Bessel function of the second kind,

$$K_\mu(z) = \frac{1}{2} \left(\frac{z}{2}\right)^\mu \int_0^{\infty} \exp\left(-t - \frac{z^2}{4t}\right) t^{-\mu-1} \, dt. \tag{2.91}$$

Use the asymptotic expression of the modified Bessel function

$$K_\mu(z) \approx \sqrt{\frac{\pi}{2z}} \, e^{-z} \quad (z \gg 1) \tag{2.92}$$

and estimate the behavior of $g(\boldsymbol{r})$ in the limit of large r with a kept finite.

---

[9] The exponent $\eta \neq 0$ introduces an *anomalous dimension* in the dimension of the order parameter $\phi(\boldsymbol{r})$ from $[\phi(\boldsymbol{r})] = L^{-(d-2)/2}$ as suggested in eqn (2.87), to $[\phi(\boldsymbol{r})] = L^{-(d-2+\eta)/2}$ as seen in the generic form, eqn (1.9). The existence of this anomalous dimension $\eta \neq 0$ is rooted on the existence of another microscopic length scale (apart from the correlation length) that needs to be included in the dimensional analysis because of ultraviolet divergences (i.e. divergences in short length scales). See, e.g., Section 3.9 for some more details.

2.10 Limit of applicability of the mean-field approximation

The mean-field approximation is valid only when fluctuations around the average of physical quantities are negligible. We therefore derive a condition for fluctuations of the magnetization to be smaller than the average in the low-temperature region $T < T_c$ in order to understand when the mean-field approximation is reliable in d dimensions.

As a measure of fluctuations, it is convenient to adopt the accumulated fluctuations of magnetization, $\delta S_r = S_r - \langle S_r \rangle$, up to the length scale of the correlation length ξ. For lengths larger than ξ the fluctuations become uncorrelated. We thus compare the following quantity with the corresponding average,

$$\sigma_m^2 \equiv \int_0^\xi \langle (S_r - \langle S_r \rangle)(S_0 - \langle S_0 \rangle) \rangle \, d\boldsymbol{r} = \int_0^\xi (\langle S_r S_0 \rangle - \langle S_r \rangle \langle S_0 \rangle) \, d\boldsymbol{r}. \tag{2.93}$$

As shown in Appendix A.2, this quantity is simply the magnetic susceptibility χ times the temperature T if the integral extends to the whole space. Actually, the integrand decreases exponentially fast as r exceeds the correlation length ξ and hence the result does not depend upon the upper limit of the integral as long as it is equal to or larger than ξ. We thus find $\sigma_m^2 = T\chi$.

This result should be compared with the square of the magnetization integrated over the same region,

$$\int_0^\xi \langle S_r \rangle \langle S_0 \rangle \, d\boldsymbol{r} \propto m^2 \xi^d. \tag{2.94}$$

If the fluctuation $\sigma_m^2 = T\chi$ is sufficiently smaller than this quantity, the mean-field approximation does not have an internal inconsistency,

$$T\chi \ll m^2 \xi^d. \tag{2.95}$$

This self-consistent condition is known as the *Ginzburg criterion*. If we rewrite the expressions for χ, m and ξ near criticality by using the critical exponents,

$$T(T_c - T)^{-\gamma} \ll (T_c - T)^{2\beta}(T_c - T)^{-\nu d}. \tag{2.96}$$

Thus, a necessary condition for consistency of the mean-field approximation is

$$\gamma < \nu d - 2\beta. \tag{2.97}$$

By inserting the mean-field values $\gamma = 1, \beta = \nu = 1/2$, we conclude $d > 4$.

The same result is derived from the following slightly different consideration. The integrand in eqn (2.93) is the connected two-point correlation function $G(r)$. This $G(r)$ takes an almost constant value for r smaller than the correlation length ξ and rapidly decreases beyond. We therefore replace $G(r)$ by its value at the correlation length $G(\xi)$ to estimate the integral,

$$\sigma_m^2 \propto G(\xi)\xi^d. \tag{2.98}$$

For internal consistency of the mean-field approximation, we then use $G(\xi)\xi^d \ll m^2 \xi^d$ instead of eqn (2.95), so that $G(\xi) \ll m^2$. The correlation function behaves like $G(\xi) \propto$

ξ^{2-d} near the critical point. To show this, we insert $t = \xi^{-2}$ and $r = \xi$ into eqn (2.85) to find

$$G(\xi) \propto \int d\mathbf{q}\, e^{iq\xi} \frac{\xi^2}{k + b(q\xi)^2}. \tag{2.99}$$

By multiplying the integration variable by $1/\xi$, we have

$$G(\xi) = \xi^{2-d} \times (\text{quantity independent of } \xi). \tag{2.100}$$

Therefore, the condition $G(\xi) \ll m^2$ implies

$$(d-2)\nu > 2\beta. \tag{2.101}$$

By using the mean-field critical exponents, we again find $d > 4$.

It is expected that, outside the critical region, fluctuations are small and the mean-field approximation should be a better theory than inside the critical region. Therefore, the close neighborhood of criticality is where we expect the mean-field approximation to fail qualitatively. We would therefore like to establish the size of the critical region where fluctuations dominate and the mean-field approximation is qualitatively incorrect. As we will see, this size is non-universal, i.e. material dependent. To this end we wish to evaluate more carefully the ratio

$$\delta g = \frac{\sigma_m^2}{\displaystyle\int_0^\xi \langle S_r \rangle \langle S_0 \rangle\, d\mathbf{r}} \tag{2.102}$$

in the mean-field approximation. The numerator ($\propto \chi$) is approximately $T_c/4k|t|$, according to eqn (2.23), while the denominator is roughly $R^d|t|^{1-d/2}$ because $m^2 \propto |t|$ and $\xi \propto R|t|^{-1/2}$. The condition $\delta g \ll 1$ then implies

$$\epsilon R^{-d} \ll |t|^{(4-d)/2}, \tag{2.103}$$

where ϵ is a number of order one. This relation determines the size of the critical region. For systems where the range of the interaction R is of the order of a microscopic length (e.g. 1 nanometer), such as normal antiferromagnets or liquid ^{4}He, the size of the critical region is large for $d < 4$ because $|t|$ should be large according to eqn (2.103). This implies that the mean-field approximation is valid only well away from the critical point. On the other hand, for type-I superconductors where R is large (of the order of the size of a Cooper pair, typically a hundred nanometers), the critical region is small, which explains why in such a case the critical region may become inaccessible experimentally. Equation (2.103) also shows that fluctuations become more relevant as the spatial dimensionality d becomes smaller.

We conclude that the mean-field approximation is reliable for $d > 4$. In fact, it is proved rigorously that the critical exponents take the mean-field values for $d > 4$. This boundary dimension $d_{uc} = 4$ is called the *upper critical dimension*. It was noted in the last part of Section 2.5 that the infinite-range model (and mean-field theories in general) predicts the critical exponents correctly in the limit of large spatial dimension. It is surprising that the mean-field approximation is already reliable as soon as d

Table 2.1 Critical exponents of the Ising model in two and three dimensions as well as typical experimental values for materials in the Ising universality class. The numbers in parentheses indicate uncertainties in the final digits. Numerical and experimental values have been taken from A. Pelissetto and E. Vicari, Phys. Rep. **368** (2002) 549.

Exponent	Mean field	$d = 3$ Ising	$d = 2$ Ising	Experiment
α	0	0.110(1)	0 (log)	$0.1105^{+0.0250}_{-0.0270}$
β	$\frac{1}{2}$	0.3265(3)	$\frac{1}{8}$	0.341(2)
γ	1	1.2372(5)	$\frac{7}{4}$	1.233(10)
δ	3	4.789(2)	15	$--$
ν	$\frac{1}{2}$	0.6301(4)	1	0.62(3)
η	0	0.0364(5)	$\frac{1}{4}$	0.042(6)

exceeds four. As was explained in Section 2.1, the mean-field approximation is a very crude approximation in which the spin variable $S_i(\pm 1)$ is separated into the average m and fluctuation δS_i and the higher-order terms of the latter are ignored, despite the fact that the difference between the two possible values of $\delta S_i = S_i - m = \pm 1 - m$, i.e. two, is clearly larger than the average m. Such a crude approximation captures the essential part of the cooperative physics of critical phenomena for d larger than four.

Our real world is three dimensional, and usually the mean-field theories do not describe critical phenomena with quantitative reliability. It is, however, very difficult to study three-dimensional problems directly, and the mean-field theories often serve as our basis to approach the realistic situation by, for example, a series expansion from four dimensions. To estimate how the mean-field theories get closer to reality, we list some critical exponents in Table 2.1. The three-dimensional values are obtained from numerical simulations and the two dimensional ones are exact results.

In some cases the upper critical dimension is different from four. For instance, the tricritical point has the mean-field exponents of $\beta = 1/4, \gamma = 1, \nu = 1/2$, which implies $d_{uc} = 3$ for the upper critical dimension according to eqn (2.97).

The mean-field approximation predicts a finite transition temperature for any spatial dimension d including the one-dimensional case. However, the lower the spatial dimension is, the more unstable ordered states are due to larger fluctuations. It actually happens that there is no finite temperature phase transition, i.e. $T_c = 0$, for d smaller than a threshold value d_{lc}, the *lower critical dimension*. For the Ising model the lower critical dimension is $d_{lc} = 1$, while the XY and Heisenberg models have $d_{lc} = 2$, as will be explained in Chapter 7.

2.11   Dynamic critical phenomena

Non-equilibrium, time-dependent quantities also show anomalous behavior near the critical point, known as *dynamic critical phenomena*. This section is an introduction to dynamic critical phenomena within the mean-field perspective. The main physical idea consists of taking the system out of equilibrium, but not very far, and studying how the system relaxes back to equilibrium when it is close to a critical point and the dynamics, bringing the system to equilibrium, is dissipative. Dynamics of phase transitions in some steady state far from equilibrium is beyond the scope of the present section. In this section t stands for time and not for the reduced temperature $(T - T_c)/T_c$.

2.11.1   Single degree of freedom

As a preparation to developing a mean-field theory for dynamic critical phenomena, we first study the simple case of a single degree of freedom (e.g. a single particle) moving with energy dissipation, i.e. a model of Brownian motion.

Suppose that a particle with instantaneous velocity v is moving in a medium under a friction of strength Γ (dissipative force), and a random time-dependent force $\zeta(t)$ due to the scattering by other particles in the medium (noise),

$$\frac{\mathrm{d}v(t)}{\mathrm{d}t} = -\Gamma v(t) + \zeta(t), \tag{2.104}$$

where we have normalized the mass to unity. Random forces are assumed to be uncorrelated at two macroscopically distinct times, and we consider it reasonable to choose $\zeta(t)$ to be a random variable with zero average and the following variance,

$$\langle \zeta(t)\zeta(t') \rangle = 2D\delta(t - t'), \tag{2.105}$$

where D is the *diffusion constant*. In other words, the random variable is chosen from a Gaussian probability distribution, a *Gaussian noise*. Equation (2.104), an example of a *stochastic differential equation* also known as the *Langevin equation*, can be solved as

$$v(t) = v(0)\mathrm{e}^{-\Gamma t} + \int_0^t \mathrm{e}^{-\Gamma(t-t_1)}\zeta(t_1)\mathrm{d}t_1. \tag{2.106}$$

The first term on the right-hand side represents the influence of the initial condition, which can be ignored after the system reaches equilibrium, and the particle moves subject to the random force of the second term only. Then, the average of the square of the velocity is

$$\langle v^2(t) \rangle = \int_0^t \mathrm{d}t_1 \mathrm{d}t_2\, \mathrm{e}^{-\Gamma(2t-t_1-t_2)}\langle \zeta(t_1)\zeta(t_2) \rangle = \frac{D}{\Gamma}\left(1 - \mathrm{e}^{-2\Gamma t}\right) \to \frac{D}{\Gamma}, \tag{2.107}$$

where the long-time limit $t \to \infty$ has been taken. According to the equipartition theorem, the left-hand side of this equation is $2 \times T/2$ in equilibrium. We therefore have *Einstein's relation*

$$D = \Gamma T, \tag{2.108}$$

a result of the interplay between fluctuations and dissipation.

If we Fourier transform the equation of motion (2.104) with respect to time using

$$v(t) = \frac{1}{2\pi} \int_{-\infty}^{\infty} d\omega \, e^{-i\omega t} \tilde{v}(\omega), \tag{2.109}$$

we find

$$-i\omega \tilde{v}(\omega) = -\Gamma \tilde{v}(\omega) + \tilde{\zeta}(\omega), \tag{2.110}$$

whose solution is

$$\tilde{v}(\omega) = \frac{\tilde{\zeta}(\omega)}{\Gamma - i\omega} \equiv \tilde{G}(\omega)\tilde{\zeta}(\omega). \tag{2.111}$$

This equation implies that the external force $\tilde{\zeta}(\omega)$ determines the system's variable $\tilde{v}(\omega)$. The coefficient of proportionality $\tilde{G}(\omega)$ is called the *response function*.[10]
Another important quantity is the correlation function

$$\langle v(t + t_0)v(t_0)\rangle \equiv C(t), \tag{2.112}$$

or its Fourier transform,

$$\langle \tilde{v}(\omega)\tilde{v}(\omega')\rangle = 2\pi\delta(\omega + \omega')\tilde{C}(\omega). \tag{2.113}$$

By inserting eqn (2.111) into this equation and using the Fourier representation of eqn (2.105)

$$\langle \tilde{\zeta}(\omega)\tilde{\zeta}(\omega')\rangle = 4\pi D\delta(\omega + \omega'), \tag{2.114}$$

we establish the relation

$$\tilde{C}(\omega) = 2D\tilde{G}(\omega)\tilde{G}(-\omega) = \frac{2\Gamma T}{\omega} \mathrm{Im}\,\tilde{G}(\omega). \tag{2.115}$$

This equation relates the correlation function, which represents fluctuations in equilibrium, and the response function, which describes the system behavior slightly away from equilibrium. The above relation (2.115) is called the *fluctuation–dissipation theorem* and is known to hold for much more general systems than a single-variable case.

2.11.2 Gaussian model

The mean-field theory of dynamic critical phenomena for multivariable systems is formulated by generalization of the single-particle case to the Gaussian model. Let us rewrite the equation of motion (2.104) using the Hamiltonian $H = v^2/2$ of a free particle of unit mass,

$$\frac{dv}{dt} = -\Gamma \frac{\partial H}{\partial v} + \zeta(t). \tag{2.116}$$

The right-hand side suggests that the motion is determined by two kinds of forces, one that tends to decrease the energy (Hamiltonian) and the other a random force.

---

[10] Strictly speaking, the response function is defined in terms of a non-random external field. Notice, however, that the randomness of $\tilde{\zeta}$ does not appear in the discussion that derives eqn (2.111).

We are therefore justified to set up the evolution equation of the time-dependent local magnetization field $\phi(\boldsymbol{r}, t)$, a generalization of $\phi(\boldsymbol{r})$ in Section 2.9, of a multivariable system as

$$\frac{\partial \phi(\boldsymbol{r}, t)}{\partial t} = -\Gamma \cdot \frac{\delta F}{\delta \phi(\boldsymbol{r}, t)} + \zeta(\boldsymbol{r}, t), \qquad (2.117)$$

which represents a non-linear generalization of the stochastic differential equation for a Brownian particle. Here, F is the Landau free energy and $\zeta(\boldsymbol{r}, t)$ is a random Gaussian variable satisfying

$$\langle \zeta(\boldsymbol{r}, t) \zeta(\boldsymbol{r}', t') \rangle = 2 T \Gamma(\boldsymbol{r} - \boldsymbol{r}') \delta(t - t'). \qquad (2.118)$$

In the absence of the random force the system relaxes to the minimum of F. Equation (2.117) is a phenomenological equation describing the time dependence of macroscopic variables and has not been derived from a microscopic starting point like the Schrödinger equation. This equation is, nevertheless, useful to analyze macroscopic dynamic phenomena and is called the *TDGL equation* (time-dependent Ginzburg-Landau equation). For a generic order parameter $O(\boldsymbol{r}, t)$ one needs to replace $\phi(\boldsymbol{r}, t)$ by $O(\boldsymbol{r}, t)$ and its corresponding Landau free energy F.

As written in eqn (2.118), Γ is a function of spatial variables and, correspondingly, the term involving Γ on the right-hand side of eqn (2.117) is an abbreviation for

$$- \int \Gamma(\boldsymbol{r} - \boldsymbol{r}') \frac{\delta F}{\delta \phi(\boldsymbol{r}', t)} \mathrm{d}\boldsymbol{r}'. \qquad (2.119)$$

The gradient (functional derivative) of the free energy at position $\boldsymbol{r}'$ affects the motion at $\boldsymbol{r}$ by the strength $\Gamma(\boldsymbol{r} - \boldsymbol{r}')$.

The mean-field theory of dynamic critical phenomena is formulated by using the Gaussian model for F. Integration by parts of the spatial derivative in eqn (2.74) yields, if we assume that boundary terms vanish,

$$F = \int \left(a \, \phi(\boldsymbol{r}, t)^2 - b \, \phi(\boldsymbol{r}, t) \nabla^2 \phi(\boldsymbol{r}, t) \right) \mathrm{d}\boldsymbol{r}. \qquad (2.120)$$

Functional variation of this expression leads to

$$\frac{\delta F}{\delta \phi(\boldsymbol{r}', t)} = 2a \, \phi(\boldsymbol{r}', t) - b \nabla^2 \phi(\boldsymbol{r}', t), \qquad (2.121)$$

indicating that the stochastic differential equation is linear in this case and can be solved by Fourier transformation. By inserting this relation into eqn (2.117) and Fourier transforming the result, we can derive

$$\frac{\partial \tilde{\phi}(\boldsymbol{q}, t)}{\partial t} = -(2a + bq^2) \tilde{\Gamma}(\boldsymbol{q}) \tilde{\phi}(\boldsymbol{q}, t) + \tilde{\zeta}(\boldsymbol{q}, t). \qquad (2.122)$$

This equation has the same form as the single-body case of eqn (2.104), and we may apply the argument of the previous section to the present problem by replacing Γ there with $(2a + bq^2) \tilde{\Gamma}(\boldsymbol{q})$.

The average of $\tilde{\phi}(\boldsymbol{q}, t)$ with respect to the probability distribution of the random variable $\tilde{\zeta}$ satisfies the following equation,

$$\frac{\partial \langle \tilde{\phi}(\boldsymbol{q}, t) \rangle}{\partial t} = -(2a + bq^2)\tilde{\Gamma}(\boldsymbol{q}) \langle \tilde{\phi}(\boldsymbol{q}, t) \rangle \tag{2.123}$$

since $\langle \tilde{\zeta} \rangle$ vanishes. Then, $\langle \tilde{\phi}(\boldsymbol{q}, t) \rangle$ decays rapidly as e^{-t/τ_q} with the following *relaxation time*,

$$\tau_q = \frac{1}{(2a + bq^2)\tilde{\Gamma}(\boldsymbol{q})}. \tag{2.124}$$

In the long-wavelength limit, $\boldsymbol{q} \to 0$, if $\tilde{\Gamma}(\boldsymbol{q})$ is finite, the relaxation time is

$$\tau_0 = \frac{1}{2a\tilde{\Gamma}(0)} \propto (T - T_{\rm c})^{-1}, \tag{2.125}$$

with a divergence inversely proportional to $T - T_{\rm c}$ because $a \propto T - T_{\rm c}$. This fact represents *critical slowing down*, in which the relaxation to equilibrium slows down near the critical point due to fluctuations of all length scales. The *dynamic critical exponent*[11] z is defined by the rate of divergence of the relaxation time written in terms of the correlation length as

$$\tau_0 \propto \xi^z. \tag{2.126}$$

The mean-field value of the dynamic critical exponent is $z = 2$ when $\tilde{\Gamma}(0) > 0$ since $\nu = 1/2$. For $\boldsymbol{q} \neq 0$ and $T \to T_{\rm c}$, a finite τ_q results.

Suppose now that the integral of the local order parameter over the whole space (i.e. the zero-wavelength limit of the Fourier component) is a conserved quantity and we wish to study its dissipative dynamics. This is indeed the case for binary alloys, in which the numbers of the two types of atoms are fixed. An explicit expression of this fact is

$$\frac{\partial \tilde{\phi}(0, t)}{\partial t} = 0. \tag{2.127}$$

Then, $\tilde{\Gamma}(\boldsymbol{q})$ approaches 0 as $\boldsymbol{q} \to 0$ according to eqn (2.123). This quantity behaves for small $\boldsymbol{q}$ (large spatial scale) as $\tilde{\Gamma}(\boldsymbol{q}) \approx cq^2$ since $\tilde{\Gamma}(\boldsymbol{q})$ is an even function due to the reflection symmetry $\Gamma(\boldsymbol{r}) = \Gamma(-\boldsymbol{r})$. The relaxation time therefore satisfies, using $a \propto (T - T_{\rm c}) \propto \xi^{-2}$,

$$\tau_q = \frac{1}{(2a + bq^2)cq^2} = \frac{\xi^4}{c(c'(\xi q)^2 + b(\xi q)^4)}, \tag{2.128}$$

where c' is a constant. If we focus our attention in the spatial region where ξq is small but non-vanishing (i.e. the lengths scales larger than ξ but finite), τ_q is proportional to the numerator ξ^4, which implies $z = 4$. The following table summarizes our conclusion:

---

[11] Not to be confused with the coordination number of the lattice z.

Exponent	Mean-field value
z	2 (non-conserved order parameter)
z	4 (conserved order parameter)

This indicates that dynamic critical phenomena may display different universality classes depending upon the dynamical process.

To develop a theory beyond the mean-field approximation, we have to take into account several points. First, the effects of the fourth-order contribution $\phi(\mathbf{r}, t)^4$ to the free energy should be evaluated, similarly to the equilibrium situations. Two additional aspects specific to non-equilibrium problems are: (1) The degrees of freedom other than the spins, like phonons, may not relax sufficiently quickly to equilibrium and may have to be treated on the same footing as the spins. (2) Non-dissipative motions such as precessions in continuous spin systems should be treated with sufficient care. The limit of applicability of the dynamic mean-field theory in the present section should be discussed under these additional conditions.

3
Renormalization group and scaling

Mean-field theory is usually taken as a first step toward understanding critical phenomena, providing an overview that reveals qualitative behavior of physical quantities. However, we have to proceed beyond the mean-field theory if we wish to better understand the situation, both qualitatively and quantitatively, when fluctuations play vital roles. Indeed, as was shown in the previous chapter, the mean-field theory loses its internal consistency for spatial dimensions less than four for the simple Ising model, and the critical exponents assume different values from the mean-field predictions. In the present chapter we explain the basic concepts of the renormalization group and scaling theory, which allow us to analyze critical phenomena with fluctuations systematically taken into account. Implementation of the renormalization group in realistic systems will be deferred to the next chapter, except for the simple one-dimensional Ising model case.

3.1 Coarse-graining and scale transformations

As described in Chapter 1, the basic concept of the renormalization group is to follow the change of physical quantities as we increase the length scale by coarse graining and rescaling, which allows us to systematically take into account fluctuations near the critical point. To quantify this idea, we start from the explanation of *coarse graining* and *(re)scaling* and their consequent influences on physical quantities.

In this and the next chapters H stands for the Hamiltonian divided by the temperature H/T, a dimensionless Hamiltonian, since the Hamiltonian appears always in this combination. The concepts of coarse graining and scaling will be best illustrated in the *real-space renormalization group*, in which we trace out a part of the microscopic degrees of freedom such as the spin variables.

Suppose that the spin degrees of freedom interact with their nearest neighbors on the square lattice as depicted on the left panel of Fig. 3.1. The model is not necessarily restricted to the ferromagnetic Ising model. Because it is usually very difficult to perform the trace over all degrees of freedom at once, we first take the trace over part of the degrees of freedom. In Fig. 3.1, tracing over the spins marked × will leave the spins marked ○ left untouched. The resulting system can be regarded as a new square lattice, though oblique by 45 degrees, in which the interactions between ○ spins have been generated by the trace-out operation on the original lattice. This operation is expressed symbolically as

$$Z_N(H) = \sum_{\{S_\circ\}} \sum_{\{S_\times\}} \mathrm{e}^{-H} = \sum_{\{S_\circ\}} \mathrm{e}^{-H'} \equiv Z_{N'}(H'). \tag{3.1}$$

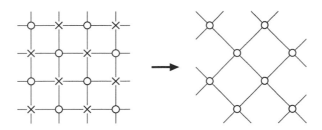

Fig. 3.1 Illustration of the real-space renormalization group. By taking the partial trace over the spin degrees of freedom marked $\times$, we are left with the other degrees of freedom (marked $\bigcirc$), which interact with each other via new, renormalized interactions.

The trace over the crossed degrees of freedom is carried out in the second equality. The symbols N and H stand for the number of the original degrees of freedom and the original Hamiltonian, respectively, and N' and H' are those after the trace operation. This latter Hamiltonian H' is formally defined by $-\log \sum_{\{S_\times\}} e^{-H}$. After the trace-out operation (coarse graining), we change the spatial scale (rescaling) and normalize the distance between the neighboring sites to the original value, unity. These two elements, coarse graining and rescaling, constitute the essence of the *renormalization group* operation. The scaling factor of space, b, is $b = \sqrt{2}$ in the present example. This operation is essentially equivalent to the numerical block-spin transformation explained in Chapter 1 since we trace out short-range fluctuations and shift our attention to longer- and longer-length behavior of the system as the operation is repeated. An unbounded repetition of these operations will reveal the critical behavior of a macroscopic system, in particular, the values of critical exponents, because we exhaust all length scales. The essence of the trick is to understand critical phenomena as the asymptotic behavior of a system in the limit of infinitely many iterations of renormalization processes only from the analysis of a single (and hence relatively simple) renormalization step.

The basis of a renormalization group analysis is the rule that shows how physical quantities are transformed in a single step. We first postulate that the partition function is invariant, as in eqn (3.1),

$$Z_{N'}(H') = Z_N(H). \tag{3.2}$$

Let us write the transformation R_b of the Hamiltonian as

$$H' = R_b(H). \tag{3.3}$$

This R_b is generally a complicated non-linear transformation. The length scale is changed by the factor $1/b$ and correspondingly the wave number is scaled by $b > 1$. The total number of degrees of freedom is reduced by b^{-d}.

$$r' = b^{-1}r, \quad q' = bq, \quad N' = b^{-d}N. \tag{3.4}$$

The free energy per degree of freedom changes, according to the invariance of the partition function (3.2) and eqn (3.4), as

$$f(H') = b^d f(H). \tag{3.5}$$

Notice that we have included the temperature factor in the free energy for simplicity and have written f for βf.

The *scaling dimension* of a physical quantity A is defined as the index x in the factor

$$A' = b^x A, \tag{3.6}$$

where A is the value before renormalization and A' after renormalization. Equation (3.4) shows that the scaling dimension of length is -1, that for wave number is 1 and the volume or the number of sites has $x = -d$. The total free energy $\beta F = -\log Z$ has scaling dimension 0 from eqn (3.2), and the free energy per degree of freedom has scaling dimension d from eqn (3.5). The magnitude of the microscopic variable, the spin, also has a scaling dimension. Let us leave its evaluation to later sections and write here a general form

$$S'(\boldsymbol{r}') = c(b)^{-1} S(\boldsymbol{r}), \tag{3.7}$$

with $c(b)$ defining the scaling of the spin field $S(\boldsymbol{r})$. The symbols $\boldsymbol{r}$ and $\boldsymbol{r}'$ denote the position vectors of the same spatial point before and after renormalization. The corresponding rule for the connected correlation function $G(\boldsymbol{r}, H) = \langle S_0 S_{\boldsymbol{r}} \rangle - \langle S_0 \rangle \langle S_{\boldsymbol{r}} \rangle$ reads

$$G(\boldsymbol{r}', H') = c(b)^{-2} G(\boldsymbol{r}, H). \tag{3.8}$$

If the original system is exactly at the critical point, the system has fluctuations of all length scales and consequently should stay essentially unchanged after many steps of renormalization. This means that the Hamiltonian reaches a fixed point H^* after many renormalization steps. The fixed point is defined by

$$H^* = R_b(H^*). \tag{3.9}$$

The critical and fixed points are closely related with each other but are not identical. If the original system with Hamiltonian H is at a critical point, H itself is not a fixed point but the renormalized Hamiltonian asymptotically approaches a fixed point H^*,

$$H^* = \lim_{n \to \infty} R_b^n(H_c), \tag{3.10}$$

where H_c is the original Hamiltonian at the critical point.

In this way, we establish a map in Hamiltonian space (or parameter space as we will see below) that mathematically defines a *semi-group* (and not a *group*) since information is erased ($b > 1$) as one traces out degrees of freedom. In other words, there is no inverse mapping (operation, R_b^{-1}, that is required for those R_b to form a group). That the set of transformations R_b forms a semi-group means that there exists an identity map when $b = 1$, and two successive mappings R_{b_1} and R_{b_2} are equivalent to a single map $R_{b_1 b_2}$. Mathematically,

$$H'' = R_{b_2}(H') = R_{b_2}.R_{b_1}(H) = R_{b_1 b_2}(H). \tag{3.11}$$

3.2   Parameter space and renormalization group equation

Let us formulate the rule of change of a Hamiltonian under the operation of renormalization group. The Hamiltonian is written as the sum of products of a parameter u_n (c-number) and operator O_n[1]

$$H = \sum_n u_n O_n, \tag{3.12}$$

and n is an integer number whose maximum value is a polynomial function of the number of elementary degrees of freedom of the system (e.g. the maximum n is a quadratic function of the total number of spins). For simplicity, the following equations are written for the Ising model case but the idea is applicable to more general cases. For example, in the Hamiltonian

$$H = -K \sum_{\langle ij \rangle} S_i S_j - h \sum_i S_i, \tag{3.13}$$

K and h are parameters (or coupling constants) and $S_i S_j$ and S_i are operators. Here again, the inverse temperature β is understood to be included in the Hamiltonian. Now, if we apply the procedure of the renormalization group, the new Hamiltonian includes operators that did not exist before renormalization. In other words, a renormalization group map (i.e. coarse graining plus rescaling) induces a change of coupling constants. In the example of the previous section, Fig. 3.1, the new Hamiltonian assumes the form

$$H' = -K' \sum_{\langle ij \rangle} S_i S_j - K_1' \sum_A S_i S_j - K_2' \sum_B S_i S_j S_k S_l + \cdots, \tag{3.14}$$

if we set $h = 0$ before renormalization. Here, A denotes the interaction between next-nearest-neighbor sites (in the direction of the diagonal of a unit square), and B is for the four-spin interaction surrounding a unit square (*plaquette* interaction). As confirmed in Exercise 3.1, the partial summation for a system with nearest-neighbor interaction only causes additional terms in the new renormalized Hamiltonian written as K', K_1', K_2' in eqn (3.14). An additional renormalization step to the latter system generates further complicated interactions. Repeated applications of this procedure may seem impracticable since more and more involved terms keep showing up. Nevertheless, we can develop a general formalism, which leads to a deep understanding of critical phenomena, by considering a very generic form of the Hamiltonian from the outset that includes all possible terms that may appear after renormalization.

> **EXERCISE 3.1** Carry out the real-space renormalization group operation in Fig. 3.1. The left panel before renormalization represents the ferromagnetic Ising model with nearest-neighbor interactions. One should take partial summation over the spins marked by crosses. Since spins marked as crosses do not have direct interactions with each other, it is allowed to consider each crossed spin separately. Let us name the cross-marked spin S_0 and the surrounding four spins marked by

---

[1] It is customary in the renormalization group theory to call the microscopic degrees of freedom *operators*, following the convention of field theory.

Fig. 3.2 The spin S_0 interacts with its four neighboring spins.

white circles S_1, S_2, S_3, S_4, see Fig. 3.2. Then, the problem is to calculate

$$\sum_{S_0=\pm1} \exp\left(KS_0(S_1 + S_2 + S_3 + S_4)\right) \tag{3.15}$$

and write the result as a function of $S_1, \cdots, S_4$. In particular, show that the result has the form of eqn (3.14).

The coarse-graining procedure is not unique. Choosing one leads to a particular renormalization group scheme. In formal terms, any renormalization group transformation can be expressed as

$$e^{-H'(S')} = \mathrm{Tr}_S P(S', S) e^{-H(S)}, \tag{3.16}$$

where $P(S', S)$ is a non-negative weight operator constructed so that the coarse-grained variables S' adopt the same values as the original S. For instance, if the original variables are Ising spins $\{S_i = \pm1\}$ each defined on lattice site i, the coarse-grained variables $\{S'_j = \pm1\}$ also represent Ising spins defined on the renormalized lattice. Clearly, $P(S', S)$ must preserve the symmetries of the original H and satisfies

$$\mathrm{Tr}_{S'} P(S', S) = 1, \tag{3.17}$$

which is equivalent to the condition (3.2). For example, in Section 1.4 we defined a block-spin transformation with an odd number of spins per block consisting of a majority rule. This amounts to the following rule

$$P(S', S) = \prod_j \delta\left(S'_j - \mathrm{sign}\left[\sum_{i \in j} S_i\right]\right), \tag{3.18}$$

where j represents a block-spin index, while i refers to a site of the original lattice.

Let us express the set of parameters corresponding to $-K', -K'_1, -K'_2, \cdots$ in eqn (3.14) as a vector $\boldsymbol{u}$ and the set of spin variables (operators) corresponding to $S_i S_j$ (the first term on the right-hand side), $S_i S_j$ (the second term), $S_i S_j S_k S_l$ (the third term) by a vector $\boldsymbol{O}$. Then, eqn (3.14) may be formally viewed as the inner product $\boldsymbol{u} \cdot \boldsymbol{O}$. We therefore write the Hamiltonians before and after renormalization as follows,

$$H = \boldsymbol{u} \cdot \boldsymbol{O}, \quad H' = \boldsymbol{u}' \cdot \boldsymbol{O}' = R_b(H). \tag{3.19}$$

The set of operators $\boldsymbol{O}$ and $\boldsymbol{O}'$ are well-defined quantities. The essential part of the renormalization group calculation is to find the rule that implements the change of the parameter sets from $\boldsymbol{u}$ to $\boldsymbol{u}'$. If we use the same symbol as in eqn (3.3), we may write the induced map as

$$\boldsymbol{u}' = R_b(\boldsymbol{u}). \tag{3.20}$$

This equation, called the *recursion relation* or *renormalization group equation*, represents the rule of change of parameters by a single step of the renormalization group operation and hence R_b involves no non-analyticities; it is an analytic transformation. Under this renormalization group map, lengths are reduced by the scale factor b, and therefore the correlation length transforms as

$$\xi[\boldsymbol{u}'] = b^{-1}\xi[\boldsymbol{u}]. \tag{3.21}$$

Successive applications of the recursion relation generates a discrete flow in the parameter space,

$$\boldsymbol{u} \to R_b(\boldsymbol{u}) \to R_b^2(\boldsymbol{u}) \to \cdots \to R_b^n(\boldsymbol{u}) \to \cdots, \tag{3.22}$$

which can be viewed as a series of points along a trajectory. The set of trajectories generated from different initial parameter values, in (infinite-dimensional, at least in principle) parameter space, is called the *renormalization group flow*. Similarly, the correlation length transforms as

$$\xi[R_b^n(\boldsymbol{u})] = b^{-n}\xi[\boldsymbol{u}], \tag{3.23}$$

eventually vanishing if $\xi[\boldsymbol{u}] < \infty$, indicating that the flow moves away from criticality. However, if $\xi[\boldsymbol{u}] = \infty$, the renormalized correlation length remains divergent.

The critical exponents that characterize the non-analyticities of physical quantities are determined by the asymptotic behavior of the parameters $\boldsymbol{u}$ that emerge in the limit of infinite repetitions of the renormalization group procedure. Singularities of physical quantities have their origin in the infinitely many applications of the renormalization group transformation, not in the function R_b itself. A *fixed point* of the renormalization group transformation is a point $\boldsymbol{u}^*$ in parameter space that is invariant

$$\boldsymbol{u}^* = R_b(\boldsymbol{u}^*), \tag{3.24}$$

and has an associated *fixed-point Hamiltonian* H^* that is also invariant under scale transformations. At the fixed point

$$\xi[R_b(\boldsymbol{u}^*)] = \xi[\boldsymbol{u}^*] = b^{-1}\xi[\boldsymbol{u}^*], \tag{3.25}$$

implying that $\xi[\boldsymbol{u}^*]$ can only take two values: 0 or ∞, and the latter case is due to a *critical fixed point*. A *trivial fixed point* is indicated by a vanishing correlation length, i.e. $\xi[\boldsymbol{u}^*] = 0$. This finding expresses the physical fact that at a fixed point there is no characteristic length scale and *scale invariance* or *self-similarity* manifests itself.

The emergence of non-analyticities has an analogy to the situation in iterative maps of classical dynamics. Only an infinite number of iterations may lead to singular behavior. Consider the dynamic equation, which corresponds to a single coupling constant,

$$\frac{\mathrm{d}u}{\mathrm{d}t} = -2u(u^2 - 1), \tag{3.26}$$

and study the behavior of $u(t)$ as a function of the initial condition u_0. The solution to this equation is $u(t) = u_0/\sqrt{u_0^2 - (u_0^2 - 1)\mathrm{e}^{-4t}}$. Clearly, for any finite t, $u(t)$ is a continuous function of u_0. It is only when $t \to \infty$ that $u(t \to \infty) = \mathrm{sign}[u_0]$ $(u_0 \neq 0)$ becomes a discontinuous function of u_0. The right-hand side of eqn (3.26) vanishes for $u = 0, \pm 1$. These three points are fixed points of eqn (3.26). In other words, if one starts $(t = 0)$ at points $u_0 > 0$, the flow is attracted toward $u^* = 1$, while for $u_0 < 0$ it is attracted to $u^* = -1$. The flow always repels $u^* = 0$, and obviously is stuck at that point only if $u_0 = 0$. It is an unstable fixed point.

As we are interested in the singular properties of physical quantities near a critical point, it is instructive to consider the departure of the parameter values slightly away from the critical point and see the behavior of the recursion relation (3.20). This would correspond to studying properties of the system with parameter values slightly away from the fixed point. We thus write the parameters before and after renormalization using the fixed-point value $\boldsymbol{u}^*$ and slight deviations from it as

$$\boldsymbol{u} = \boldsymbol{u}^* + \delta\boldsymbol{u}, \quad \boldsymbol{u}' = \boldsymbol{u}^* + \delta\boldsymbol{u}'. \tag{3.27}$$

Although the recursion relation $\boldsymbol{u}' = R_b(\boldsymbol{u})$ is in general a non-linear transformation, we expand the right-hand side of this equation around the fixed point and keep the first-order term

$$\boldsymbol{u}' = \boldsymbol{u}^* + \delta\boldsymbol{u}' = R_b(\boldsymbol{u}^* + \delta\boldsymbol{u}) = R_b(\boldsymbol{u}^*) + \left.\frac{\partial R_b}{\partial \boldsymbol{u}}\right|_{\boldsymbol{u}^*} \cdot \delta\boldsymbol{u} + \cdots, \tag{3.28}$$

because we are only interested in the vicinity of the fixed point. A Taylor expansion is possible since R_b has no singularities, i.e. it is analytic. The linearized recursion relation is then written as

$$\delta\boldsymbol{u}' = T_b(\boldsymbol{u}^*) \cdot \delta\boldsymbol{u}, \tag{3.29}$$

where $T_b(\boldsymbol{u}^*) = \left.\dfrac{\partial R_b}{\partial \boldsymbol{u}}\right|_{\boldsymbol{u}^*}$ is a real matrix, with components given by

$$[T_b(\boldsymbol{u}^*)]_{ij} = \left.\frac{\partial u_i'}{\partial u_j}\right|_{\boldsymbol{u}^*}, \tag{3.30}$$

not necessarily symmetric. We are interested in those situations where T_b is diagonalizable with real eigenvalues. Critical phenomena will turn out to be characterized by the eigenvalues and eigenvectors of this linear transformation T_b.

Let us recall that T_b is a function of the rescaling factor b. An eigenvalue of T_b is in general expressed as a power of b,

$$\lambda_i(b) = b^{y_i}. \tag{3.31}$$

The reason is as follows. A successive operation of two renormalization group transformations of rescaling factors b_1 and b_2 is equivalent to a single transformation of factor $b_1 b_2$ (i.e. its semi-group property $T_{b_2} \cdot T_{b_1} = T_{b_1 b_2}$). The eigenvalue of the latter $\lambda_i(b_1 b_2)$ coincides with the product of the former $\lambda_i(b_1)\lambda_i(b_2)$ due to the linearity of T_b: The first transformation of b_1 multiplies the eigenvector by $\lambda_i(b_1)$ and the second by $\lambda_i(b_2)$. The resulting relation $\lambda_i(b_1 b_2) = \lambda_i(b_1)\lambda_i(b_2)$ is satisfied only by a power of b, $\lambda_i(b) = b^{y_i}$.

It is instructive to expand $\delta \boldsymbol{u}$ and $\delta \boldsymbol{u}'$ by the set of eigenvectors $\{\boldsymbol{\phi}_i\}$ of T_b, $T_b \cdot \boldsymbol{\phi}_i = \lambda_i(b)\boldsymbol{\phi}_i$ as,

$$\boldsymbol{u} = \boldsymbol{u}^* + \sum_i g_i \boldsymbol{\phi}_i, \quad \boldsymbol{u}' = \boldsymbol{u}^* + \sum_i g_i' \boldsymbol{\phi}_i, \tag{3.32}$$

implying after eqn (3.29) that g_i' and g_i are related as $g_i' = b^{y_i} g_i$. These $g_1, g_2, \cdots$ are very important quantities that characterize the properties of the parameter space $\boldsymbol{u}$ near the fixed point and are called *scaling fields*. The description of the behavior of the system by the renormalization group has thus been reduced to the study of the properties of the fixed point, the exponents $y_1, y_2, \cdots$ of eigenvalues of the linearized transformation T_b and the scaling fields $g_1, g_2, \cdots$. The following table summarizes the main steps in a renormalization group analysis

$$\boxed{\text{Renormalization group procedure}}$$

i— Coarse graining and rescaling as represented by

$$P(S', S) \quad \text{and} \quad b.$$

ii— Write down the renormalization group equations:

$$\boldsymbol{u}' = R_b(\boldsymbol{u}).$$

iii— Solve the renormalization group equations iteratively:

$$\boldsymbol{u} \rightarrow \boldsymbol{u}' \rightarrow \boldsymbol{u}'' \rightarrow \cdots$$

iv— Determine the phase diagram from the flow diagram and fixed points $\boldsymbol{u}^*$.

v— Linearize $R_b \rightarrow T_b$ close to the critical fixed point $\boldsymbol{u}^*$.

vi— Determine the eigenvalues and eigenvectors of $T_b(\boldsymbol{u}^*)$:

$$\{\lambda_i(b) = b^{y_i}\} \quad \text{and} \quad \{\boldsymbol{\phi}_i\}.$$

vii— Extract the exponents and scaling fields:

$$\{y_i, g_i\}.$$

One can formally write the renormalization group equation $\boldsymbol{u}' = R_{\eta b}(\boldsymbol{u})$ as a coupled set of non-linear differential equations by considering an infinitesimal rescaling $\eta b = (1 + \epsilon)b$, with $\epsilon \ll 1$

$$\frac{\mathrm{d}\boldsymbol{u}}{\mathrm{d}b} = \lim_{\epsilon \to 0} \frac{\boldsymbol{u}' - \boldsymbol{u}}{\epsilon b} = \frac{1}{b}\beta(\boldsymbol{u}) , \tag{3.33}$$

or

$$\frac{\mathrm{d}\boldsymbol{u}}{\mathrm{d}\tau} = \beta(\boldsymbol{u}) \tag{3.34}$$

with $\tau = \log b$. The zeros of the *beta function* $\beta(\boldsymbol{u})$, $\beta(\boldsymbol{u}^*) = 0$, define the fixed points of the transformation. This formulation elucidates the relation to eqn (3.26) for the dynamical systems.

3.3   Renormalization group flow near a fixed point and universality

The exponent y_i appearing in the eigenvalue of the linear transformation T_b is an important quantity characterizing the parameter flow near the fixed point. If y_i is positive, the eigenvalue b^{y_i} is larger than unity ($b > 1$), and the scaling field g_i becomes amplified by the factor b^{y_i} after each application of the renormalization group transformation. The parameter therefore moves away from the fixed point. For negative y_i, on the other hand, the parameter converges to the fixed-point value. It is therefore concluded that all the scaling fields g_i with positive exponent $y_i > 0$ should be tuned to 0 for the system parameters to be attracted to the fixed point. Since the adjustment of the value of g_i has a decisive effect on the system behavior, a scaling field with positive exponent $y_i > 0$ is called a *relevant* variable or a relevant field. A scaling field with negative exponent $y_i < 0$ rapidly diminishes toward 0 as the renormalization step proceeds and has no essential effects on the critical properties. Such scaling fields are thus called *irrelevant* variables. The intermediate case of $y_i = 0$ is said to be *marginal*. Marginal scaling fields are associated with logarithmic corrections to scaling. Note that the notions of relevance, irrelevance or marginality are relative to a particular fixed point. A relevant variable at one fixed point may be irrelevant at another.

Figure 3.3 illustrates the renormalization group flow of scaling fields g_i and g_j with $y_i > 0$ and $y_j < 0$. The flow of g_j with $y_j < 0$ converges to the fixed point F at $g_i = g_j = 0$ along the horizontal line $g_i = 0$. If, however, g_i is non-vanishing, g_i rapidly diverges away from the fixed-point value of 0 since $y_i > 0$. Even in this case g_j approaches 0.

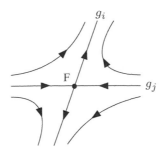

Fig. 3.3 Schematic illustration of the parameter flow when one of the exponents is positive and the other negative, $y_i > 0$ and $y_j < 0$. The fixed point F has $g_i = g_j = 0$.

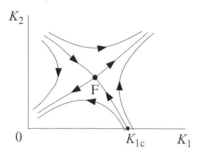

Fig. 3.4 Schematic diagram of the renormalization group flow in the space of interactions in the absence of external field. The nearest-neighbor interaction is written as K_1 and the next-nearest-neighbor interaction as K_2.

Critical phenomena are observed in a ferromagnetic system only when the two parameters T and h are adjusted to critical values, $T = T_c$ and $h = 0$, as shown in Fig. 1.3. Slight deviations of these values from the critical point will drive the system away from the critical point by renormalization group operations, as mentioned in Section 1.4. This fact leads us to the identification of T and h as two quantities related to relevant variables in typical critical phenomena. The scaling fields corresponding to T and h, g_1 and g_2, will have positive exponents, $y_1 > 0, y_2 > 0$. The others are negative, $y_3 < 0, y_4 < 0, \cdots$. The scaling fields are derived from the function realizing the transformation $R_b(\cdot)$, an analytic function, and hence g_i should be analytic with respect to T and h. We may therefore infer that g_1 is proportional to $t = (T - T_c)/T_c$ and g_2 proportional to h near the fixed point. Variables other than the temperature and external field, represented by the other scaling fields $g_3, g_4, \cdots$, do not affect the essential features of critical phenomena, typically the critical exponents. Details of the system properties other than the values of temperature and external field have no influence on the critical exponents. This is the statement of *universality* from the standpoint of renormalization group. We hereafter write g_t, y_t for g_1, y_1 and g_h, y_h for g_2, y_2.

Figure 3.4 is a flow diagram under the condition $h = 0$, from which we learn important lessons on the significance of scaling fields. Various types of interactions emerge as the renormalization step proceeds, and Fig. 3.4 represents a projection of such a multidimensional space onto the two-dimensional plane of K_1 and K_2, where K_1 is the nearest-neighbor interaction divided by temperature and K_2 stands for the next-nearest-neighbor interaction divided by temperature. Each parameter changes as the renormalization proceeds. If, for instance, there are only nearest-neighbor interactions initially ($K_2 = 0$), there appear next-nearest-neighbor interactions ($K_2 > 0$) after a single step of renormalization. Since only g_t is a relevant scaling field when $h = 0$, there is just a single direction along which the system is attracted toward the fixed point F at $g_t = g_h = g_3 = \cdots = 0$. The g_t axis corresponds to the direction along which the renormalization flow moves away from the fixed point. The other directions correspond to g_3 or other irrelevant scaling fields, along which the flow is attracted to the fixed point.

It should now be clear that the scaling field g_t should not be directly identified with the parameter $K_1 - K_{1c}$ of the initial Hamiltonian with $K_2 = 0$, the so-called *bare parameter*. We, nevertheless, observe that the scaling field g_t, which expresses a small deviation from the fixed point, is proportional to the deviation of the initial parameter value from the critical value $(K_1 - K_{1c})$ as long as the system is close to the fixed point. Both are proportional to $t = (T - T_c)/T_c$.

The set of fixed points is an important characteristic of a renormalization group study. Each fixed point $\boldsymbol{u}^*$ has a set of points in parameter space, known as a *basin of attraction*, which flow into the fixed point $\boldsymbol{u}^*$ under renormalization. The basin of attraction of a critical fixed point is called a *critical surface* or a *critical manifold*. Points on a critical manifold are attracted by renormalization to the fixed point F in Figs. 3.3 and 3.4. As seen in Fig. 3.4, the transition point (critical point) lies on the critical surface but should be distinguished from the fixed point itself. Moreover, points $\boldsymbol{u}$ on the critical surface have infinite correlation length since $\xi[\boldsymbol{u}] = b^n \xi[R_b^n[\boldsymbol{u}]]$, and in the limit $n \to \infty$, $\xi[\boldsymbol{u}] = \infty$ since in that limit $\xi[\boldsymbol{u}^*] = \xi[R_b^n[\boldsymbol{u}]] = \infty$.

Fixed points are not necessarily isolated points in parameter space. They may represent lines or generic surfaces, and one may classify them according to their *codimension*. Points with codimension 0 (known as *sinks*) have no relevant direction associated to them and represent bulk phases. Fixed points with codimension 1 can be discontinuity fixed points (corresponding to first-order transitions, as will be discussed in Exercise 3.3) or they may represent a stable bulk phase. A critical point has codimension 2 since two relevant fields are involved. Those fixed points organize the parameter space into regions with qualitatively different physical behavior. A phase diagram, which has an associated fixed-point structure, summarizes the global structure of the renormalization group flow.

3.4   Scaling law and critical exponents

We next relate the positive exponents y_t and y_h of eigenvalues of the linearized renormalization group transformation T_b with the critical exponents. This is the quintessence of the renormalization group theory and yet is formulated in a surprisingly simple and elegant manner. The central idea lies in the analysis of the free energy under a renormalization group transformation. This explains how the renormalization group theory accounts for the origin of scaling law, as will be detailed toward the end of this section.

According to eqn (3.5) and the discussions in Section 3.2, the free energy is transformed in the following manner by a renormalization group transformation,

$$f(g_t, g_h, g_3, \cdots) = b^{-d} f(g_t', g_h', g_3', \cdots), \qquad (3.35)$$

where g_t is proportional to the deviation of temperature from the critical value t, and g_h is proportional to the external field h. Hereafter, we understand that f denotes the singular part of the free energy. Then, as will be illustrated later in Section 3.6.3, the right-hand side should include an additional non-singular term, rigorously speaking,

$$f(g_t, g_h, g_3, \cdots) = b^{-d} f(g_t', g_h', g_3', \cdots) + w(g_t, g_h, g_3, \cdots). \qquad (3.36)$$

The last term w, however, does not play a crucial role in the determination of the critical point or critical exponents, thus we omit it in this book and use eqn (3.35). Notice, however, that this regular term is important to determine the total free energy.

Let us drop $g_3, g_4, \cdots$ as they represent irrelevant fields. We also ignore the constants of proportionality between g_t and t and between g_h and h since these play no roles. Then, eqn (3.35) becomes, after n steps of renormalization,

$$f(t, h) = b^{-nd} f(b^{n y_t} t, b^{n y_h} h). \tag{3.37}$$

For $t \neq 0$ we choose the number n such that the first argument of the right-hand side reduces to unity, $b^{n y_t} t = 1$.[2] The physical idea is that we repeat the renormalization group transformation many times so that the effective temperature is pushed away from the critical region. As a result, the critical condition $|t| \ll 1$ is replaced by $t' = b^{n y_t} t = 1$ and the system has a high (or low) effective temperature. Panels (d) of Figs. 1.5 and 1.9 correspond to this circumstance. Then, by inserting $b^n = t^{-1/y_t}$ in the right-hand side of eqn (3.37), we find a very important relation known as the *scaling law*,

$$f(t, h) = t^{d/y_t} f(1, h t^{-y_h/y_t}) \equiv t^{d/y_t} \Psi(h t^{-y_h/y_t}). \tag{3.38}$$

The free energy originally has two independent variables, the temperature and external field. Nevertheless, as expressed in the last part of the above equation, it has effectively become a single-variable function as far as the critical phenomena are concerned. The function $\Psi(\cdot)$ is called the *scaling function*. As we will see below, the scaling law implies functional relations among critical exponents.

From a mathematical standpoint, the scaling law above asserts that the (singular part of the) free energy is a *generalized homogeneous function*. For n variables, those functions transform as

$$f(\lambda^{\alpha_1} g_1, \lambda^{\alpha_2} g_2, \cdots, \lambda^{\alpha_n} g_n) = \lambda f(g_1, g_2, \cdots, g_n), \tag{3.39}$$

where λ, α_i are arbitrary numbers. It is clear that homogeneous functions can always be written in terms of scaling functions such as the case of two variables in eqn (3.38). Consider a homogeneous function $f(g_1, g_2)$ and perform a scale transformation $\lambda^{\alpha_1} = 1/g_1$. Then, $f(\lambda^{\alpha_1} g_1, \lambda^{\alpha_2} g_2) = g_1^{-1/\alpha_1} \Psi(g_1^{-\alpha_2/\alpha_1} g_2)$, where $\Psi(z) = f(1, z)$.

EXERCISE 3.2 Show that the *lattice constant a*, the distance between neighboring lattice sites, is an irrelevant variable. It will be useful to generalize eqn (3.37) to include a as an additional variable.

Comment: We conclude from this result that the value of the lattice constant has no influence on the critical behavior. In particular, continuous field theories obtained in the limit $a \to 0$ are often used for the evaluation of critical exponents. However, care must be exercised since in some cases irrelevant variables affect

---

[2] For $T < T_c$, we choose $b^{n y_t} |t| = 1$ as t is negative. For simplicity of notation, we often write t for $|t|$ even when $T < T_c$ in the present book. In this case, the first argument of the second expression of eqn (3.38) is -1 instead of 1.

scaling. Those variables are known as *dangerous irrelevant variables*, as will be discussed in Section 4.2.1.

The scaling law is useful since it relates the exponents y_t and y_h to the critical exponents. For instance, the specific heat is the second-order derivative of the free energy with respect to the temperature, usually under the condition $h = 0$. Setting $h = 0$ in eqn (3.38) and taking the derivative, we have

$$C(t, 0) \propto \frac{\partial^2 f(t, 0)}{\partial t^2} \propto t^{d/y_t - 2}. \tag{3.40}$$

This should be proportional to $t^{-\alpha}$, from which we find $\alpha = 2 - d/y_t$.[3] The exponent y_t, which describes the rate of amplification of the scaling field $g_t \propto t$ by the renormalization group operation, determines the critical exponent α of the specific heat. Next, to find the critical exponent β, we differentiate eqn (3.38) with respect to h and then set $h = 0$,

$$m(t, 0) \propto \left. \frac{\partial f(t, h)}{\partial h} \right|_{h=0} \propto t^{(d - y_h)/y_t}, \tag{3.41}$$

which means $\beta = (d - y_h)/y_t$. Similarly for the critical exponent γ: By twice differentiating eqn (3.38) with respect to h and setting $h = 0$, we obtain the susceptibility χ,

$$\chi(t, 0) \propto \left. \frac{\partial^2 f(t, h)}{\partial h^2} \right|_{h=0} \propto t^{(d - 2y_h)/y_t}. \tag{3.42}$$

We thus conclude $\gamma = (2y_h - d)/y_t$. As for the critical exponent δ, we set $t = 0$ in eqn (3.37) and differentiate the resulting expression with respect to h,

$$m(0, h) \propto \frac{\partial f(0, h)}{\partial h} = b^{-nd + ny_h} f_2(0, b^{ny_h} h), \tag{3.43}$$

where f_2 is the partial derivative of f with respect to the second argument. If we choose n such that $b^{ny_h} h = 1$ is satisfied, the h dependence of the right-hand side becomes $h^{(d - y_h)/y_h}$, and $\delta = y_h/(d - y_h)$ follows. Those results are summarized as follows.

$$\alpha = 2 - \frac{d}{y_t}, \quad \beta = \frac{d - y_h}{y_t}, \quad \gamma = \frac{2y_h - d}{y_t}, \quad \delta = \frac{y_h}{d - y_h}. \tag{3.44}$$

We have used the relations $b^n = t^{-1/y_t}$ and $b^n = h^{-1/y_h}$ in the above argument. These relations are interpreted as describing the system in a gradually coarse-grained manner in renormalization group since smaller t or h (i.e. closer to the critical point) needs larger steps of renormalization n according to $b^n = t^{-1/y_t}$ and $b^n = h^{-1/y_h}$.

The behaviors of the scaling function $\Psi(x)$ in eqn (3.38) in the limits of $x \to 0$ and $x \to \infty$ are determined from the following discussion. When $h = 0$, f satisfies

---

[3] As mentioned before, the right-hand side of eqn (3.40) represents the singular term of the specific heat. Additional contributions from non-singular terms must be added to recover the full temperature dependence of the specific heat. This is particularly important when the critical exponent α is negative, since then the magnitude of the non-singular terms is much larger than the singular contribution. The same comment applies to all the following relations.

$f(t, 0) \approx t^{2-\alpha}$ and therefore $\Psi(0)$ should be a non-vanishing constant according to the relation $d/y_t = 2 - \alpha$. To see the properties in the other limit, we notice the simple fact that $f(t, h)$ is a function only of h for $t \to 0$. Then, the t dependence of the right-hand side of eqn (3.38) must be canceled by the following behavior of $\Psi(x)$ for $x \to \infty$,

$$\Psi(ht^{-y_h/y_t}) \approx \left(ht^{-y_h/y_t}\right)^{d/y_h} = h^{d/y_h} t^{-d/y_t}. \tag{3.45}$$

Hence, for $x \gg 1$, $\Psi(x) \approx x^{d/y_h}$. From the above result we find in the limit $t \to 0$

$$f(0, h) \approx h^{d/y_h}, \tag{3.46}$$

which implies $m \approx h^{d/y_h - 1}$, leading to $\delta = y_h/(d - y_h)$. This is consistent with the previous result.

We have derived important formulas to relate the exponents of eigenvalues to critical exponents. These formulas are at the core of a theoretical framework to evaluate critical exponents from the linearized renormalization group equation and its eigenvalues. It is worth remembering here that there are usually only two relevant scaling fields, i.e. positive exponents y_t and y_h, which implies that the four critical exponents $\alpha, \beta, \gamma, \delta$ are not completely independent. Knowledge of two of them is sufficient to determine the remaining two. The *scaling relations* are explicit expressions of this fact, which can be derived by eliminating y_t and y_h from eqn (3.44),

$$\boxed{\alpha + 2\beta + \gamma = 2, \quad \gamma = \beta(\delta - 1).} \tag{3.47}$$

Since these are derived from thermodynamic quantities, they represent *thermodynamic scaling relations*. The critical exponents of the mean-field theories of critical and tricritical points satisfy these scaling relations.

We note in passing that the following inequality, known as *Rushbrooke's inequality*, holds for the critical exponents $\alpha_-, \beta, \gamma_-$,

$$\alpha_- + 2\beta + \gamma_- \geq 2, \tag{3.48}$$

where α_- and γ_- are the exponents α and γ, respectively, for the low-temperature side of the critical point.[4] Rushbrooke's inequality can be proved rigorously using thermodynamics, as shown in Appendix A.3.

3.5 Scaling law for correlation functions and hyperscaling

In the previous section we have expressed the critical exponents $\alpha, \beta, \gamma, \delta$ in terms of y_t and y_h from the scaling law of the free energy. An additional scaling law for the correlation function is necessary in order to relate the critical exponents ν and η of the correlation function to y_t and y_h. Let us assume that $h = 0$. We identify the scaling field g_t with the temperature deviation from the critical point t as before. The connected correlation function $\langle S(0)S(\boldsymbol{r})\rangle - \langle S(0)\rangle\langle S(\boldsymbol{r})\rangle$ will be written as $G(\boldsymbol{r}, t)$, a

---

[4] Remember that β is defined only on the low-temperature side (ordered phase). The other critical exponents are believed to have the same values below and above the critical point, which has, however, been proved only in limited cases. An example will be given in Chapter 10.

function of distance and temperature. Assume that $G(\boldsymbol{r},t)$ depends on $\boldsymbol{r}$ only through its absolute value $r = |\boldsymbol{r}|$, independent of the direction, which is very plausible near the critical point where short-scale properties (such as the lattice anisotropy) are irrelevant.

According to the transformation rule of the spin variable S, eqn (3.7), the correlation function $G(r,t)$ acquires the factor $c(b)^2$ after a single step of a renormalization group operation since the correlation function is the average of the product of two spin variables,

$$G(r,t) = c^2(b)G(b^{-1}r, b^{y_t}t). \tag{3.49}$$

Another piece of information comes from the scaling law of the magnetization m, which is obtained by differentiation of eqn (3.37), for $n = 1$, with respect to h and then setting $h = 0$,

$$m(t,0) = b^{-d+y_h}m(b^{y_t}t, 0). \tag{3.50}$$

A comparison of this equation with a relation similar to eqn (3.49),

$$m(t,0) = c(b)m(b^{y_t}t, 0) \tag{3.51}$$

reveals that $c(b) = b^{-d+y_h}$. Then, by writing eqn (3.51) as

$$m(b^{y_t}t, 0) = b^{d-y_h}m(t,0), \tag{3.52}$$

we find the scaling dimension of the spin variable to be $d - y_h$ according to the definition of scaling dimension, eqn (3.6).

Therefore, eqn (3.49) reduces to

$$G(r,t) = b^{-2d+2y_h}G(b^{-1}r, b^{y_t}t). \tag{3.53}$$

By renormalizing n times, we obtain a relation where b is replaced by b^n in this equation. For $t \neq 0$, we can choose n such that $b^{ny_t}t = 1$ holds as in the previous section, and then the following scaling law for the correlation function is derived,

$$G(r,t) = t^{2(d-y_h)/y_t}\Phi(rt^{1/y_t}) \quad (T \neq T_{\mathrm{c}}). \tag{3.54}$$

If we fix t to a small but finite value and let r increase, the correlation function should decay exponentially as $\mathrm{e}^{-r/\xi}$. Since the correlation length ξ diverges proportionally to $t^{-\nu}$, the exponent r/ξ in $\mathrm{e}^{-r/\xi}$ should be proportional to rt^ν. Comparison of this fact with eqn (3.54) indicates that r appears as a product with a power of t, rt^ν in $\mathrm{e}^{-r/\xi}$ and rt^{1/y_t} in eqn (3.54). Since both of these represent the same function, $1/y_t$ should coincide with ν. We therefore conclude $\nu = 1/y_t$.

The critical exponent η determines the power of decay of the correlation function $r^{-d+2-\eta}$ exactly at the critical point. Setting $t = 0$ and $b = r$ in eqn (3.53), we have

$$G(r,0) \propto r^{-2d+2y_h} \quad (T = T_{\mathrm{c}}), \tag{3.55}$$

from which we conclude $\eta = d - 2y_h + 2$. To summarize,

$$\nu = \frac{1}{y_t}, \quad \eta = d - 2y_h + 2. \tag{3.56}$$

Notice that the space dimensionality d appears explicitly in the scaling law for the correlation function. Several scaling relations follow naturally,

$$\alpha = 2 - d\nu, \ \beta = \frac{\nu(d - 2 + \eta)}{2}, \ \gamma = \nu(2 - \eta), \ \delta = \frac{d + 2 - \eta}{d - 2 + \eta}. \tag{3.57}$$

These equations relate the exponents for singularities of the free energy, $\alpha, \beta, \gamma, \delta$, with those for the correlation function, ν and η, and are called *hyperscaling relations*. The first relation in eqn (3.57) is sometimes referred to as the *Josephson scaling relation*. In contrast to the (ordinary or thermodynamic) scaling relations (3.47), hyperscaling relations are sometimes violated. For example, the mean-field exponents do not satisfy the first relation of eqn (3.57) for $d > 4$. As will be discussed in the next chapter, hyperscaling relations may be violated when a variable named 'dangerous irrelevant variable' affects the behaviors of the free energy and correlation function in different ways.

> **EXERCISE 3.3** Let us study a first-order phase transition from the perspective of the renormalization group methodology. The ferromagnetic Ising model below the critical temperature $T < T_c$ goes through a first-order transition at $h = 0$ as one changes h across $h = 0$ at a fixed temperature. For $h > 0$, the spins align upwards and the magnetization is positive ($m > 0$), whereas for $h < 0$ down-pointing spins dominate and $m < 0$, see the left-most panel of Fig. 1.3. As explained in Section 1.4, the Ising model with $h = 0$ approaches the fixed point at $T = 0$ if the initial temperature satisfies $T < T_c$, which implies that $T = 0, h = 0$ is a fixed point stable along the temperature axis. This point determines the properties of the first-order transition and is called a *discontinuity fixed point*.
>
> Show that the exponent of renormalization y of the external field h at the discontinuity fixed point is equal to the spatial dimension d. As a hint, the correlation length does not diverge at a first-order transition, in contrast to a second-order phase transition. This means that the correlation function does not decay as a power law, from which one can infer the scaling dimension.

3.6 A simple example: One-dimensional Ising model

It is instructive to illustrate the general ideas using a simple example. We can solve the one-dimensional Ising model exactly without recourse to the renormalization group, as will be shown in Chapter 9. Nevertheless, it is well worth studying this model by the renormalization group because this is one of the rare examples in which the renormalization steps can be carried out exactly. Another reason is that we can compare the result with the exact solution to check if the prescription of the renormalization group works as expected.

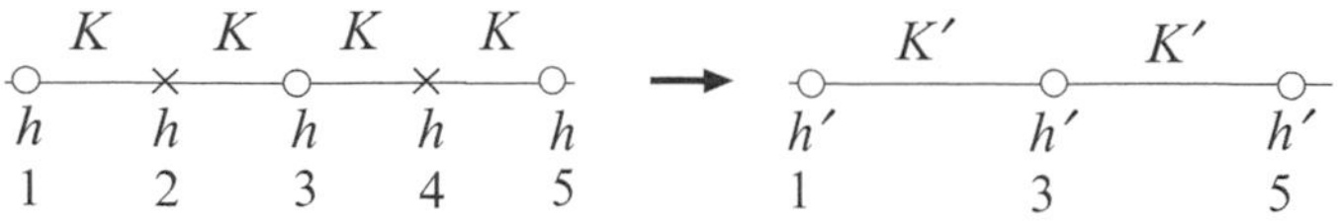

Fig. 3.5 Spin variables on even-numbered sites are traced out as a realization of the real space renormalization group.

3.6.1 Recursion relation

Let us take the sum over half of the spin degrees of freedom, that is, over the spins on even-numbered sites, as a simple realization of the renormalization group for the one-dimensional Ising model. This is one of the simplest cases of the *real-space renormalization group*. This method is also called *decimation*. The starting Hamiltonian is

$$H = -K \sum_{i=1}^{N} S_i S_{i+1} - h \sum_{i=1}^{N} S_i. \tag{3.58}$$

We assume that the inverse temperature $1/T$ is included in the Hamiltonian, and a periodic boundary condition is used, $S_{N+1} = S_1$. A transformation of scale factor $b = 2$ is realized by the summation over spin variables on even-numbered sites as in Fig. 3.5. For example, the summation over S_2 proceeds as

$$\sum_{S_2=\pm1} \exp\left(K S_2(S_1 + S_3) + h S_2\right) = e^{K(S_1+S_3)+h} + e^{-K(S_1+S_3)-h}. \tag{3.59}$$

This result can be expressed as a function of S_1 and S_3,

$$A \exp\left(K' S_1 S_3 + h_1(S_1 + S_3)\right). \tag{3.60}$$

The reason is as follows. Since eqn (3.59) is a function of S_1 and S_3, the exponent in eqn (3.60) should have the form $g(S_1, S_3)$. The expansion of this $g(S_1, S_3)$ in powers of S_1 and S_3 does not include terms of order higher than S_1, S_3 and $S_1 S_3$ (such as S_1^2 or $S_1 S_3^2$) because of the identity $S_1^2 = S_3^2 = 1$.

We defer the actual evaluation of K' and h_1 for a while and perform similar manipulations at all even-numbered sites to find

$$\sum_{S_2,S_4,\cdots} e^{-H} = \tilde{A} \exp\left(K'(S_1 S_3 + S_3 S_5 + \cdots) + (h + 2h_1)(S_1 + S_3 + \cdots)\right), \tag{3.61}$$

where $\tilde{A}$ is the product of A appearing in eqn (3.60). The coefficient $h + 2h_1$ of $(S_1 + S_3 + \cdots)$ is the renormalized field h'. The factor two in front of h_1 comes from the fact that, for example, $h_1 S_3$ emerges twice from the traces over S_2 and S_4. The remaining task is to find the forms of K' and h' as functions of K and h explicitly and to estimate the eigenvalues of the linearized recursion relation around a fixed point. Notice that in this problem the renormalization group mapping preserves the form of the Hamiltonian with renormalized parameters. In other words, the mapping does not generate new couplings as in the two-dimensional case of eqn (3.14).

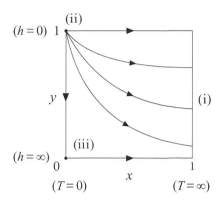

Fig. 3.6 Renormalization flow, or trajectories, of the one-dimensional Ising model.

To obtain K' and h', we equate the right-hand side of eqn (3.59) to eqn (3.60) and write four relations resulting from the four combinations, $(S_1 = \pm 1, S_3 = \pm 1)$. Actually, $(S_1 = 1, S_3 = -1)$ and $(S_1 = -1, S_3 = 1)$ give the same relation, and we find three equations, which are necessary and sufficient to determine K', h' and A as functions of K and h. Details are left to Exercise 3.4, and we just quote the result,

$$e^{4K'} = \frac{\cosh(2K + h)\cosh(2K - h)}{\cosh^2 h} \tag{3.62}$$

$$e^{2h'} = \frac{e^{2h}\cosh(2K + h)}{\cosh(2K - h)} \tag{3.63}$$

$$A^4 = 16\cosh^2 h \,\cosh(2K + h)\cosh(2K - h). \tag{3.64}$$

These are the recursion relations of the renormalization group transformation with scale factor $b = 2$ for the one-dimensional Ising model.

EXERCISE 3.4 Derive the recursion relations (3.62) to (3.64).

3.6.2   Fixed points and eigenvalues

It is necessary to find fixed points of the recursion relations (3.62) and (3.63) and evaluate the eigenvalues of the linearized relations. It will be useful to use the variables $x = e^{-4K}$ and $y = e^{-2h}$ instead of K and h since the critical temperature is at absolute zero $(K \to \infty)$ in the present one-dimensional system. Equations (3.62) and (3.63) are then expressed as

$$x' = \frac{x(1 + y)^2}{(x + y)(1 + xy)} = f_1(x, y) \,, \qquad y' = \frac{y(x + y)}{1 + xy} = f_2(x, y) \tag{3.65}$$

or in a vector notation

$$\begin{pmatrix} x' \\ y' \end{pmatrix} = \begin{pmatrix} f_1(x, y) \\ f_2(x, y) \end{pmatrix}. \tag{3.66}$$

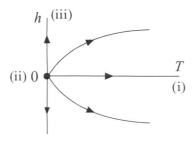

Fig. 3.7 Figure 3.6 redrawn in terms of the conventional variables as axes.

Figure 3.6 illustrates how x and y transform as $(x,y) \to (x',y') \to (x'',y'') \to \cdots$ by the above recursion relation. We find three fixed points (x^*, y^*):

(i) $x^* = 1$ and arbitrary y^* (attractive line of fixed points). This corresponds to the high-temperature limit ($K = 0$) and has no important physical significance. It represents the disordered phase.

(ii) $x^* = 0, y^* = 1$, i.e. $T = 0, h = 0$. This is a critical point (repulsive or unstable fixed point). As shown in the phase diagram of Fig. 3.7, which is drawn in terms of the conventional variables T and h instead of x and y, the ordered phase of the one-dimensional Ising model is restricted to a single point at $T = 0, h = 0$. Figure 3.6 shows that the fixed point (ii) is unstable against infinitesimal introduction of finite values of T and h. Thus, these two variables are relevant at this fixed point.

(iii) The last fixed point is at $x^* = 0$ and $y^* = 0$ or $T = 0$ and $h = \infty$. This is also uninteresting similarly to (i).

We now linearize the recursion relation (3.66) around the fixed point (ii) and evaluate the eigenvalues. Linearization near $x^* = 0, y^* = 1$ amounts to the approximation that drops second and higher orders of x and $\epsilon = 1 - y$,

$$x' \approx 4x, \quad \epsilon' = 2\epsilon. \tag{3.67}$$

Eigenvalues within this linear approximation are clearly $\lambda_t = 4, \lambda_h = 2$. Using the scaling factor $b = 2$, we obtain $y_t = 2$ and $y_h = 1$. Thus, the critical exponents are found to be $\alpha = 3/2, \beta = 0, \gamma = 1/2, \delta \to \infty, \nu = 1/2$ and $\eta = 1$. The exponent β is not actually well defined since the critical point is $T = 0$. The magnetization at $T = 0$ is $m = 1$ for any positive h, which may be considered consistent with $\delta \to \infty$ in $m \propto h^{1/\delta}$. Similarly, the correlation function is fixed to $G(r) = 1$ at the critical point $T = 0$ and is consistent with $d = 1, \eta = 1$ in $G(r) \propto r^{-d+2-\eta}$. The scaling dimension of spins is $x_h = d - y_h = 0$ since $d = y_h = 1$. This is again in agreement with $G(r) \propto r^{-2x_h} = 1$. The vanishing scaling dimension of spin variables $x_h = 0$ is a special feature of the one-dimensional system in which the magnetization jumps to a finite value at the critical point.

3.6.3 Singularities in physical quantities

Let us rewrite x and y in terms of the original variables to see the behavior of physical quantities around the critical point $T = 0, h = 0$. The magnetic susceptibility is, from $\gamma = 1/2$,

$$\chi \propto x^{-\gamma} = (\mathrm{e}^{-4K})^{-1/2} = \mathrm{e}^{2K}. \tag{3.68}$$

This is in agreement with the exact solution derived in Section 9.1.1. The specific heat diverges as

$$C \propto x^{-\alpha} = (\mathrm{e}^{-4K})^{-3/2} = \mathrm{e}^{6K}, \tag{3.69}$$

in contrast to the vanishing behavior $C \propto \mathrm{e}^{-2K}$ as $T \to 0$ of the exact solution of Section 9.1.1. The reason for this discrepancy is in the replacement $t \to x$ in $C \propto \partial^2 f/\partial t^2$ in the discussions of the scaling law of the free energy. The scaling field in the present section is $x = \mathrm{e}^{-4K}$, but the specific heat is the second order derivative of the free energy with respect to t, and not x. Hence, the correction factor $x^2 = \mathrm{e}^{-8K}$ is necessary in $\partial^2 f/\partial x^2$ to correctly reproduce the specific heat, which, in conjunction with e^{6K} in eqn (3.69), gives the correct temperature dependence $C \propto \mathrm{e}^{-2K}$.

EXERCISE 3.5 Explain why the susceptibility result (3.68) agrees with the exact solution without a correction factor, whereas the specific heat (3.69) should be supplemented by e^{-8K}.

A similar consideration holds for a general value of the scaling factor b. Suppose that we take the trace over spins, one among b of them. Assume that there is no external field, $h = 0$, for simplicity. The renormalization group equation then becomes

$$u' = u^b, \tag{3.70}$$

where $u = \tanh K$.

EXERCISE 3.6 Derive the recursion relation (3.70). Hint: Manipulations that generalize eqn (3.59) to the scaling factor b will have to be carried out, that is to take the trace over $S_2, S_3, \cdots, S_b$ to find the effective coupling constant between S_1 and S_{b+1}. It will be useful to take the trace over S_2 first, S_3 next, and so on.

Equation (3.70) shows that $T = 0$ ($u = 1$) is a fixed point. For $T > 0$, u is smaller than 1 and decreases as the renormalization process of eqn (3.70) proceeds. This means that the fixed point $u = 1$ (point (ii) Figs. 3.6 and 3.7) is unstable, as expected. The correlation length should be transformed as

$$\xi(u') = \frac{1}{b}\xi(u). \tag{3.71}$$

From $u' = u^b$, we have $\xi(u^b) = \xi(u)/b$, which is satisfied by the inverse of the logarithmic function,

$$\xi(u) = \frac{\mathrm{const}}{|\log u|} = \frac{\mathrm{const}}{|\log \tanh K|}. \tag{3.72}$$

The temperature dependence of the correlation length has thus been worked out. The above expression agrees with the exact solution to be derived in Section 9.1.1. In the low-temperature limit $K \to \infty$, the correlation length diverges exponentially due to the expansion $\tanh K \approx 1 - 2\mathrm{e}^{-2K}$,

$$\xi \approx \mathsf{const} \cdot \mathrm{e}^{2K}. \tag{3.73}$$

We have shown that we can carry out the renormalization-group calculations without approximations for the one-dimensional Ising model.

Singularities in physical quantities near the critical point usually emerge as powers of $T - T_\mathrm{c}$, as has been discussed at length in the previous sections. It often happens, however, that such a power-law singularity is replaced by an exponential law at the lower critical dimension. Since the lower critical dimension of the ferromagnetic Ising model is 1, the exponential divergences of the specific heat and magnetic susceptibility reflect this feature of the lower critical dimension. The two-dimensional XY model to be discussed in Chapter 7 also shows exponential singularities for the same reason.

It is useful to recall that the partition function acquires the factor $\tilde{A}(K, h)$ appearing in eqn (3.61) after a single step of the renormalization process. The logarithm of $\tilde{A}(K, h)$ corresponds to the regular part w in eqn (3.36). Repeating the renormalization group calculations leads to successive multiplications of this factor, $\tilde{A}(K, h)\tilde{A}(K', h') \cdots$, which reflects the change of parameters. After taking the trace over all spin variables by very many steps of renormalization group calculations, we are left with this factor only, which is simply the partition function of the whole system. This argument applies not just to the one-dimensional Ising model but to any case. The multiplicative factor does not affect critical exponents but should not be forgotten when one wishes to know the value of the free energy, the logarithm of the partition function.

3.7 Mean-field theory and scaling law

It is instructive here to analyze the mean-field theory from the standpoint of scaling and renormalization and derive explicit forms of scaling functions.

Let us consider again the Landau theory for the Ising universality class. Minimization of the Landau free energy gives a relation between m and h, the so-called mean-field equation of state, eqn (2.20),

$$2am + 4bm^3 = h. \tag{3.74}$$

To solve this equation for m, we divide both sides by $t^{3/2}$ and write $a = kt$ to find

$$\frac{m}{\sqrt{t}} + c_1 \left(\frac{m}{\sqrt{t}}\right)^3 = c_2 \cdot \frac{h}{t^{3/2}} \tag{3.75}$$

with some constants c_1 and c_2. This is a cubic equation for $m/\sqrt{t}$ and the solution has the form

$$m = \sqrt{t}\, g\left(\frac{h}{t^{3/2}}\right). \tag{3.76}$$

Now, according to the general theory developed in the previous sections, the h-derivative of eqn (3.38) yields the scaling law of magnetization,

$$m = t^{\beta} \Psi'(ht^{-\beta\delta}), \tag{3.77}$$

where use has been made of $\beta = (d - y_h)/y_t$ and $\delta = y_h/(d - y_h)$. These two equations (3.76) and (3.77) coincide if we use the mean-field exponents $\beta = 1/2, \delta = 3$. Hence, the mean-field theory satisfies the scaling law, and the scaling function $\Psi'(\cdot)$ is a solution to the cubic equation (3.75).

A similar analysis reveals the scaling law for the magnetic susceptibility. The Landau expression of the magnetic susceptibility (2.21) becomes, after using the above relation for the magnetization (3.76),

$$\chi = \frac{1}{t\left(c_3 + c_4 g^2 (ht^{-3/2})\right)}, \tag{3.78}$$

with constants c_3 and c_4. The second-order derivative of the free energy (3.38) with respect to h gives the susceptibility in its scaling form,

$$\chi = t^{-\gamma} \Psi''(ht^{-\beta\delta}). \tag{3.79}$$

These two equations (3.78) and (3.79) become compatible when the mean-field exponents $\gamma = 1, \beta = 1/2$, and $\delta = 3$ are used.

It can also be confirmed that the Landau free energy before differentiation also satisfies the scaling law (Exercise 3.7). We conclude that the mean-field theory (Landau theory) is consistent with the scaling law, as far as the free energy is concerned.

EXERCISE 3.7 Show that the Landau free energy can be written in the form of the scaling law of eqn (3.38).

EXERCISE 3.8 Is it possible to rewrite the equation of state of the mean-field theory $m = \tanh \beta(Jmz + h)$ in a form that satisfies the scaling law? If not, mention the physical reason.

3.8 Scaling dimension and scaling law

The two-point correlation function decays algebraically at the critical point with a power that is twice the scaling dimension of the spin variables $x_h \equiv d - y_h$ as in eqn (3.55),

$$G(r, 0) \propto r^{-2x_h}. \tag{3.80}$$

Such algebraic behavior is observed also in other operators. Let us consider the example of the local energy. The internal energy is given by the temperature derivative of the free energy. In the case of the ferromagnetic Ising model with only nearest-neighbor interactions, the internal energy E is proportional to the average of the local energy operator $E_{nn}(\boldsymbol{x}) \equiv S(\boldsymbol{x})S(\boldsymbol{x} + \boldsymbol{\delta})$, i.e. the product of neighboring spin operators. Here,

δ is the vector to a neighboring site. Therefore, the t-derivative of the relation (3.37) with $n = 1$ gives the scaling law of the local energy,

$$\langle E_{\mathrm{nn}}(\boldsymbol{x}) \rangle = b^{-d+y_t} \langle E_{\mathrm{nn}}(b^{y_t} \boldsymbol{x}) \rangle. \tag{3.81}$$

This equation shows that the scaling dimension of the local energy operator is $x_t = d - y_t$. This result allows us to derive the following asymptotic behavior of the energy–energy correlation function at the critical point in a manner similar to the usual spin–spin correlation function,

$$G_E(r) \equiv \langle E_{\mathrm{nn}}(\boldsymbol{x}) E_{\mathrm{nn}}(\boldsymbol{x} + \boldsymbol{r}) \rangle \propto r^{-2x_t}. \tag{3.82}$$

More generally, when the operator ψ_i corresponds to the scaling field g_i and the exponent of the eigenvalue is y_i, the scaling dimension of ψ_i is $x_i = d - y_i$. This relation is understood by g_i differentiation of the free-energy scaling law with an explicit g_i dependence,

$$f(t, h, g_i) = b^{-d} f(b^{y_t} t, b^{y_h} h, b^{y_i} g_i). \tag{3.83}$$

The relation between g_i and the average ψ_i is

$$\bar{\psi}_i \equiv \langle \psi_i \rangle = \frac{\partial f}{\partial g_i}, \tag{3.84}$$

in the same way that the internal energy relates to t through the temperature derivative of the free energy or the magnetization relates to h. We thus find

$$\bar{\psi}_i(t, h, g_i) = b^{-d+y_i} \bar{\psi}_i(b^{y_t} t, b^{y_h} h, b^{y_i} g_i), \tag{3.85}$$

which indicates that $x_i = d - y_i$. Exactly at the critical point, the correlation function decays as a power

$$\langle \psi_i(\boldsymbol{x}) \psi_j(\boldsymbol{x} + \boldsymbol{r}) \rangle \propto r^{-x_i - x_j}. \tag{3.86}$$

These discussions are valid even when g_i and g_j represent irrelevant variables.

The concept of scaling dimension makes it possible to derive scaling relations, which connect critical exponents, through dimensional analysis. Let us assume that singularities of physical quantities are caused essentially by the divergence of the correlation length $\xi \propto t^{-\nu}$. This is a reasonable assumption since the correlation length is the most important characteristic length scale, the divergence of which should affect all physical quantities. Therefore, by considering ξ to be the only fundamental physical quantity with a dimension of length, we can derive scaling relations. For instance, the free energy per degree of freedom, f, has the scaling dimension d, as was mentioned in Section 3.1, and the correlation length ξ, the standard of distance, has the same scaling dimension -1 as the length. These facts lead us to the relation between f and ξ,

$$f \propto \xi^{-d} \propto t^{\nu d}, \tag{3.87}$$

from which the hyperscaling relation $2 - \alpha = d\nu$ results. Notice here that eqn (3.87) means that the most singular term in the free energy is proportional to ξ^{-d}. The whole free energy, including the regular part, should not be considered to be directly proportional to ξ^{-d}.

3.9   Scaling and anomalous dimensions

We have already introduced the concept of scaling dimension x of a physical quantity in Section 3.1, and in the last section we related it to the scaling laws. We remind the reader that the scaling dimension defines the behavior of a physical quantity under a scale transformation. In many situations the scaling dimension of a physical quantity is simply determined by dimensional analysis. For example, consider the dimensionless quantity that is a part of the Gaussian model (2.74) with the prefactor b chosen to be 1,[5]

$$\left[\int \mathrm{d}\boldsymbol{r}(\nabla\phi(\boldsymbol{r}))^2 \right] = 1. \tag{3.88}$$

The notation $[\cdots]$ indicates the dimension of the quantity inside the square bracket, and the fact that it is dimensionless is indicated with 1 on the right-hand side. Since the dimensions of the quantities inside the integral are

$$[\mathrm{d}\boldsymbol{r}] = L^d \quad , \quad [\nabla] = L^{-1}, \tag{3.89}$$

eqn (3.88) implies that

$$L^d \cdot L^{-2} \cdot [\phi(\boldsymbol{r})]^2 = 1 \iff [\phi(\boldsymbol{r})] = L^{-(d-2)/2}, \tag{3.90}$$

where L is the unit of length (associated to the correlation length ξ). This equation defines the dimension of the physical quantity $\phi(\boldsymbol{r})$. In this way, one determines a dimension from the dimensional analysis, also known as the *canonical dimension*, that we name $d_\phi = 1 - d/2$ in this particular example. It turns out that in the Gaussian model of Section 2.9 the scaling dimension of the magnetization field $\phi(\boldsymbol{r})$ is identical to the dimension obtained from the dimensional analysis, i.e. $x_\phi = d_\phi$. As we will explain below, this is why the exponent η of the Gaussian model is zero, as in the mean-field case.

In general, the scaling dimension *is not* identical to the canonical dimension obtained from the dimensional analysis. The difference between the two is proportional to what is known as the *anomalous dimension*. For example, in a theory whose two-point correlation function of scalar fields has non-vanishing η, the anomalous dimension means that $d_\phi - x_\phi = (1 - d/2) - (d - y_\phi) = \eta/2 \neq 0$.[6] This may also happen to other critical exponents such as ν but in those cases it is customary not to give any special name or symbol to the difference.

We have already seen in Chapter 2 that mean-field theories (such as the Landau theory) share, regardless of the details, the same set of critical exponents typically represented in terms of rational numbers. An important observation is that the exponents of any mean-field theory can be obtained from considerations of simple dimensional analysis. This fact explains why those exponents are not irrational numbers. Therefore, a central task of the theory of critical phenomena is to explain and determine those anomalous dimensions, for example, by using the renormalization group method. After

---

[5] Typically, the Hamiltonian as defined in eqn (3.1) is dimensionless, i.e. $[H] = 1$, since it appears in the exponent of an exponential.

[6] Notice that y_ϕ here is identical to y_h.

all, critical behavior is dominated by fluctuations at all length scales, including lengths smaller than ξ. The argument that close to the critical point the microscopic length a may be ignored since $\xi \gg a$ is not always accurate, as will be shown below. In other words, if there were no anomalous dimensions, the whole field of critical phenomena would be quite boring since all critical exponents would acquire the mean-field values.

The reason behind the existence of anomalous dimensions can be heuristically attributed to the importance of irrelevant fields (or variables, or operators) in establishing scaling, and thus modifying exponents. For example, in Exercise 3.2 one showed that the lattice constant a is an irrelevant variable, and we said in the paragraph above that critical phenomena are dominated by fluctuations at all length scales, including the microscopic (minimal) length a or, equivalently, a given short-wavelength cutoff Λ^{-1} in the terminology of field theory (see Chapter 5). Then, it may happen that the correlation length ξ is insufficient to establish the right scaling and one also needs to take into account a, which we show below. In the two-point correlation function of eqn (3.53) the effect of the microscopic length a has been ignored since $a \ll \xi$. At the critical point, $t = 0$, $G(r, 0) \propto r^{-2d+2y_h}$. From considerations of simple dimensional analysis, on the other hand, we have seen above that $G(r, 0)$ should scale as $G(r, 0) \propto r^{-d+2}$, with the result that $\eta = 0$. Indeed, this is the exact scaling behavior of the Gaussian model, i.e. eqn (2.87), and of mean-field theories in general. However, in most of the interesting cases, $\eta \neq 0$.

Consider the modified scaling relation, which includes the effect of a, at criticality

$$G(r, a) = b^{-d+2} G(b^{-1} r, b^{-1} a), \tag{3.91}$$

and choose $b = r$. Then,

$$G(r, a) = r^{-d+2} G(1, a/r). \tag{3.92}$$

The condition $a \ll r$ does not imply that the function $G(1, x)$ is non-vanishing near $x = 0$. Close to $x = 0$ it can certainly behave as

$$G(1, x) \propto x^{\eta}, \tag{3.93}$$

with an exponent $\eta \neq 0$, implying that the two-point correlation function scales as

$$G(r, a) \propto a^{\eta} r^{-d+2-\eta}, \tag{3.94}$$

thus heuristically explaining the origin of anomalous dimensions.

3.10 Data analysis by scaling law and finite-size scaling

Scaling laws are useful to estimate critical exponents from experimental or numerical data. Let us first explain how one can extract the values of critical exponents β and δ out of experimental data of magnetization m measured near the critical temperature. A naive fit of magnetization data for very small external fields to the definition of β, $m \approx |t|^{\beta}$, is better replaced by a more systematic application of the scaling laws.

We first rewrite the scaling law for magnetization, eqn (3.77), as

$$t^{-\beta} m(t, h) = \Psi'(h t^{-\beta\delta}). \tag{3.95}$$

This equation implies that, if we know the values of β and δ, the plot of data with abscissa $ht^{-\beta\delta}$ and ordinate $t^{-\beta}m(t,h)$ will give a single curve for any h and t. More practically, for a fixed h, we scan t and plot the values of $t^{-\beta}m(t,h)$ with the above-mentioned abscissa and ordinate. Then, the same process is repeated for another fixed value of h. If the scaling law does not apply, the second plot will give a different curve from the first. However, due to the scaling law (3.95), the values of $t^{-\beta}m(t,h)$ coincide for different hs for the same value of $ht^{-\beta\delta}$. The above procedure is repeated for various values of h. In practice, we have no precise knowledge of β and δ beforehand, and it is necessary to guess these values to proceed by trial and error to find the appropriate β and δ, by which a single curve is obtained to a satisfactory precision.

Next, we show how to extract critical exponents from numerical data for the magnetic susceptibility. Critical phenomena take place in macroscopic systems, and physical quantities show singular behavior near a critical point theoretically only when the system size is infinite. We can, however, carry out numerical computations only for finite-size systems. It is therefore necessary to estimate critical exponents, which characterize singularities of infinite systems, from the data for finite-size systems. The idea of *finite-size scaling* is a well-established method for this purpose. We will see later in this section that finite-size scaling can also be interpreted as a crossover phenomenon in terms of the size of the system.

Suppose that we perform the process of renormalization group for a system on a hypercubic lattice with linear size L. The parameters of the system should be carefully tuned for the system to be at the critical point, i.e. $t = h = 0$. This condition only applies to the infinite-size system, $L \to \infty$. We interpret this last fact as the condition that the parameter L^{-1} must be tuned to 0 in addition to $t = h = 0$ to keep the system at the critical point, which implies that L^{-1} is a relevant variable. We thus include L^{-1} in the argument of the free energy and write

$$f(t,h,L^{-1}) = b^{-d}f(b^{y_t}t, b^{y_h}h, bL^{-1}). \tag{3.96}$$

It is seen that L^{-1} is a relevant variable with exponent $y_L = 1$.

To derive the finite-size scaling of the magnetic susceptibility, we differentiate eqn (3.96) twice with respect to h and set $h = 0$ to find

$$\chi(t,0,L^{-1}) = b^{2y_h-d}f_2(b^{y_t}t, 0, bL^{-1}), \tag{3.97}$$

where f_2 is the second-order partial derivative of $f(t,h,L^{-1})$ with respect to the second argument. By choosing $b = L$, we obtain the following equation for some function $\tilde{\Psi}(\cdot)$, analytic for finite L,

$$\chi(t,0,L^{-1}) = L^{2-\eta}\tilde{\Psi}(tL^{1/\nu}), \tag{3.98}$$

where we have used $2y_h - d = 2 - \eta = \gamma/\nu$ and $y_t = 1/\nu$. The scaling function $\tilde{\Psi}(\cdot)$ depends generally on the boundary conditions.

The analysis of numerical data proceeds as follows. One first fixes L and plots the data by changing t with abscissa $tL^{1/\nu}$ and ordinate $L^{\eta-2}\chi$ and repeats it for other values of L. If the presumed values of η and ν are appropriate, these plots fall on the same curve. An example is given in Fig. 3.8. As in the case of experimental data, one

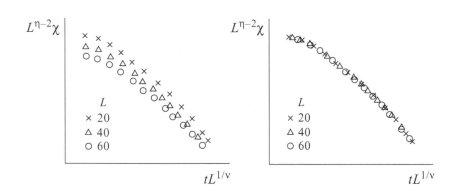

Fig. 3.8 Schematic illustration of data analysis by finite-size scaling. With inappropriate values of the critical exponents and critical points, the data for different sizes L do not lie on a single curve (left), whereas they collapse on a common curve with the correct values (right). Both axes are drawn in logarithmic scales.

adjusts the working values of η and ν by trial and error to find the best possible single curve.

Another parameter to be found in practice is the critical point T_c. Consider the behavior of the correlation length under a renormalization group transformation

$$\xi(t, L^{-1}) = b\,\xi(b^{y_t}t, bL^{-1}) = L\Phi(tL^{1/\nu}), \qquad (3.99)$$

where we chose $b = L$ and defined the scaling function $\Phi(x)$ that is regular when $x \to 0$. Then, the Taylor expansion of the scaling function Φ about $t = 0$ results in

$$\frac{\xi(t, L^{-1})}{L} = \Phi(0) + \Phi'(0)\,tL^{1/\nu} + \cdots . \qquad (3.100)$$

This relation shows that the critical temperature T_c (or $t = 0$) can be determined as the intersection point of curves $\xi(t, L^{-1})/L$ versus t for different sizes L since $\xi(t = 0, L^{-1})/L$ has no L dependence.

We notice a couple of points in relation to finite-size scaling. The magnetic susceptibility diverges at the critical point in the thermodynamic limit $L \to \infty$ but remains finite for finite-size systems. Then, where does the peak of the susceptibility appear in a finite-size system? According to eqn (3.98), the peak of the susceptibility in a finite-size system as a function of temperature coincides with the peak of the scaling function $\tilde{\Psi}(x)$, which may or may not be located at $x = 0$, i.e. $t = 0$. If the peak of $\tilde{\Psi}(x)$ is located at $x = c(\neq 0)$, the peak of the susceptibility as a function of temperature T is at $t = cL^{-1/\nu}$ (the sign of c depends on the boundary conditions). This implies that the peak is shifted from $t = 0$ for the infinite-size system to $cL^{-1/\nu}$. We sometimes call $1/\nu$ the *shift exponent* for this reason. Another point to be noticed is the height of the peak of the finite-size susceptibility, which is proportional to $L^{2-\eta}$. This observation allows us to estimate $2 - \eta$ from numerical data for the peak value.

3.11   Crossover phenomena

In some cases we should tune the values of more than two parameters to observe richer critical phenomena. An example is the tricritical point explained in Section 2.4 in the context of the Blume–Capel model. In such a case there are more than two relevant variables. In the present section we elucidate the concept of *crossover* between two different types of critical phenomena that typically takes place when three or more relevant variables compete. In other words, the phenomenon of crossover happens when more than one critical fixed point appears in the phase diagram.

Suppose that a system has two or more relevant variables. For example, the Heisenberg model with a uniaxial anisotropy (e.g. due to crystal fields)

$$H = -J \sum_{\langle ij \rangle} \boldsymbol{S}_i \cdot \boldsymbol{S}_j - D \sum_i (S_i^z)^2 \tag{3.101}$$

shows critical behavior of the Heisenberg universality type for $D = 0$. At high temperatures the system is in a paramagnetic (disordered) phase, and as one lowers the temperature the system orders. For $D \neq 0$ the anisotropy explicitly breaks the global rotational symmetry of the pure Heisenberg model in which simultaneous rotation of all spins keeps the inner product $\boldsymbol{S}_i \cdot \boldsymbol{S}_j$ invariant. Now, if $D > 0$, the anisotropy energy of the second term on the right-hand side of eqn (3.101) is lower for larger $|S_i^z|$. In particular, in the limit $D \to \infty$ the system reduces to the Ising model with $S_i^z = \pm 1$ under the normalization condition $(\boldsymbol{S}_i)^2 = 1$. For finite $D(> 0)$ the critical phenomena are essentially of the Ising type. The reason is that repeated applications of the renormalization group enhance the effective value of D and the parameters flow toward an Ising-type fixed point, the point marked 'I' in Fig. 3.9. If $D < 0$, a smaller $|S_i^z|$ is stable and hence the spin variables tend to be restricted to the XY surface. Then, the system is attracted toward another, XY-type, fixed point, which controls

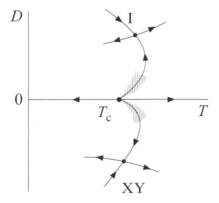

Fig. 3.9 Renormalization group flow of the anisotropic Heisenberg model with single-ion anisotropy. The Heisenberg fixed point on the $D = 0$ axis is unstable along both the t and D axes. Notice that critical and fixed points are not distinguished in this schematic diagram.

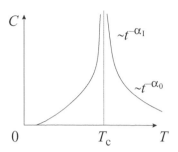

Fig. 3.10 Crossover in the specific-heat divergence when $D > 0$. The critical exponent looks different depending on the temperature range; α_0 for the Heisenberg model is observed a little away from the critical point and α_{I} for the Ising model is seen when the parameter T is very close to the critical point T_{c}. Note that the transition between regimes is not sharp.

the critical behavior for all negative D. This discussion indicates that the Heisenberg (higher symmetry) fixed point with $D = 0$ has two relevant variables $t \propto T - T_{\mathrm{c}}$ and D. The external field is of course also relevant.

When $D > 0$, the critical properties are governed by a fixed point that describes the Ising model ($\mathbb{Z}_2$ symmetry), while the critical behavior of the system with $D < 0$ will be determined by the fixed point of the XY model. In practice, however, when $|D|$ is small, these non-Heisenberg behaviors do not show up unless $|t|$ is very small.

Let us write the scaling law of the free energy near the Heisenberg fixed point for $h = 0$ as follows,

$$f(t, D) = b^{-d} f(b^{y_t} t, b^{y_D} D). \tag{3.102}$$

Both variables are relevant, $y_t > 0$ and $y_D > 0$. We may choose b to satisfy $b^{y_t} t = 1$,

$$f(t, D) = t^{d/y_t} f(1, Dt^{-y_D/y_t}) \equiv t^{2-\alpha_0} \Psi(Dt^{-\phi}), \tag{3.103}$$

where we have written α_0 for the specific-heat critical exponent of the Heisenberg model. The ratio of the two relevant exponents $\phi = y_D/y_t > 0$ is called the *crossover exponent* of anisotropy D. If D vanishes exactly, critical behavior of the Heisenberg model is observed,

$$f(t, 0) = t^{2-\alpha_0} \Psi(0). \tag{3.104}$$

For small D satisfying $|Dt^{-\phi}| \ll 1$, $\Psi(Dt^{-\phi})$ should be approximately equal to $\Psi(0)$ and then the above equation (3.104) would be a good approximation. This suggests that, for fixed D, the singularity of the specific heat would look like $t^{-\alpha_0}$ in the temperature range satisfying $|Dt^{-\phi}| \ll 1$ (i.e. t not too close to 0), see Fig. 3.10. For t smaller, $|t| < |D|^{1/\phi} = t_{\mathrm{cross}}$ with t_{cross} the *crossover temperature*, the asymptotic properties of the function $\Psi(x)$ for large x determine the behavior of $f(t, D)$. The Ising critical behavior appears for $D > 0$ and the XY-like properties dominate when $D < 0$. This is the crossover of the critical region.

The crossover exponent ϕ determines the parameter range $Dt^{-\phi} \approx 1$, where crossover takes place. The size of the crossover region is not universal. The exponent for the temperature field y_t is usually smaller than the exponent for anisotropy y_D and consequently $\phi > 1$. A small deviation from zero for D changes the critical exponent to a new value, which is observed in a temperature range around the critical point $|Dt^{-\phi}| \gg 1$. Out of this range, the critical exponent looks unaffected.

Crossover is observed generally between two relevant variables, not just between the temperature and anisotropy D. As an example, the external field h and temperature t are two prominent relevant variables and the crossover exponent is $\phi = y_h/y_t = \beta\delta$. The mean-field value of this crossover exponent is $\phi = 3/2$. Thus, zero-field ($h = 0$) critical phenomena are observed in the presence of a very small field (which can be a residual field experimentally) as long as we observe the temperature range $|ht^{-\phi}| \ll 1$. A difference from the case of anisotropy is that the external field h totally eliminates critical phenomena, and no new critical phenomena are observed.

As already mentioned, the finite-size scaling may also be regarded as a kind of crossover phenomenon. Equation (3.98) shows that the scaling function $\tilde{\Psi}(\cdot)$ should behave as $\tilde{\Psi}(tL^{1/\nu}) \approx (tL^{1/\nu})^{(\eta-2)\nu}$ for sufficiently large $tL^{1/\nu}$ because then the L dependence of the right-hand side cancels and the critical behavior $\chi \approx t^{(\eta-2)\nu} = t^{-\gamma}$ of the infinite-size system is recovered. When $tL^{1/\nu}$ is not very large, finite-size effects show up and $\chi(t, 0, L^{-1})$ has L dependence. The condition $tL^{1/\nu} \gg 1$ that finite-size effects disappear is reduced to $L \gg \xi$ because of $\xi \propto t^{-\nu}$. This is a very reasonable criterion that the system size should be much larger than the correlation length for the system to be regarded as infinitely large. The crossover exponent from finite- to infinite-size system is $\phi = \nu$ (Exercise 3.9). Crossover takes place at $|Dt^{-\phi}| \approx 1$ in the case of anisotropy. Correspondingly, the crossover region for finite-size effects is $|L^{-1}t^{-\nu}| \approx 1$, i.e. $L \approx \xi$, in agreement with the above conclusion.

EXERCISE 3.9 Show that the crossover exponent for finite-size effects is $\phi = \nu$.

3.12   Dynamic scaling law

The concept of scaling also applies to non-equilibrium systems near a critical point. It is convenient first to rewrite the scaling law of the correlation function in equilibrium (3.53) in terms of its spatial Fourier transformation,

$$\langle \tilde{S}(\boldsymbol{q})\tilde{S}(-\boldsymbol{q}) \rangle = b^{2-\eta} \langle \tilde{S}(b\boldsymbol{q})\tilde{S}(-b\boldsymbol{q}) \rangle. \tag{3.105}$$

Here, we have used the fact that the argument r on the right-hand side of eqn (3.53) has a factor b^{-1}, which leads to the additional factor of b^d after Fourier transformation, in conjunction with the relation $2y_h - d = 2 - \eta$. We have omitted the scaling field proportional to $(T - T_c)/T_c$ because the symbol t may be confused with time in the present non-equilibrium problem. Now, let us introduce time t and generalize the above relation to the *dynamic correlation function*

$$\langle \tilde{S}(\boldsymbol{q},t)\tilde{S}(-\boldsymbol{q},0)\rangle = b^{2-\eta}\langle \tilde{S}(b\boldsymbol{q},b^{-z}t)\tilde{S}(-b\boldsymbol{q},0)\rangle, \qquad (3.106)$$

where z is the scaling dimension of time t. If we write the time Fourier transformation of the left-hand side as $\tilde{C}(\boldsymbol{q},\omega)$ following the notation of Section 2.11, we have

$$\tilde{C}(\boldsymbol{q},\omega) = b^{2-\eta+z}\tilde{C}(b\boldsymbol{q},b^z\omega). \qquad (3.107)$$

The right-hand side carries the additional factor of b^z due to the factor b^{-z} in front of t on the right-hand side of eqn (3.106).

The fluctuation–dissipation theorem (2.115) indicates that the scaling behavior of the response function is different from that of the dynamic correlation function due to the factor ω.[7] The scaling law of the response function therefore reads

$$\tilde{G}(\boldsymbol{q},\omega) = b^{2-\eta}\tilde{G}(b\boldsymbol{q},b^z\omega). \qquad (3.108)$$

If we choose $b = \xi$, reducing the unwritten variable $b^{y_t}(T-T_c)/T_c$ to a constant, the *dynamic scaling law* results,

$$\tilde{G}(\boldsymbol{q},\omega) = \xi^{2-\eta}\Phi(\xi\boldsymbol{q},\xi^z\omega). \qquad (3.109)$$

This equation suggests that the typical time scale, the *relaxation time* τ_q, is given as

$$\tau_q = \xi^z g(\xi q). \qquad (3.110)$$

The mean-field relations (2.124) and (2.128) satisfy this equation and their dynamic critical exponents are confirmed to be $z = 2$ and $z = 4$, respectively.

At the critical point, we set $b = q^{-1}$ in eqn (3.108) to obtain

$$\tilde{G}(\boldsymbol{q},\omega) = q^{\eta-2}\Phi(q^{-z}\omega). \qquad (3.111)$$

The relaxation time is proportional to a power of the wave number, $\tau_q \propto q^{-z}$.

---

[7] Although eqn (2.115) was derived for a system with a single degree of freedom, results of dimensional analysis from such a system remain valid for many-body systems.

4

Implementation of the renormalization group

We have formulated the general framework of the renormalization group theory. It has been shown that the eigenvalues of a linearized recursion relation around the fixed point determine the critical exponents. It has also been elucidated that the free energy and related functions satisfy scaling laws, which are useful to analyze experimental and numerical data. We next derive explicit forms of the recursion relation and find its fixed points and eigenvalues. This chapter will discuss these topics. In contrast to the elegant general theory of the previous chapter, actual computations of fixed points and eigenvalues usually involve approximations, often crude ones, except for a very limited number of simple cases such as the one-dimensional Ising model of the previous chapter. There are established methods to systematically improve precision, but they usually need a large number of (often numerical) calculations. The scope of the present chapter is modest as we limit ourselves to basic examples. It is often difficult to present a general prescription directly applicable to practical problems that the reader may have at hand. Nevertheless, studies of well-known instances will help better understand the renormalization group in general and may provide hints to attack novel unsolved problems.

4.1 Real-space renormalization group for arbitrary dimensions

We have already discussed the simple example of the one-dimensional Ising model. When the spatial dimension of the system is two or higher, the partial trace operation generates various complicated interactions, and it rapidly becomes difficult to explicitly implement the renormalization group in terms of a tractable number of parameters. It thus becomes essential to introduce certain approximations. Unfortunately, in real-space renormalization group theory, there are no generic prescriptions applicable to arbitrary problems of practical interest to systematically improve the degree of approximation. We therefore have to devise a clever approximate method to treat the problem of interest, taking full advantage of the specific features of the problem. The following are some representative examples.

4.1.1 Partial sums and the two-dimensional Ising model

What happens if we naively trace over a part of the spin variables on the square lattice as in Fig. 3.1? Implementation of the trace operation in this situation has been carried out in Exercise 3.1 and represents a particular coarse-graining scheme. The result shows that three types of interactions are generated after the partial trace when only nearest-neighbor interactions exist initially and $h = 0$. More explicitly, if we denote the initial nearest-neighbor interaction by K and the renormalized nearest-neighbor, next-nearest-neighbor and the four-spin interaction around a plaquette (unit square) by K', K_1' and K_2', respectively, we find

$$K' = \frac{1}{4} \log \cosh 4K, \tag{4.1}$$

$$K_1' = \frac{1}{8} \log \cosh 4K, \tag{4.2}$$

$$K_2' = \frac{1}{8} \log \cosh 4K - \frac{1}{2} \log \cosh 2K. \tag{4.3}$$

We have assumed that there is no external field. It is difficult to carry out another step of renormalization that starts from these three kinds of interactions, which would lead to additional complicated interactions. A crude approximation is to consider only the nearest-neighbor interactions and neglect the next-nearest-neighbor and four-body interactions. This approximation amounts to regarding eqn (4.1) as the renormalization group equation for nearest-neighbor interactions and finding its fixed point and eigenvalue. Equations (4.2) and (4.3) are simply ignored. A justification may be that there is only one relevant variable in the absence of external field and the nearest-neighbor interactions are the most important ones to be treated seriously.

Unfortunately, this idea does not work since the fixed points are located only at $K = 0$ and $K \to \infty$. This result is in contradiction to the existence of a critical point in the two-dimensional Ising model. Note that eqn (4.1) is similar to eqn (3.62) for the one-dimensional Ising model at zero field where we know that there is no finite-temperature phase transition. We must conclude that the approximation has been too crude. It may be worth trying to ignore only the four-body interaction and keep the nearest K' and next-nearest-neighbor K_1' interactions. Both couplings K' and K_1' have the same sign and produce the net effect of aligning the spins. It turns out that some further approximations should be introduced to perform calculations that lead to renormalization group equations for these two kinds of interactions. A high-temperature assumption, that these two interactions are small and higher-order terms in the Taylor expansion are ignorable, makes it possible to explicitly write the renormalization group equations. Let us mention only the consequence of this approximation because a detailed account of the crude approximation is not very productive. The result is that the critical point is at $K_c = 0.333$ with the critical exponent $\nu = 0.64$, which are not too far away from the exact solutions of $K_c = \log(1 + \sqrt{2})/2 = 0.441, \nu = 1$ but are not particularly impressive.[1]

Another very crude and easy way of estimating the effective renormalization group equation is to note that the number of nearest- and next-nearest neighbors is the same on the square lattice. In the fully aligned situation they provide the same contribution

---

[1] From the exact exponents $y_t = 1$ and $y_h = 15/8$ one can determine the other critical exponents by using, for example, the scaling relations eqns (3.44) and (3.56) with $d = 2$.

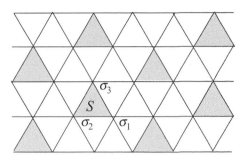

Fig. 4.1 Three Ising spins on a shaded triangle $\sigma_1, \sigma_2, \sigma_3$ ($\sigma_i = \pm 1$) are grouped into a single block spin S whose value is determined by the majority rule of eqn (4.7).

to the energy, and then one can define the effective nearest-neighbor coupling from eqns (4.1) and (4.2)

$$\tilde{K}' = K' + K_1' = \frac{3}{8} \log \cosh 4K, \tag{4.4}$$

leading to the non-trivial (unstable) fixed point $K_c = 0.507$. From

$$\left. \frac{\mathrm{d}\tilde{K}'}{\mathrm{d}K} \right|_{K_c} = 1.449 = b^{y_t}, \tag{4.5}$$

with $b = \sqrt{2}$, it results that $\nu = 1/y_t = 0.935$.

Our conclusion is that direct partial sums are generally not a method of choice for low-dimensional systems except for the one-dimensional case.

4.1.2 Block-spin transformation

The technique of *block-spin transformation* is an analytic realization of the numerical renormalization group of Figs. 1.5, 1.8 and 1.9 in Section 1.4,[2] for example, by a majority rule for three neighboring spins on the triangular lattice. In this section we are interested in applying this technique to the Ising model on the triangular lattice

$$H = -K \sum_{\langle ij \rangle} S_i S_j - h \sum_i S_i, \tag{4.6}$$

where $\langle ij \rangle$ represents bonds on the triangular lattice (see Fig. 4.1). For notational convenience, in the following, the spins of the original lattice will be denoted as σ_i instead of S_i. As shown in Fig. 4.1, we first group sites on the triangular lattice into triples on triangles and then coarse grain the system by choosing a representative single spin value for each of the triples forming a new triangular lattice with the scale factor $b = \sqrt{3}$.

---

[2] We have seen in Chapter 3 that the renormalization group mapping induces a flow in the *parameter space*. In Figs. 1.5, 1.8 and 1.9 we show the associated flow in the *configuration space* for the two-dimensional Ising model on the square lattice with $b = 3$.

If we denote the three spins to be grouped as σ_1, σ_2 and σ_3, the new block spin S takes the value according to the majority rule,

$$S = \text{sign}(\sigma_1 + \sigma_2 + \sigma_3) \left(= \frac{\sigma_1 + \sigma_2 + \sigma_3 - \sigma_1 \sigma_2 \sigma_3}{2} \right). \tag{4.7}$$

Each value of the block spin S arises from four spin-σ configurations. The set of σ configurations in the block I is denoted as $\sigma_I = \{\sigma_1, \sigma_2, \sigma_3\}$. Then, the transformation

$$\sigma_I^+ = \{-++, +-+, ++-, +++\} \rightarrow S = 1 \tag{4.8}$$

$$\sigma_I^- = \{+--, -+-, --+, ---\} \rightarrow S = -1 \tag{4.9}$$

realizes a partition of the block configuration space into two sectors each with a well-defined value of the block spin S. In the language of the previous chapter (Section 3.2) this renormalization group scheme corresponds to a weight operator $P(S, \sigma)$ like the one in eqn (3.18).

One can rewrite the Hamiltonian H in terms of intrablock, H_0, and interblock, V, interactions, $H = H_0 + V$, where

$$H_0 = -K \sum_I \sum_{\langle ij \rangle \in I} \sigma_i \sigma_j - h \sum_I \sum_{i \in I} \sigma_i, \tag{4.10}$$

$$V = -K \sum_{I \neq J} \sum_{i \in I, j \in J} \sigma_i \sigma_j, \tag{4.11}$$

with letters I and J representing block indices. Our goal is to determine the renormalized Hamiltonian $H' = R_b(H)$, and the idea is to accomplish this by dealing with V in a perturbative manner.

For a given value of S in the block (e.g. $S = 1$), let us write the interaction within a block as $H_0(S, \sigma)$ and the interaction between blocks as $V(S, \sigma)$. Then, the interaction among renormalized (block) spins $H'(S)$ is calculated as

$$e^{-H'(S)} = \sum_{\{\sigma_I^S\}} e^{-H_0(S,\sigma)} e^{-V(S,\sigma)} = \left\langle e^{-V(S,\sigma)} \right\rangle_0 \sum_{\{\sigma_I^S\}} e^{-H_0(S,\sigma)}. \tag{4.12}$$

Here, $\langle \cdots \rangle_0$ is the expectation value with respect to the weight $e^{-H_0(S,\sigma)}$,

$$\left\langle e^{-V(S,\sigma)} \right\rangle_0 = \frac{\sum_{\{\sigma_I^S\}} e^{-H_0(S,\sigma)} e^{-V(S,\sigma)}}{\sum_{\{\sigma_I^S\}} e^{-H_0(S,\sigma)}}, \tag{4.13}$$

and the sums are performed over all configurations realizing the chosen block-spin value S in all blocks, i.e. $\{\sigma_I^S\} = \sigma_1^S, \sigma_2^S, \cdots, \sigma_I^S, \cdots, \sigma_M^S$, with M the total number of blocks in the system.

The evaluation of the denominator in eqn (4.13) is straightforward after realizing that the term H_0 does not connect blocks and it simply represents a sum of independent block terms

$$\sum_{\{\sigma_I^S\}} e^{-H_0(S,\sigma)} = \prod_{I=1}^{M} Z_{\text{block}}(S_I). \tag{4.14}$$

Let us compute the partition function of a single block I

$$Z_{\text{block}}(S) = \sum_{\sigma_I^S} e^{K \sum_{\langle ij \rangle \in I} \sigma_i \sigma_j + h \sum_{i \in I} \sigma_i} = 3\, e^{-K+hS} + e^{3K+3hS}, \tag{4.15}$$

where we have used the configurations of eqns (4.8) and (4.9).

Our goal is to evaluate the transformation rule of the renormalization group (4.12) by using the coarse-graining scheme of eqn (4.7). This is actually difficult due to the non-trivial couplings between many σs. The key approximation is then to assume that the interblock interactions are weak and to only take into account the first-order contributions of V in the evaluation of the expectation value of eqn (4.13). To first order in a cumulant expansion, as described in Appendix A.4, we have

$$\langle e^{-V(S,\sigma)} \rangle_0 \approx e^{-\langle V \rangle_0}, \tag{4.16}$$

meaning that the renormalized Hamiltonian can be written as

$$H'(S) = -\sum_I \log Z_{\text{block}}(S_I) + \langle V \rangle_0 + \mathcal{O}(V^2). \tag{4.17}$$

We now consider the interblock interactions

$$V = \sum_{I \neq J} V_{IJ}, \quad V_{IJ} = -K \sum_{i \in I, j \in J} \sigma_i \sigma_j. \tag{4.18}$$

The interaction between two blocks $I = 1$ and $J = 2$ (see Fig. 4.2) turns out to be

$$\langle V_{IJ} \rangle_0 = -K \langle \sigma_{11} \sigma_{22} \rangle_0 - K \langle \sigma_{13} \sigma_{22} \rangle_0$$

$$= -2K \langle \sigma_{11} \sigma_{22} \rangle_0 = -2K \langle \sigma_{11} \rangle_0 \langle \sigma_{22} \rangle_0, \tag{4.19}$$

the latter equality resulting from the fact that H_0 does not couple different blocks. To keep track of the block index I we have replaced σ_j by the variable σ_{Ii} (where $i = 1, 2, 3$). The expectation values appearing in eqn (4.19) involve only quantities within a block and can be evaluated relatively easily. This is an advantage of the

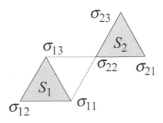

Fig. 4.2 When we take into account the interaction between blocks by a first-order perturbation calculation, the couplings among σ_{11}, σ_{13} and σ_{22} (the first index is the block index, while the second represents the position inside the block) determine the effective interblock interaction between S_1 and S_2.

first-order approximation. Here, $\langle\sigma_{11}\rangle_0$, $\langle\sigma_{13}\rangle_0$, and $\langle\sigma_{22}\rangle_0$ are expectation values with fixed values of S_1 and S_2,

$$\langle\sigma_{Ii}\rangle_0 = \frac{1}{Z_{\mathsf{block}}(S_I)} \sum_{\sigma_I^{S_I}} \sigma_{Ii}\, \mathrm{e}^{K(\sigma_{I1}\sigma_{I2}+\sigma_{I1}\sigma_{I3}+\sigma_{I2}\sigma_{I3})+h(\sigma_{I1}+\sigma_{I2}+\sigma_{I3})}. \tag{4.20}$$

As an example, let us show the explicit formula for $\langle\sigma_{11}\rangle_0$. When $S_1 = 1$, the allowed configurations for the block are σ_1^+ of eqn (4.8), which leads to

$$\langle\sigma_{11}\rangle_0 = \frac{\mathrm{e}^{-K+h} + \mathrm{e}^{3K+3h}}{3\,\mathrm{e}^{-K+h} + \mathrm{e}^{3K+3h}}. \tag{4.21}$$

The other case of $S_1 = -1$ requires the configurations σ_1^-. The end result can be summarized as

$$\langle\sigma_{11}\rangle_0 = S_1 \frac{\mathrm{e}^{-K+hS_1} + \mathrm{e}^{3K+3hS_1}}{3\,\mathrm{e}^{-K+hS_1} + \mathrm{e}^{3K+3hS_1}}. \tag{4.22}$$

Similarly, $\langle\sigma_{12}\rangle_0 = \langle\sigma_{11}\rangle_0 = \langle\sigma_{13}\rangle_0$ because of the equivalence of three spins σ_{11}, σ_{12} and σ_{13} within the block.

Equation (4.22) needs some additional algebraic manipulations since the renormalized block spin S_1 appears in the exponent. One would like to write $\langle\sigma_{Ii}\rangle_0$ as some expression linear in the block spin, i.e. $\langle\sigma_{Ii}\rangle_0 = A + BS_I$, and this is always possible. Simple algebra shows that eqn (4.22) can be written in this way with

$$A = \frac{1}{2}\left(\langle\sigma_{11}^+\rangle_0 + \langle\sigma_{11}^-\rangle_0\right) \tag{4.23}$$

$$B = \frac{1}{2}\left(\langle\sigma_{11}^+\rangle_0 - \langle\sigma_{11}^-\rangle_0\right), \tag{4.24}$$

where $\langle\sigma_{11}^\pm\rangle_0$ represents $\langle\sigma_{11}\rangle_0$ for $S_I = \pm 1$. Similarly, one needs to rewrite the partition function of the block, eqn (4.15), in the form $Z_{\mathsf{block}}(S_I) = \mathrm{e}^{C+DS_I}$ with

$$C = \frac{1}{2}(\log Z_{\mathsf{block}}(+) + \log Z_{\mathsf{block}}(-)) \tag{4.25}$$

$$D = \frac{1}{2}(\log Z_{\mathsf{block}}(+) - \log Z_{\mathsf{block}}(-)). \tag{4.26}$$

One can then write the resulting renormalized Hamiltonian of eqn (4.17) to linear order in V as

$$H'(S) = -MC - D\sum_I S_I - 2K\sum_{\langle IJ\rangle}(A+BS_I)(A+BS_J), \tag{4.27}$$

and from this expression obtain the renormalization group equation

$$\begin{pmatrix} K' \\ h' \end{pmatrix} = \begin{pmatrix} 2K(B(K,h))^2 \\ D(K,h) + 12KA(K,h)B(K,h) \end{pmatrix}, \tag{4.28}$$

where the number 12 has been derived as $12 = 2 \times z$ with $z = 6$ the coordination number of the triangular lattice.[3]

For $h = 0$ the effective coupling K' between blocks reads

$$K' = 2K \left(\frac{e^{-K} + e^{3K}}{3 \, e^{-K} + e^{3K}} \right)^2 . \tag{4.29}$$

The critical fixed point is at $K^* = \frac{1}{4} \log(1 + 2\sqrt{2}) \approx 0.336$, which is not very close to the exact critical point for the Ising model on the triangular lattice, $K_c(= 1/T_c) = \frac{1}{4} \log 3 \approx 0.275$, but is also not too far away. Linearization of the renormalization group equation (4.28) around the fixed point and the fact that

$$\left. \frac{\mathrm{d}K'}{\mathrm{d}h} \right|_{K^*,h^*} = \left. \frac{\mathrm{d}h'}{\mathrm{d}K} \right|_{K^*,h^*} = 0, \tag{4.30}$$

lead to the eigenvalues of the relevant operators,

$$\left. \frac{\mathrm{d}K'}{\mathrm{d}K} \right|_{K^*,h^*} = b^{y_t} \, , \quad \left. \frac{\mathrm{d}h'}{\mathrm{d}h} \right|_{K^*,h^*} = b^{y_h}, \tag{4.31}$$

with $h^* = 0$, giving finally the critical exponent $\nu = 1.13$ as shown in Exercise 4.1. This value is fairly close to the exact solution $\nu = 1$. One also obtains the exponent $y_h = 2.034$, which is to be compared to the exact $y_h = 15/8$. We may regard this result as a relatively satisfactory one in consideration of the crude approximations involved in the above manipulations as well as the relative compactness of the calculations in comparison with the derivation of the exact solution to be discussed in Section 9.5.

> **EXERCISE 4.1** Compute the fixed point and eigenvalue around the fixed point for the renormalization group equation (4.29). Confirm that the critical exponent ν is 1.13.

Three comments are in order. The first emphasizes the fact that both of the Ising models on the square and triangular lattices share the same set of exact critical exponents, which means that they belong to the same two-dimensional Ising universality class. The critical temperature, on the other hand, is a non-universal quantity. The exact critical temperature (in units of J) for the triangular lattice is higher than that for the square lattice, i.e. $T_c^{\triangle} = 4/\log 3 > T_c^{\square} = 2/\log(1 + \sqrt{2})$, which is consistent with the fact that the coordination number z of the triangular lattice is larger than the one for the square lattice and so in agreement with the mean-field situation: Remember that the mean-field value in units of J is $T_c = z$. The second comment concerns the systematic improvement of this cumulant expansion approach to renormalization. One could proceed to the next order of approximation by including the second cumulant, and then higher-order cumulants. The convergence in general is non-uniform. Finally, we mention that a simple way to proceed to lead to quite accurate results consists of dealing with larger blocks known as the cluster method. For example, we could have chosen as our block two triangles (six spins) instead of a single one.

---

[3] Do not confuse this z with the dynamic critical exponent of Chapter 2.

4.1.3   Migdal–Kadanoff renormalization group

Another approximate real-space renormalization group method often used in practice is the *Migdal–Kadanoff renormalization group* (MKRG). A main technical problem in the real-space renormalization group method is the difficulty in taking partial sums except for the one-dimensional case. To avoid this problem, let us adopt an approximation that ignores part of the interactions, as illustrated in Fig. 4.3 for $b = 2$. To partly compensate for possible errors caused by this approximation, we multiply the remaining interaction by the factor $b(= 2)$. Then, the spins denoted by crosses in the middle panel of Fig. 4.3 can be traced out easily as in the one-dimensional case. The remaining spins shown in black dots have only nearest-neighbor interactions. The type of interactions, nearest-neighbor only, is kept intact and only the strength of the interactions has been renormalized from K to K'. Then, we are able to repeat the renormalization group calculations for the single variable K to find the fixed point and eigenvalues.

Let us show explicit formulas to realize the MKRG for $b = 2$. The remaining interactions in the middle panel of Fig. 4.3 have a strength of $2K$. The partial trace over the spins shown as crosses can be carried out as in the one-dimensional case, see eqn (3.70), and $\tanh 2K$ is squared,

$$\tanh K' = (\tanh 2K)^2, \tag{4.32}$$

where K' is the renormalized coupling. Using the notation $u = \tanh K, u' = \tanh K'$, the above equation is expressed as

$$u' = \left(\frac{2u}{1 + u^2} \right)^2. \tag{4.33}$$

A non-trivial (critical) fixed point u_0 is easily found at $u_0 = 0.296$, namely $T_c = 3.28$. The exact critical point for the ferromagnetic Ising model on the square lattice is $T_c = 2/\log(1 + \sqrt{2}) = 2.269$, and the result $T_c = 3.28$ is not very impressive. To evaluate the critical exponent, we linearize eqn (4.33) around the fixed point, $u - u_0 = \epsilon \ll 1$.

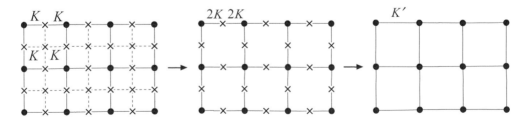

Fig. 4.3 Steps to realize the MKRG for $b = 2$. The interactions shown as dashed lines in the left panel are ignored (set to zero) and the other interaction strengths (couplings) are doubled (middle panel). Then, the spins shown as crosses are traced out to yield the renormalized coupling K' (right panel).

From

$$u_0 + \epsilon' = \left(\frac{2(u_0 + \epsilon)}{1 + (u_0 + \epsilon)^2} \right)^2, \tag{4.34}$$

we find

$$\epsilon' = \frac{8u_0(1 - u_0^2)}{(1 + u_0^2)^3} \epsilon = 2^{y_t} \epsilon. \tag{4.35}$$

Then, the critical exponent is $y_t = 0.747$ or $\nu = 1.338$. The exact value is $\nu = 1$.

For general b, eqn (4.32) generalizes to

$$\tanh K' = (\tanh bK)^b. \tag{4.36}$$

If we could carry out renormalization group calculations without approximations, the results on the fixed point and critical exponents would not depend on the scaling factor b. In practice, approximations lead to b-dependent answers. In the MKRG, approximations are considered less crude for b closer to unity because the number of ignored interactions is smaller when b is closer to 1. Thus, it seems a good approximation to set $b = 1 + \epsilon$ in eqn (4.36) and drop orders higher than the first in ϵ to find

$$K' = K - \epsilon\beta(K), \quad -\beta(K) = K + \cosh K \sinh K \log(\tanh K). \tag{4.37}$$

If we set $db = \epsilon$, the above equation is rewritten as a differential equation,

$$\frac{dK}{db} = -\beta(K). \tag{4.38}$$

The function $\beta(K)$ on the right-hand side is the beta function of the renormalization group. This name 'beta function' is used also for generic cases, not just for the MKRG as we saw in Section 3.2. As depicted in Fig. 4.4, the zero of the beta function $\beta(K) = 0$ is identified with the fixed point. The coupling K increases in the range above the fixed point and decreases below. The fixed point of eqn (4.37), $K^* = 0.4407$, turns out to coincide with the exact solution derived in Sections 9.5 and 10.1, not just as a number but as the analytical expression.

The eigenvalue b^{y_t} of the linearized renormalization group mapping is the derivative of the recursion relation at the fixed point as one sees in eqns (4.33)–(4.35). In terms of the present notation, it is the derivative of $-\beta(K)\epsilon$ at the fixed point. Since $-\beta'(K^*)\epsilon$ corresponds to ϵy_t in $b^{y_t} = (1 + \epsilon)^{y_t} \approx 1 + \epsilon y_t$, we find

$$y_t = -\left. \frac{d\beta}{dK} \right|_{K^*} = 0.7535. \tag{4.39}$$

This means $\nu = 1/y_t = 1.327$. The infinitesimal MKRG for the square lattice gives the exact critical point but the associated critical exponent turns out to be approximate.

Although the MKRG may appear to be an unsophisticated approximation, it is known that this method with integer b actually gives exact solutions for a special class of lattices called *hierarchical lattices*.

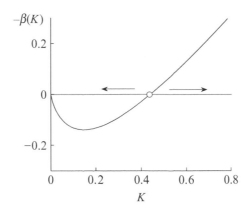

Fig. 4.4 The beta function of the infinitesimal MKRG on the square lattice. The small circle denotes the fixed point. The arrows indicate the directions of the renormalization flow of K.

In general, a problem with the real-space renormalization group method is that the limit of applicability of the approximations involved is not clear in most cases. We therefore are not certain beforehand how far we may trust the result. Thus, we should be careful in applying the real-space renormalization group to problems for which we do not have an idea about the correct result derived by other methods such as numerical simulations or physical intuition. Nevertheless, it is sometimes important to have a way to reach a solution, albeit approximate, when other methods are not easily employed for whatever reasons.

4.2 Momentum-space renormalization group: $\epsilon = 4 - d$ expansion

The accuracy of the block-spin transformation deteriorates as the spatial dimension d increases. For instance, imagine that neighboring 3×3 spins on the square lattice are represented by a single block spin, assuming values ± 1, determined by the majority rule. The same block-spin operation in three dimensions involves $3^3 = 3 \times 3 \times 3$ spins. In a general d-dimensional case a block spin having values ± 1 replaces 3^d spins by the simple majority rule. The range of the sum for 3^d spins, $-3^d, -3^d + 2, -3^d + 4, \cdots, 3^d$, increases with d, and the approximation to assign ± 1 to a block spin becomes more and more inappropriate as d increases. This consideration suggests to use continuous-spin variables (i.e. fields), as in the Landau theory, defined over a continuous space. These spin variables can in principle take any value with more probable values around ± 1. We therefore discuss the properties of a continuous model near four dimensions in the present section. More details about field-theoretic descriptions of statistical systems will be developed in Chapter 5.

4.2.1 Gaussian fixed point

A standard model with continuous spin variables (fields) in continuous space is the ϕ^4 *model* or *the Landau–Ginzburg–Wilson model,*

$$H = \int d\boldsymbol{r} \left\{ \left(\nabla\phi(\boldsymbol{r})\right)^2 + t\phi(\boldsymbol{r})^2 + u\phi(\boldsymbol{r})^4 - h\phi(\boldsymbol{r}) \right\}, \tag{4.40}$$

obtained from eqn (2.74) in Section 2.9 with an additional quartic term and an external field.[4] We will show in Chapter 5 that this model serves as an effective field theory for the critical Ising model near four dimensions. The Gaussian model of Section 2.9 appropriately describes the disordered phase when $T > T_c$, but fails to describe the ordered phase for $T < T_c$. To describe the latter phase one needs to have $u \neq 0$.

Let us write the Hamiltonian renormalized by the scale factor b as

$$H' = \int d\boldsymbol{r}' \left\{ \left(\nabla'\phi'(\boldsymbol{r}')\right)^2 + t'\phi'(\boldsymbol{r}')^2 + u'\phi'(\boldsymbol{r}')^4 - h'\phi'(\boldsymbol{r}') \right\}. \tag{4.41}$$

We now check the consequences of the scale invariance of the system, i.e. the equivalence of eqns (4.41) and (4.40). It is useful to recall the transformation rules,

$$\boldsymbol{r}' = b^{-1}\boldsymbol{r}, \quad \nabla' = b\nabla, \quad \phi' = b^{d-y_h}\phi, \quad t' = b^{y_t}t, \quad u' = b^{y_u}u, \quad h' = b^{y_h}h. \tag{4.42}$$

If we express the first term on the right-hand side of eqn (4.41) by the original variables using eqn (4.42), the factor $b^{-d+2+2d-2y_h}$ appears. The requirement of scale invariance then suggests $2 + d - 2y_h = 0$, or $y_h = d/2 + 1$. Similarly, scale invariance of the second and third terms leads to $y_t = 2$ and $y_u = 4 - d = \epsilon$. The fourth term automatically remains invariant. Hence, scaling fields transform as

$$t' = b^2 t, \quad u' = b^{4-d}u, \quad h' = b^{d/2+1}h. \tag{4.43}$$

An immediate consequence is that the quartic term is irrelevant for $d > 4$ (and becomes relevant for $d < 4$). We are thus justified to ignore the u term and discuss the Gaussian model if $d > 4$. The above recursion relations can also be written in a differential form

$$\frac{dt}{d\tau} = 2t, \quad \frac{dh}{d\tau} = \left(1 + \frac{d}{2}\right)h, \tag{4.44}$$

with $\tau = \log b$.

EXERCISE 4.2 Show that the sixth-order term $v\phi(\boldsymbol{r})^6$, if added to the ϕ^4 model, would be more irrelevant than the quartic term for $d > 4$ in the sense that it decreases more rapidly than the quartic term.

Setting $u = 0$ in the renormalization group equation (4.43), we find the *Gaussian fixed point* at $t^* = h^* = 0$. Therefore, the fixed-point Hamiltonian is

$$H^* = \int d\boldsymbol{r} \left(\nabla\phi(\boldsymbol{r})\right)^2. \tag{4.45}$$

The critical exponents around this fixed point are derived from the eigenvalues $y_t = 2, y_h = d/2 + 1$ as

$$\alpha = 2 - \frac{d}{2}, \quad \beta = \frac{d-2}{4}, \quad \gamma = 1, \quad \delta = \frac{d+2}{d-2}, \quad \nu = \frac{1}{2}, \quad \eta = 0. \tag{4.46}$$

[4] The symbol F in Chapter 2 and the present H represent the same quantity.

The values for the exponents γ, ν and η in this list agree with the mean-field prediction, but the others α, β, and δ coincide with the mean-field theory only at $d = 4$. The Gaussian fixed point indeed describes critical phenomena of the mean-field type for $d > 4$. The reason for the superficial deviation of α, β and δ, all of which include d in eqn (4.46), from the mean-field values is now explained.

Let us write the scaling law of magnetization (3.52), when $h = 0$, with the variable u (irrelevant for $d > 4$) explicitly included as

$$m(t, u) = b^{1-d/2} m(b^2 t, ub^{4-d}). \tag{4.47}$$

We may choose b arbitrarily, thus set $b = t^{-1/2}$ to reduce the first argument of the right-hand side to unity,

$$m(t, u) = t^{(d-2)/4} m(1, ut^{(d-4)/2}). \tag{4.48}$$

If we simply ignore u since it is irrelevant for $d > 4$, we have

$$m(t, 0) = t^{(d-2)/4} m(1, 0) \propto t^{(d-2)/4}, \tag{4.49}$$

which gives the same β as in eqn (4.46). The problem is that $m(1, u)$ actually behaves like $u^{-1/2}$ as $u \to 0$, which precludes us from simply ignoring $m(1, 0)$ as an unimportant constant. We check this last fact as follows.

The minimization condition of the Landau free energy, $2tm + 4um^3 = 0$, for the case of a spatially uniform order parameter $\phi(\boldsymbol{r}) = m$, leads to $m \propto u^{-1/2}$. We then insert

$$m(1, ut^{(d-4)/2}) \propto \left(ut^{(d-4)/2} \right)^{-1/2} \tag{4.50}$$

into eqn (4.48) for small u to find

$$m(t, u) \propto t^{(d-2)/4} u^{-1/2} t^{-(d-4)/4} = u^{-1/2} t^{1/2}, \tag{4.51}$$

which shows the correct mean-field value $\beta = 1/2$. We can verify $\alpha = 0$ and $\delta = 3$ from a similar argument. A variable u of this nature is called a *dangerous irrelevant variable* when $d > 4$ and $T < T_{\rm c}$.

EXERCISE 4.3 Show that the critical exponent δ assumes the mean-field value 3 if we take into account the dangerous irrelevant variable.

Let us expose the origin of these dangerous irrelevant variables. In eqn (3.37) we assumed that the irrelevant fields $g_3, g_4, \cdots$ can be dropped from the scaling analysis. Implicit in that fact was the assumption that the (singular part of the) free energy can be Taylor expanded in those irrelevant variables near the critical fixed point $t^* = 0, h^* = 0$. Consider only one irrelevant field g_3, with exponent $y_3 < 0$, then

$$f(t, h, g_3) = b^{-d} f(b^{y_t} t, b^{y_h} h, b^{y_3} g_3). \tag{4.52}$$

If one chooses $b^{y_t}t = 1$, it leads to

$$f(t, h, g_3) = t^{d/y_t} f(1, ht^{-y_h/y_t}, g_3 t^{|y_3|/y_t}) \equiv t^{d/y_t} \Psi(ht^{-y_h/y_t}, g_3 t^{|y_3|/y_t})$$

$$= t^{d/y_t} \left(\Psi(ht^{-y_h/y_t}, 0) + \Psi'(ht^{-y_h/y_t}, 0)g_3 t^{|y_3|/y_t} + \cdots \right), \qquad (4.53)$$

where $\Psi'(\cdot, \cdot)$ stands for the partial derivative with respect to the second argument. This equation shows that the main effect of the irrelevant field is to provide *correction terms to scaling*. However, there are situations, e.g. the variable u in the ϕ^4 model, where such a Taylor expansion is impossible and the field g_3 becomes a dangerous irrelevant variable. In our present ϕ^4 model the dangerous irrelevant variable affects the scaling law of the free energy and its derivatives but not the correlation functions. This leads to the breakdown of hyperscaling like $\alpha = 2 - d\nu$ that connects exponents related to the free energy with those for the correlation function. This is an exceptional situation and, while we may usually ignore irrelevant variables, care must always be exercised.

The reader may wonder if it is legitimate to use the Landau theory to derive the asymptotic form $u^{-1/2}$ of $m(1, u)$ as $u \to 0$ because it amounts to using the Landau (mean-field) theory itself to show that the critical exponent is of the mean-field type, $\beta = 1/2$. It therefore makes sense to verify that the magnetization behaves as $u^{-1/2}$ for small u without recourse to the Landau theory. Since magnetization is the expectation value of the spin variable $\phi(\boldsymbol{x})$, magnetization for the ϕ^4 theory is

$$m = \frac{\displaystyle\int \left(\prod_{\boldsymbol{y}} d\phi(\boldsymbol{y}) \right) \phi(\boldsymbol{x}) \exp\left(-\int d\boldsymbol{z} \left\{ (\nabla\phi)^2 + t\phi(\boldsymbol{z})^2 + u\phi(\boldsymbol{z})^4 - h\phi(\boldsymbol{z}) \right\} \right)}{\displaystyle\int \left(\prod_{\boldsymbol{y}} d\phi(\boldsymbol{y}) \right) \exp\left(-\int d\boldsymbol{z} \left\{ (\nabla\phi)^2 + t\phi(\boldsymbol{z})^2 + u\phi(\boldsymbol{z})^4 - h\phi(\boldsymbol{z}) \right\} \right)}. \qquad (4.54)$$

We change the integration variable as $\phi \to u^{-1/2}\phi$ to obtain

$$m = \frac{\displaystyle\int \left(\prod_{\boldsymbol{y}} d\phi(\boldsymbol{y}) \right) \phi(\boldsymbol{x}) \exp\left(-\frac{1}{u}\int d\boldsymbol{z} \left\{ (\nabla\phi)^2 + t\phi(\boldsymbol{z})^2 + \phi(\boldsymbol{z})^4 - u^{1/2}h\phi(\boldsymbol{z}) \right\} \right)}{u^{1/2} \displaystyle\int \left(\prod_{\boldsymbol{y}} d\phi(\boldsymbol{y}) \right) \exp\left(-\frac{1}{u}\int d\boldsymbol{z} \left\{ (\nabla\phi)^2 + t\phi(\boldsymbol{z})^2 + \phi(\boldsymbol{z})^4 - u^{1/2}h\phi(\boldsymbol{z}) \right\} \right)}.$$

$$(4.55)$$

Apparently, we may drop the external-field term in the limit $u \to 0$, while h is kept infinitesimally small but positive. Then, the asymptotic form of the magnetization as $u \to 0$ is determined by the saddle-point method due to the large factor $1/u$. The extremal values of the exponential parts of the integrands of the denominator and numerator cancel out. Since the saddle-point equation is written only in terms of the variable ϕ without u, the saddle-point value of $\phi(\boldsymbol{x})$ is independent of u. This means that the saddle-point value of $\phi(\boldsymbol{x})$ in front of the exponential function in the numerator does not depend on u. Consequently, the asymptotic form of m as $u \to 0$ is not influenced by the ratio of the integrals, implying the desired result $m \propto u^{-1/2}$.

4.2.2   Expansion from four dimensions

The quartic term is relevant below four dimensions. We therefore have to find a non-Gaussian fixed point and study the renormalization flow around it. A standard method to study critical phenomena is the ϵ *expansion*, in which we expand critical exponents around four dimensions as a power series in terms of $\epsilon = 4 - d$, i.e. near the upper critical dimension.

When the quartic term is relevant, the renormalization group recursion relation of the ϕ^4 model, for t and u, has a fixed point with $t^* \neq 0$ and $u^* \neq 0$. The recursion relation should have a little different form from the corresponding one near the Gaussian fixed point (4.43). We may, nevertheless, expect that the effects of u are not too large near four dimensions and thus a cumulant expansion in the quartic term would be useful. Actual manipulations to derive the explicit recursion relations are delegated to Appendix A.5. The result is written in a differential form,

$$\frac{dt}{db} = 2t + \frac{3c}{t + \Lambda^2} u \tag{4.56}$$

$$\frac{du}{db} = \epsilon u - \frac{9c}{(t + \Lambda^2)^2} u^2, \tag{4.57}$$

for infinitesimal $b - 1 = db$ with Λ and c representing positive constants, the former being the cutoff in the momentum-space integration. The Gaussian fixed point, $t^* = u^* = 0$, continues to be a trivial fixed point of these equations. But now a non-Gaussian fixed point emerges, also determined from the zeros of the beta function (which is analytic), and satisfies

$$t^* = -\frac{3c}{2(t^* + \Lambda^2)} u^*, \quad u^* = \frac{(t^* + \Lambda^2)^2}{9c} \epsilon. \tag{4.58}$$

Insertion of the second relation into the first leads to

$$t^* = -\frac{t^* + \Lambda^2}{6} \epsilon, \tag{4.59}$$

from which we obtain

$$t^* = -\frac{\epsilon}{6 + \epsilon} \Lambda^2 \approx -\frac{\Lambda^2}{6} \epsilon, \tag{4.60}$$

the last relation being valid for small ϵ. We also find

$$u^* = \frac{4\epsilon}{(6 + \epsilon)^2} \frac{\Lambda^4}{c} \approx \frac{\Lambda^4}{9c} \epsilon \tag{4.61}$$

by ignoring $\mathcal{O}(\epsilon^2)$ terms. Note that the non-trivial fixed point depends on the cutoff Λ. However, we will see below that the critical exponents are independent of the cutoff and only depend upon ϵ.

Linearization of eqns (4.56) and (4.57) by writing $t = t^* + \delta t$, $u = u^* + \delta u$ results in, to lowest order in ϵ,

$$\frac{\mathrm{d}(\delta t)}{\mathrm{d}b} = 2(t^* + \delta t) + 3c(u^* + \delta u) \cdot \Lambda^{-2} \cdot \left(1 - \frac{t^* + \delta t}{\Lambda^2}\right)$$

$$= \left(2 - \frac{\epsilon}{3}\right)\delta t + \frac{c}{\Lambda^2}\left(3 + \frac{\epsilon}{2}\right)\delta u + \mathcal{O}(\epsilon^2, \delta t \delta u) \tag{4.62}$$

$$\frac{\mathrm{d}(\delta u)}{\mathrm{d}b} = \epsilon(u^* + \delta u) - 9c(u^* + \delta u)^2 \cdot \Lambda^{-4} \cdot \left(1 - 2\frac{t^* + \delta t}{\Lambda^2}\right)$$

$$= -\epsilon \delta u + \mathcal{O}(\epsilon^2, \epsilon \delta t \delta u, (\delta u)^2). \tag{4.63}$$

These equations imply $y_t = 2 - \epsilon/3$ and $y_u = -\epsilon$ because, for example, $\delta u' = b^{y_u}\,\delta u$ for $b = 1 + \mathrm{d}b$ leads to

$$\frac{\mathrm{d}(\delta u)}{\mathrm{d}b} = y_u\,\delta u, \tag{4.64}$$

to be compared with eqn (4.63). Notice that we may ignore the off-diagonal term $c(3 + \epsilon/2)\delta u/\Lambda^2$ in eqn (4.62) since this term does not contribute to the eigenvalues of the linearized transformation. Remember that for the Gaussian fixed point $y_t = 2$ and $y_u = \epsilon$, as can be verified from the equations above when one uses $t^* = u^* = 0$.

The result $t^* = -(\Lambda^2/6)\epsilon$, $u^* = (\Lambda^4/9c)\epsilon$ means that the Gaussian fixed point destabilizes below four dimensions and a non-Gaussian fixed point $t^* < 0$, $u^* > 0$ emerges. Figure 4.5 illustrates this situation in the t–u plane. The eigenvalue corresponding to the external field can be evaluated in essentially the same manner, and

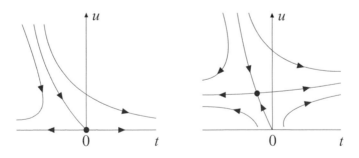

Fig. 4.5 The left panel is for the case $d > 4$ and has a set of points (critical surface) that are attracted to the Gaussian fixed point at $t^* = u^* = 0$. All points on this critical surface (represented as a line in the figure) flow toward the Gaussian fixed point and have the standard critical exponents of the mean-field theory. This is the universality of critical exponents stated in terms of the renormalization group. The situation changes in the right panel when $d < 4$, in which a non-Gaussian fixed point attracts points on the critical surface. The external field h is set to zero in both cases and the system has only one relevant variable corresponding to the temperature. When $d < 4$ the Gaussian fixed point is unstable along the two directions.

the resulting exponent is $y_h = 3 - \epsilon/2$ to leading order in ϵ. The critical exponents are then written explicitly to first order in ϵ as, for instance, $\nu = 1/(2 - \epsilon/3) \approx 1/2 + \epsilon/12$ and, to $\mathcal{O}(\epsilon)$, $\eta = d + 2 - 2y_h = d + 2 - (6 - \epsilon) = 0$, which means that $\eta = \mathcal{O}(\epsilon^2)$.

The above discussions apply to systems with a one-component $(n = 1)$ order parameter, the Ising model being the typical example. A similar theory has been developed for multicomponent cases like the XY and Heisenberg models with their critical exponents evaluated below four dimensions as series expansions in powers of ϵ.

Summarizing, we have sketched a theoretical framework to systematically derive the deviations of critical exponents from their mean-field values in powers of $\epsilon = 4 - d$, i.e. near the upper critical dimension. As long as the trivial fixed-point constitutes a reasonable starting point and one has enough information about it, the ϵ expansion represents a useful tool. These expansions are actually not Taylor-series expansions but asymptotic expansions. According to the properties of the asymptotic expansion explained in Appendix A.1, we may expect to have good estimates of the exponents for small ϵ by truncating the expansions at an appropriate order. Indeed we find good agreement between the values obtained by the ϵ expansion to second order (with $\epsilon = 1$) and those estimated from direct methods like numerical simulations, high-temperature expansions and experiments.

The following table summarizes the expansions of α, β and γ to $\mathcal{O}(\epsilon^2)$ as functions of the number of components n of the basic operator, $n = 1$ for the Ising model and $n = 2$ for the XY model, for example. The other exponents can be estimated by scaling relations from these values.

$$\alpha = -\frac{n - 4}{2(n + 8)}\epsilon - \frac{(n + 2)^2(n + 28)}{4(n + 8)^3}\epsilon^2$$

$$\beta = \frac{1}{2} - \frac{3}{2(n + 8)}\epsilon + \frac{(n + 2)(2n + 1)}{2(n + 8)^3}\epsilon^2$$

$$\gamma = 1 + \frac{n + 2}{2(n + 8)}\epsilon + \frac{(n + 2)(n^2 + 22n + 52)}{4(n + 8)^3}\epsilon^2.$$

For example, the exponent γ of the Ising model $(n = 1)$ to first and second order in $\epsilon(= 1)$ is 1.167 and 1.244, respectively. Numerical simulations indicate that $\gamma = 1.240$ for the three-dimensional Ising model in good agreement with the second-order result. It should, however, be pointed out that a simple inclusion of higher-order terms in the ϵ expansion leads to deteriorated results. This reflects a characteristic of asymptotic expansions, and a special caution is needed to make use of the results of ϵ expansions.

4.3   Real-space renormalization group for a quantum system

We have so far considered a renormalization group framework that is suited to study the critical phenomena in problems of classical statistical physics. One of the reasons is that critical phenomena in macroscopic systems involve a large number of degrees of freedom, which masks quantum effects at finite temperatures. At very low temperatures, quantum effects cannot be ignored in general and *quantum phase transitions* may occur. At zero temperature, a quantum phase transition between two quantum states with different correlations may take place as a parameter of the *quantum* Hamiltonian is varied. Note that this parameter, which would represent some physical interaction, plays the role of temperature in classical phase transitions in that it is the 'knob' (the relevant variable to be tuned) that drives the transition. It is not within the scope of this book to expand on quantum phase transitions. However, it is illuminating to show a simple case that exemplifies the extension of the real-space renormalization group framework to quantum systems.

4.3.1 Quantum phase transition in the transverse-field Ising model

Consider the spin system with a quantum Hamiltonian

$$H = -J \sum_{j=1}^{N-1} \sigma_j^z \sigma_{j+1}^z - h \sum_{j=1}^{N} \sigma_j^x, \qquad (4.65)$$

where free boundary conditions are assumed. The spin-1/2 quantum operators $S_j^{x,y,z} = \sigma_j^{x,y,z}/2$ (σ^x, σ^y, and σ^z are the Pauli matrices) are represented by the following matrices ($\hbar = 1$)

$$S^x = \begin{pmatrix} 0 & \frac{1}{2} \\ \frac{1}{2} & 0 \end{pmatrix}, \quad S^y = \begin{pmatrix} 0 & -\frac{i}{2} \\ \frac{i}{2} & 0 \end{pmatrix}, \quad S^z = \begin{pmatrix} \frac{1}{2} & 0 \\ 0 & -\frac{1}{2} \end{pmatrix}, \qquad (4.66)$$

satisfying the algebraic relations $S_j^z S_j^x = -S_j^x S_j^z$, $S_i^z S_j^x = S_j^x S_i^z$ ($i \neq j$) and similarly for the x, y and y, z components. This model is known as the one-dimensional *Ising model in a transverse field* or the *transverse-field Ising model*. Let us denote the eigenvectors of the spin operator S^z as

$$S^z |\uparrow\rangle = \frac{1}{2} |\uparrow\rangle, \quad S^z |\downarrow\rangle = -\frac{1}{2} |\downarrow\rangle. \qquad (4.67)$$

Then, the spin operator S^x flips the states as

$$S^x |\uparrow\rangle = \frac{1}{2} |\downarrow\rangle, \quad S^x |\downarrow\rangle = \frac{1}{2} |\uparrow\rangle. \qquad (4.68)$$

The eigenvectors of S^x can be determined in terms of $|\uparrow\rangle$ and $|\downarrow\rangle$ as

$$S^x |+\rangle = \frac{1}{2} |+\rangle, \quad |+\rangle = \frac{|\uparrow\rangle + |\downarrow\rangle}{\sqrt{2}} \qquad (4.69)$$

$$S^x |-\rangle = -\frac{1}{2} |-\rangle, \quad |-\rangle = \frac{|\uparrow\rangle - |\downarrow\rangle}{\sqrt{2}}. \qquad (4.70)$$

The factor $1/\sqrt{2}$ is for normalization, i.e. $\langle +|+\rangle = \langle -|-\rangle = 1$.

This model has a quantum phase transition in the ground state as a function of h/J, as depicted in Fig. 4.6. On the one hand, when $h/J = \infty$ ($h > 0, J = 0$), the spins have a lower energy if they align parallel to the external magnetic field (x-direction), and the ground state (the equilibrium state at zero temperature) is given by $|\Psi_0\rangle_+ = |+++\cdots+\rangle$ (assuming $h > 0$). On the other hand, when $h/J = 0$ ($h = 0, J > 0$),

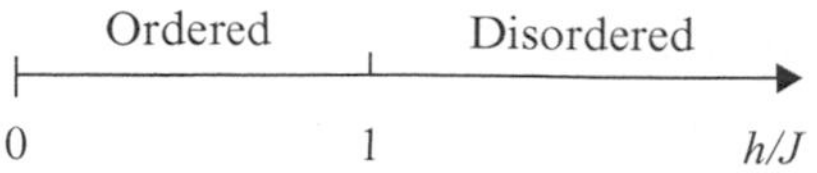

Fig. 4.6 The one-dimensional Ising model in a transverse field has a spontaneous magnetization along the z-direction for small values of the transverse field h but not for large h in the ground state. A quantum phase transition takes place at $(h/J)_c = 1$.

the spins align in the ground state in the z-direction, $|\Psi_0\rangle_\uparrow = |\uparrow\uparrow \cdots \uparrow\rangle$ or $|\Psi_0\rangle_\downarrow = |\downarrow\downarrow \cdots \downarrow\rangle$, which are doubly degenerate. As will be shown in Chapter 9, Exercise 9.8 in particular, the system undergoes a quantum phase transition at $(h/J)_c = 1$ between a ferromagnetic phase (ordered state, which approaches $|\Psi_0\rangle_\uparrow$ or $|\Psi_0\rangle_\downarrow$ as $h/J \to 0$) and a paramagnetic phase (disordered state from the viewpoint of ordering along the z-axis, which approaches $|\Psi_0\rangle_+$ as $h \to \infty$). This latter state is expanded as

$$|\Psi_0\rangle_+ = 2^{-N/2} \prod_{j=1}^{N} \left(|\uparrow\rangle_j + |\downarrow\rangle_j \right)$$

$$= 2^{-N/2} \left(|\uparrow\uparrow \cdots \uparrow\rangle + |\downarrow\uparrow \cdots \uparrow\rangle + \cdots + |\downarrow\downarrow \cdots \downarrow\rangle \right), \qquad (4.71)$$

which indicates that this state is completely disordered from the viewpoint of the z-axis. For $h/J < 1$, long-range order in the z-direction develops and the system has a spontaneous magnetization in the z-direction, i.e. a finite expectation value of the operator $m_z = (\sum_j S_j^z)/N$. The same quantity vanishes for $h/J > 1$. This is one of the simplest examples of quantum phase transitions.

4.3.2   Real-space renormalization group

We now develop a real-space renormalization group procedure to study this quantum phase transition. To this end we divide the N-site lattice into M blocks of $n = N/M$ spins each. Let us rewrite the Hamiltonian in eqn (4.65) as

$$H = \sum_{j \in \text{odd}} H_{b,j} + \sum_{j \in \text{even}} H_{b,j} = H_b + H_c, \qquad (4.72)$$

where the *block Hamiltonian* is given by

$$H_{b,j} = -J\sigma_j^z \sigma_{j+1}^z - h\sigma_j^x. \qquad (4.73)$$

Obviously, for the partition in eqn (4.72), $n = 2$ and the rescaling of the lattice spacing is $b = 2$. This is a most symmetric way to decimate the Hamiltonian, where the intrablock Hamiltonian H_b and the interblock Hamiltonian H_c share the same functional form. We could have chosen other ways still preserving the form of the lattice. For example, we could have partitioned

$$H = \sum_{j \in \text{odd}} H_{b',j} + \sum_{j \in \text{even}} H_{c,j}, \qquad (4.74)$$

with

$$H_{b',j} = -J\sigma_j^z\sigma_{j+1}^z - h(\sigma_j^x + \sigma_{j+1}^x), \quad H_{c,j} = -J\sigma_j^z\sigma_{j+1}^z. \tag{4.75}$$

However, the particular partition of eqns (4.72) and (4.73) satisfies a form-preserving transformation property, called self-duality (see Section 10.4), which leads to a better approximation.

Once we have partitioned the system, we have to diagonalize the intrablock Hamiltonian $H_{b,j}$ ($j \in$ odd). In our example, the latter is represented by a $2^2 \times 2^2$ matrix

$$\begin{pmatrix} -J & 0 & -h & 0 \\ 0 & J & 0 & -h \\ -h & 0 & J & 0 \\ 0 & -h & 0 & -J \end{pmatrix} \tag{4.76}$$

in the S^z-orthonormal basis $\{|\uparrow\uparrow\rangle, |\uparrow\downarrow\rangle, |\downarrow\uparrow\rangle, |\downarrow\downarrow\rangle\}$. The eigenvectors are given by

$$|1\rangle = \frac{1}{\sqrt{h^2 + (\sqrt{J^2 + h^2} + J)^2}} \left[(\sqrt{J^2 + h^2} + J)|\uparrow\uparrow\rangle + h|\downarrow\uparrow\rangle\right] \tag{4.77}$$

$$|2\rangle = \frac{1}{\sqrt{h^2 + (\sqrt{J^2 + h^2} - J)^2}} \left[(\sqrt{J^2 + h^2} - J)|\uparrow\downarrow\rangle + h|\downarrow\downarrow\rangle\right] \tag{4.78}$$

$$|3\rangle = \frac{1}{\sqrt{h^2 + (\sqrt{J^2 + h^2} + J)^2}} \left[-(\sqrt{J^2 + h^2} + J)|\uparrow\downarrow\rangle + h|\downarrow\downarrow\rangle\right] \tag{4.79}$$

$$|4\rangle = \frac{1}{\sqrt{h^2 + (\sqrt{J^2 + h^2} - J)^2}} \left[(-\sqrt{J^2 + h^2} + J)|\uparrow\uparrow\rangle + h|\downarrow\uparrow\rangle\right], \tag{4.80}$$

with corresponding eigenvalues $\epsilon_1 = \epsilon_2 = -\sqrt{J^2 + h^2}$ and $\epsilon_3 = \epsilon_4 = \sqrt{J^2 + h^2}$. Notice that each eigenvalue is doubly degenerate, which implies that there is a symmetry under the exchange of two degenerate states, as was the case in the original system.

The central idea of the present renormalization group is to keep the two lowest-lying eigenstates of each block, $|1\rangle$ and $|2\rangle$, and ignore $|3\rangle$ and $|4\rangle$. This is an approximation expected to be effective to study the ground state. Then, the new renormalized block operators can be defined as

$$\tilde{1} = |1\rangle\langle1| + |2\rangle\langle2| \,, \quad \tilde{\sigma}^z = |1\rangle\langle1| - |2\rangle\langle2| \,, \quad \tilde{\sigma}^x = |1\rangle\langle2| + |2\rangle\langle1|, \tag{4.81}$$

and a new renormalized Hamiltonian is defined on a lattice with $N/2$ sites. The operator $\tilde{1}$ is the projection onto the subspace spanned by $|1\rangle$ and $|2\rangle$, and $\tilde{\sigma}^x$ exchanges these two states as σ^x did in the original Hamiltonian. The operator $\tilde{\sigma}^z$ gives 1 for $|1\rangle$ and -1 for $|2\rangle$,

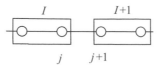

Fig. 4.7 The index j belongs to the block I and $j+1$ to $I+1$ for j even.

To write the renormalized Hamiltonian, one needs to construct the projector onto the coarse-grained system,

$$\mathbb{P} = \mathbb{P}_1 \otimes \mathbb{P}_2 \cdots \otimes \mathbb{P}_{N/2} \,, \quad \mathbb{P}_I = (|1\rangle\langle 1| + |2\rangle\langle 2|)_I = \tilde{1}_I, \tag{4.82}$$

where I represents a block index that will become a site index in the next iteration. The renormalized intrablock Hamiltonian ($j \in$ odd) is trivially a diagonal operator

$$\mathbb{P}_I H_{b,j} \mathbb{P}_I = \epsilon_1 \tilde{1}_I, \tag{4.83}$$

because $H_{b,j}|1\rangle = \epsilon_1|1\rangle$ and $H_{b,j}|2\rangle = \epsilon_1|2\rangle$.

The site indices j and $j+1$ define the block index I. The interactions in the new coarse-grained lattice are dictated by the corresponding projection of the interblock Hamiltonian that connects two blocks with indices I and $I+1$. The index j ($j \in$ even) is related to the block index I, while $j+1$ belongs to the index $I+1$ (see Fig. 4.7). Therefore,

$$\left(\mathbb{P}_I \otimes \mathbb{P}_{I+1}\right) H_{b,j} \left(\mathbb{P}_I \otimes \mathbb{P}_{I+1}\right)$$
$$= -J \left(\mathbb{P}_I \sigma_j^z \mathbb{P}_I\right) \otimes \left(\mathbb{P}_{I+1} \sigma_{j+1}^z \mathbb{P}_{I+1}\right) - h \left(\mathbb{P}_I \sigma_j^x \mathbb{P}_I\right) \otimes \tilde{1}_{I+1}. \tag{4.84}$$

We are thus left with the task of computing the projection of the original Pauli spin operators. To evaluate the projections in eqn (4.84), the following relations will be useful, which can be verified by using eqns (4.77) to (4.80),

$$_I\langle 1|\sigma_j^z|1\rangle_I = -_I\langle 2|\sigma_j^z|2\rangle_I = 1 \tag{4.85}$$

$$_I\langle 1|\sigma_j^z|2\rangle_I = _I\langle 2|\sigma_j^z|1\rangle_I = 0 \tag{4.86}$$

$$_{I+1}\langle 1|\sigma_{j+1}^z|1\rangle_{I+1} = -_{I+1}\langle 2|\sigma_{j+1}^z|2\rangle_{I+1} = \frac{J}{\sqrt{J^2 + h^2}} \tag{4.87}$$

$$_{I+1}\langle 1|\sigma_{j+1}^z|2\rangle_{I+1} = _{I+1}\langle 2|\sigma_{j+1}^z|1\rangle_{I+1} = 0 \tag{4.88}$$

$$_I\langle 2|\sigma_j^x|1\rangle_I = _I\langle 1|\sigma_j^x|2\rangle_I = \frac{h}{\sqrt{J^2 + h^2}} \tag{4.89}$$

$$_I\langle 1|\sigma_j^x|1\rangle_I = _I\langle 2|\sigma_j^x|2\rangle_I = 0. \tag{4.90}$$

Then, simple algebraic manipulations lead to

$$\mathbb{P}_I \sigma_j^z \mathbb{P}_I = \tilde{\sigma}_I^z \tag{4.91}$$

$$\mathbb{P}_I \sigma_j^x \mathbb{P}_I = \frac{h}{\sqrt{J^2 + h^2}} \, \tilde{\sigma}_I^x \tag{4.92}$$

$$\mathbb{P}_{I+1} \sigma_{j+1}^z \mathbb{P}_{I+1} = \frac{J}{\sqrt{J^2 + h^2}} \, \tilde{\sigma}_{I+1}^z. \tag{4.93}$$

Notice the asymmetry in the projections between sites j and $j+1$. This asymmetry can be traced back to the way we partitioned the lattice in eqn (4.73), where sites j and $j+1$ are non-equivalent.

We can finally collect all these pieces together and write the full renormalized Hamiltonian after one step of renormalization,

$$\mathbb{P} H \mathbb{P} = \tilde{H} = \epsilon_1 \sum_{I=1}^{N/2} \tilde{1}_I - \frac{J^2}{\sqrt{J^2 + h^2}} \sum_{I=1}^{N/2-1} \tilde{\sigma}_I^z \tilde{\sigma}_{I+1}^z - \frac{h^2}{\sqrt{J^2 + h^2}} \sum_{I=1}^{N/2} \tilde{\sigma}_I^x. \tag{4.94}$$

It is interesting to note that, apart from a constant, our choice of renormalization group transformation preserves the form of the Hamiltonian with no additional couplings. Had we chosen a different partition, the form of the Hamiltonian would not have been preserved. We have thus generated the following renormalization group equation

$$\begin{pmatrix} J' \\ h' \end{pmatrix} = \begin{pmatrix} \dfrac{J^2}{\sqrt{J^2 + h^2}} \\ \dfrac{h^2}{\sqrt{J^2 + h^2}} \end{pmatrix}, \tag{4.95}$$

or equivalently $k' = h'/J' = (h/J)^2 = k^2$, which must be iterated many times.

The resulting fixed-point equation is $k^* = (k^*)^2$. This recursion relation has two trivial (stable) fixed points. One is $k^* = 0$ and corresponds to the ordered ferromagnetic phase and the other is $k^* = \infty$, which characterizes the disordered paramagnetic phase. These trivial fixed points are separated by a non-trivial (unstable) fixed point, which is critical, $k^* = k_c = 1$. This critical value is exact.

To determine the values of the critical exponents, we linearize the renormalization group equation $k' = k^2$ close to the critical fixed point $k_c = 1$ and obtain the appropriate eigenvalue

$$\lambda_k = b^{y_k} = \left. \frac{\mathrm{d}k'}{\mathrm{d}k} \right|_{k_c} = 2 = b^1, \tag{4.96}$$

which implies $y_k = 1$. Since the correlation length should diverge as $\xi \sim (k - k_c)^{-\nu}$, and $\xi' = \xi/b$, the linearized recursion relation

$$k' - k_c = 2(k - k_c) \tag{4.97}$$

implies that $\nu = 1/y_k = 1$. This is indeed the exact critical exponent for the correlation length of the present system because the one-dimensional transverse-field Ising model is equivalent to the two-dimensional classical Ising model in a sense that will be elucidated in Exercise 9.8 and Section 10.4.2.

The critical exponent β^* of the magnetization of the left-most (boundary) spin with $j = 1$ is calculated from the recursion relation of eqn (4.93),

$$\langle \sigma_1^z \rangle = \frac{J}{\sqrt{J^2 + h^2}} \langle \tilde{\sigma}_1^z \rangle = \frac{1}{\sqrt{1 + k^2}} \langle \tilde{\sigma}_1^z \rangle. \tag{4.98}$$

This is to be compared with the generic form of eqn (3.50) to give

$$b^{-x} = \frac{1}{\sqrt{1 + k_c^2}} = \frac{1}{\sqrt{2}}, \tag{4.99}$$

which implies $x = 1/2$. Then, according to eqn (3.44), we conclude $\beta^* = 1/2$. This is the exact value for the surface (boundary) magnetization. One should notice, however, that the right-most spin with $j = N$ has $\beta^* = 0$ from the same argument using eqn (4.91), which is inconsistent with $\beta^* = 1/2$ for $j = 1$. Hence, the present method is an approximation, which happens to give a part of the exact values for the critical point and critical exponents.

5
Statistical field theory

We have seen that statistical-mechanical systems often involve discrete elementary degrees of freedom such as spins in the Ising model. Field theories, on the other hand, have continuous fields, defined over the whole space-time or part of it, as fundamental degrees of freedom. These two seemingly different descriptions of physical phenomena can be related close to the critical point. The physical idea behind this is that close to the critical region some correlation length diverges and the behavior of the correlations between degrees of freedom over long distances is independent of the microscopic details of the theory. This is true for both discrete variables and fields alike, and what conceptually connects the two representations is the hypothesis of universality underlying the renormalization group framework. In the present chapter we summarize how the description by continuous fields emerges from discrete degrees of freedom in a more systematic manner than in previous chapters. The important roles of symmetry and topology are also elucidated in some detail.

5.1 From bits to fields

At its most fundamental level, the microscopic description of matter is in terms of its elementary degrees of freedom, such as spins $\{S_j\}$ in the case of a magnet, or positions and momenta $\{q_j, p_j\}$ in the case of an atomic gas. Typically, the time evolution of those degrees of freedom is governed by a set of equations of motion derivable from a Hamiltonian, H, or a Lagrangian density, $\mathcal{L}$, that encodes the interactions between those elementary degrees of freedom, or by master equations in the case of open systems (i.e. systems that are not isolated but coupled to some external environment) as described in Chapter 11. One can imagine that solving these equations for many degrees of freedom (say $N = 10^{23}$) is a daunting task that not only involves great complexity but also is prone to failure. It is, in general, an intractable problem, except for limited cases, as discussed in Chapter 9.

To attack this problem, one of the common procedures consists in performing some sort of averaging, i.e. coarse graining, over many degrees of freedom with the expectation that the system still retains its main physical properties, and at the same time, the problem becomes manageable. The averaged degrees of freedom are no longer discrete but turn into slowly varying continuous *fields*, thus eliminating the short-wavelength (short-distance) modes. A field represents an infinite number of degrees of freedom and it is, in general, a tensor-valued function of the coordinates r (or space-time). It happens that this methodology is more accurate when the relevant physics one is trying to describe is regulated by the *collective* behavior of those elementary degrees of freedom, where long wavelengths and long times are involved. As indicated in previous chapters, it is precisely at a critical region where fluctuations have long wavelengths and are correlated over distances of the order of the correlation length, $\xi \gg a$, with a denoting some microscopic distance, typically the lattice constant. Therefore, the local connectivity of the lattice, e.g. whether it is the square lattice or the triangular lattice, is irrelevant from the standpoint of critical properties. It is only relevant to determine non-universal properties such as the specific value of the critical temperature. These correlated fluctuations involve many elementary degrees of freedom and, thus, critical phenomena represent one of those problems where a description by *statistical field theory* seems appropriate.

The equivalence between the original model described in terms of discrete variables and the field theory is not usually realized by an exact algebraic mapping, like in the one-dimensional quantum XY and free Fermion models in Section 9.4. It should rather be regarded as an equivalence in the sense that both models share the same critical behavior and thus belong to the same universality class. One may conjecture that, as long as the statistical-mechanical system displays universal behavior at criticality, there should be a corresponding statistical field theory that describes the same physics of long wavelength.

Indeed, we have already seen a few examples of statistical field theories, e.g. the Gaussian model in Sections 2.9 and 4.2. In this chapter we summarize more systematic approaches to the foundation of statistical field theories. The present exposition will also serve as a bridge to the conformal field theory described in the next chapter. Also explained are the roles of symmetry and topology, in particular the concepts of symmetry breaking, long-range order and topological defects, which are essential to the deep understanding of phase transitions and critical phenomena.

5.2 Continuum limit and field theory

Before we proceed with the derivation of effective field theories for models with discrete variables, let us illustrate the passage to the continuum with the simple case of a linear chain of identical torsion pendulums that are coupled through bars of elastic constant κ (Fig. 5.1). Let us denote the moment of inertia and the length of the pendulum as

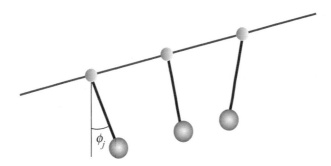

Fig. 5.1 One-dimensional series of torsion pendulums.

I and l, respectively. The mass of the weight is m. Each pendulum is constrained to move in a plane perpendicular to the elastic bars with an angle ϕ_j from the vertical direction. Then, one can write the Lagrangian of the system of $2N + 1$ pendulums as $(\partial_t \phi_j = \mathrm{d}\phi_j/\mathrm{d}t)$

$$L = \frac{I}{2} \sum_{j=-N}^{N} (\partial_t \phi_j)^2 - \Big(\frac{\kappa}{2} \sum_{j=-N}^{N-1} (\phi_j - \phi_{j+1})^2 + mgl \sum_{j=-N}^{N} (1 - \cos \phi_j) \Big). \tag{5.1}$$

The equilibrium situation corresponds to $\phi_j = 0$ $(\forall j)$, and g is the gravitational acceleration. The Lagrangian L in eqn (5.1) is the difference between the kinetic energy, the first term, and the potential energy.

We are interested in describing the system in the continuum limit. This limit is obtained by letting the spacing between two pendulums, a, tend to zero and $N \to \infty$. In this limit, by making the associations $(\partial_x \phi = \mathrm{d}\phi/\mathrm{d}x)$

$$aj \to x \ , \ \ \phi_j \to \phi(x,t) \ , \ \ \frac{\phi_{j+1} - \phi_j}{a} \to \partial_x \phi(x,t) \ , \ \ a \sum_j \to \int \mathrm{d}x, \tag{5.2}$$

one obtains the following expression for L, known as the *sine-Gordon Lagrangian*,

$$L = \frac{1}{2} \int \mathrm{d}x \Big(\lambda (\partial_t \phi(x,t))^2 - Y(\partial_x \phi(x,t))^2 - 2G(1 - \cos \phi(x,t)) \Big), \tag{5.3}$$

in terms of the *scalar field* $\phi(x,t)$ and its derivatives. Here, $\lambda = I/a$ is the density of the moment of inertia, $Y = \kappa a$ is the Young modulus, and $G = \rho gl$ with $\rho = m/a$. In the limit that the gravitational force vanishes, the above Lagrangian becomes

$$L = \frac{1}{2} \int \mathrm{d}x \Big(\lambda (\partial_t \phi(x,t))^2 - Y(\partial_x \phi(x,t))^2 \Big). \tag{5.4}$$

Formally, this is the same Lagrangian as the one that describes a harmonic crystal whose elementary excitations are sound waves. The physical differences, though, are that in the crystal case there are no pendulums but only masses connected by elastic springs in a linear chain, and the angle variables are replaced by longitudinal displacements of those masses from their equilibrium positions. The elasticity theory, a phenomenological approach to studying the elastic properties of a crystal, is the archetypal example of a field theory.

Since we are interested in studying the thermodynamic aspects of field theories, i.e. statistical field theories, we would like to compute their generating functional Z. This latter quantity corresponds to the partition function of statistical mechanics and is defined in terms of an action $S[\phi]$, which plays a similar role as the Hamiltonian in classical statistical mechanics. More precisely, one analytically continues time as $t \to -it$, which changes the Minkowski space with infinitesimal distance $(\mathrm{d}r)^2 - (\mathrm{d}t)^2$ to the Euclidean space with $(\mathrm{d}r)^2 + (\mathrm{d}t)^2$. By this correspondence, the Lagrangian density of the linear chain of pendulums and the corresponding action are

$$\mathcal{L}(\phi, \partial_\mu \phi) = \frac{\lambda}{2} (\partial_{r_1} \phi(r))^2 + \frac{Y}{2} (\partial_{r_2} \phi(r))^2, \tag{5.5}$$

and

$$S[\phi] = \int d^d r \, \mathcal{L}(\phi, \partial_\mu \phi), \tag{5.6}$$

respectively, where $r = (r^1, r^2)$.[1] The first component of the coordinate, r^1, corresponds to imaginary time. This *Euclidean Lagrangian density* can be identified with the Hamiltonian density in classical statistical mechanics. The expression of the action of eqn (5.6) is generic and we shall now discuss general properties not restricted to the system of torsion pendulums.

The generating functional $Z[J]$ is an equivalent of the partition function,

$$Z[J] = \int \mathcal{D}\phi(r) \, W[\phi(r)], \tag{5.7}$$

where the statistical weight of a field configuration is

$$W[\phi(r)] = \exp\left(-S[\phi] + \int d^d r \, J(r)\phi(r)\right), \tag{5.8}$$

with $J(r)$ being a *generating current* or a *source term* linearly coupled to the field $\phi(r)$. The *functional integral* in eqn (5.7) will be defined below.

The source term plays the role of a probe to determine n-point correlation functions by functional differentiation of the generating functional,

$$\langle \phi(r_1) \cdots \phi(r_n) \rangle = \frac{1}{Z[J]} \frac{\delta^n Z[J]}{\delta J(r_1) \cdots \delta J(r_n)}. \tag{5.9}$$

This is a continuum analog of the logarithmic derivative of the partition function by local fields in classical statistical mechanics,

$$\langle S_1 \cdots S_n \rangle = \frac{1}{Z(h)} \frac{\partial^n Z(h)}{\partial(\beta h_1) \cdots \partial(\beta h_n)}, \tag{5.10}$$

where, for example, the partition function is given by

$$Z(h) = \sum_{\{S_i\}} \exp\left(\beta \sum_{\langle ij \rangle} J_{ij} S_i S_j + \beta \sum_i h_i S_i\right). \tag{5.11}$$

It should be clear from this analogy that we may call the action $S[\phi]$ an effective Hamiltonian $\tilde{H}$

$$S[\phi] \to \tilde{H}(\phi), \tag{5.12}$$

where the inverse temperature $\beta = 1/T$ has been included in the effective Hamiltonian. We use both names interchangeably for the same quantity in this and the next chapters.

---

[1] The change $t \to -it$ in eqn (5.4) yields the opposite sign for $\mathcal{L}$. This does no harm if we choose an appropriate sign in exponentiating the action to define the statistical weight, as will be done shortly. Also, we write explicitly the superscript d for the differential of the variable of integration to emphasize the important role of dimensionality.

In eqn (5.7) the generating functional $Z[J]$ has been given by a *functional integral* obtained as a limit of discrete integrals over all allowed configurations of the variables. The passage to the continuum from a d-dimensional lattice model with scalar degrees of freedom is accomplished by a simple extension of the relations in eqn (5.2) with a functional measure defined as an appropriate limit

$$\int \mathcal{D}\phi(\boldsymbol{r}) = \lim_{a \to 0, N \to \infty} \int \prod_{j=1}^{N} \mathrm{d}\phi_j, \qquad (5.13)$$

where, without loss of generality, we assumed that there are N points defining the discretized lattice of constant spacing a and infinitesimal volume a^d.

In field theories one often encounters unphysical divergences originating in the continuum nature of fields or the infinite degrees of freedom as seen in the limits in eqn (5.13). In such cases, discrete lattice models (before taking the limits $N \to \infty$ and $a \to 0$ on the right-hand side of eqn (5.13)) provide a natural *regularization*, a way to remove divergences, and thus supply the precise mathematical meaning to the functional integrals and other tricks of statistical field theories. The small length scale a, the lattice constant, is called a *cutoff* in this context.

Clearly, the continuum limit of the mechanical system described in eqn (5.1) is a well-defined field theory. There are several important differences between this simple mechanical system and statistical field theories. First, the Hamiltonian of the Ising model, for example, is not a genuine mechanical Hamiltonian since there are no intrinsic dynamics and conjugate variables related by Poisson brackets. Secondly, one needs to distinguish between lattice systems with continuous variables from those with discrete degrees of freedom. In models of statistical mechanics the elementary degrees of freedom are often realized by discrete variables, e.g. Ising spins. In those cases, the way a continuous field is generated requires some care since there is a constraint in the allowed values for the variables. Finally, most of the statistical field theories require a physical cutoff to be mathematically well defined to avoid infinite integrals. These cutoffs determine the limits of integration. An *infrared cutoff* refers to a long-distance (or small-momentum or low-energy) cutoff, while an *ultraviolet cutoff* alludes to a short-distance (or large-momentum or high-energy) cutoff.

5.3 Hubbard–Stratonovich transformation

In this section we illustrate a process to start from a microscopic lattice model and derive its mapping to a functional integral over continuous fields. The Ising model on a d-dimensional hypercubic lattice with lattice constant a is taken as an example. The technique is known as the *Hubbard–Stratonovich transformation*, or the *Gaussian transformation*, and is a generalization of the completing-the-square method in the standard Gaussian integration.

The Ising Hamiltonian and its partition function are given by

$$H = -\sum_{i,j} J_{ij} S_i S_j \ , \ \ Z = \frac{1}{2^N} \sum_{\{S_i = \pm 1\}} \exp\left(\sum_{i,j} K_{ij} S_i S_j\right), \qquad (5.14)$$

with $K_{ij} = \beta J_{ij}$ a matrix of coupling constants, not necessarily of nearest-neighbor type. A trivial prefactor 2^{-N} has been given to Z for later simplicity of notation. For example, the uniform coupling $J_{ij} = J/2N$ defines the infinite-range Ising model of Section 2.5, while $J_{ij} = J/2$ for nearest-neighboring sites and zero otherwise represents the usual Ising model with short-range interactions. We assume that the system is translationally invariant.

Define the N-component vector $\boldsymbol{S} = (S_1, S_2, \cdots, S_N)$, and an $N \times N$ symmetric coupling matrix $\tilde{K}_{ij} = K_0 \mathbb{1}_{ij} + K_{ij}$, where $\mathbb{1} = \mathrm{diag}(1, \cdots, 1)$, with $K_0 \geq \max_i \sum_j K_{ij}$ so that $\tilde{\boldsymbol{K}}$ is a positive definite matrix. Then,

$$Z = \frac{e^{-K_0 N/2}}{2^N} \sum_{\{S_i = \pm 1\}} \exp\left(\frac{1}{2} \boldsymbol{S} \cdot \tilde{\boldsymbol{K}} \cdot \boldsymbol{S}\right). \tag{5.15}$$

According to a formula of multivariable Gaussian integral, eqns (A.253) and (A.258), the exponential on the right-hand side of eqn (5.15) is expressed as

$$\exp\left(\frac{1}{2} \boldsymbol{S} \cdot \tilde{\boldsymbol{K}} \cdot \boldsymbol{S}\right) = e^{-\mathcal{A}} \int_{\mathbb{R}^N} \left(\prod_{j=1}^N d\sigma_j\right) \exp\left(-\frac{1}{2} \boldsymbol{\sigma} \cdot \tilde{\boldsymbol{K}}^{-1} \cdot \boldsymbol{\sigma} - \boldsymbol{\sigma} \cdot \boldsymbol{S}\right), \tag{5.16}$$

where $e^{\mathcal{A}} = (2\pi)^{N/2} (\det \tilde{\boldsymbol{K}})^{1/2}$. This is the Hubbard–Stratonovich transformation and is a multivariable generalization of eqn (2.36). If we realize that

$$2^{-N} \sum_{\{S_i = \pm 1\}} \exp\left(-\boldsymbol{\sigma} \cdot \boldsymbol{S}\right) = \exp\left(\sum_{i=1}^N \log \cosh \sigma_i\right), \tag{5.17}$$

the partition function of the original Ising model becomes

$$Z = e^{-(\mathcal{A} + K_0 N/2)} \int_{\mathbb{R}^N} \left(\prod_{j=1}^N d\sigma_j\right) \exp\left(-\frac{1}{2} \boldsymbol{\sigma} \cdot \tilde{\boldsymbol{K}}^{-1} \cdot \boldsymbol{\sigma} + \sum_{i=1}^N \log \cosh \sigma_i\right). \tag{5.18}$$

We would like to emphasize at this point that the mapping of the partition function of eqn (5.14) into eqn (5.18) is exact. There are no approximations involved, and we have simply transformed the original discrete variables into continuous variables.

The first term in the exponential of eqn (5.18) should provide terms leading to derivatives in the field theory since it is the one containing spatial variations,

$$\boldsymbol{\sigma} \cdot \tilde{\boldsymbol{K}}^{-1} \cdot \boldsymbol{\sigma} = 2 \sum_{i>j} \sigma_i \tilde{K}_{ij}^{-1} \sigma_j + \sum_i \tilde{K}_{ii}^{-1} \sigma_i^2. \tag{5.19}$$

The second term gives a potential, local, contribution

$$\log \cosh \sigma_i = \frac{\sigma_i^2}{2} - \frac{\sigma_i^4}{12} + \mathcal{O}(\sigma_i^6). \tag{5.20}$$

Notice that $\tilde{\boldsymbol{K}}^{-1}$ can be written as

$$\tilde{\boldsymbol{K}}^{-1} = (K_0 \mathbb{1} + \boldsymbol{K})^{-1} = K_0^{-1} \mathbb{1} - K_0^{-2} \boldsymbol{K} + K_0^{-3} \boldsymbol{K}^2 - K_0^{-4} \boldsymbol{K}^3 + \mathcal{O}(\boldsymbol{K}^4) \tag{5.21}$$

in terms of the original coupling constants. Assume now a nearest-neighbor Ising model. Then $\boldsymbol{K}$ includes non-vanishing elements to connect neighboring sites, $\boldsymbol{K}^2$ for second nearest neighbors, $\boldsymbol{K}^3$ for third neighbors, and so on. If nearest neighbors are the dominant, most relevant, parts of interactions, we may eliminate $\boldsymbol{K}^2$ and higher orders in the expansion of eqn (5.21). Then, the leading contribution in eqn (5.19) can be written in the form $((\sigma_i - \sigma_j)/a)^2$ for nearest neighboring i, j. The reason is that the nearest-neighbor term (proportional to $-\boldsymbol{\sigma} \cdot \boldsymbol{K} \cdot \boldsymbol{\sigma}$) coming from $-K_0^{-2}\boldsymbol{K}$ in eqn (5.21) has a contribution proportional to $-\sigma_i\sigma_j$, which can be combined with part of the simple quadratic terms of eqn (5.19) to yield the desired expression $-2\sigma_i\sigma_j + \sigma_i^2 + \sigma_j^2$. In the continuum limit, where $a \to 0$ and $N \to \infty$,

$$a\, i \to \boldsymbol{r} \ , \ \ \sigma_i \to \phi(\boldsymbol{r}) \ , \ \ \left(\frac{\sigma_i - \sigma_j}{a}\right)^2 \to (\nabla\phi(\boldsymbol{r}))^2, \tag{5.22}$$

and the scalar field theory belonging to the Ising universality class is

$$Z \sim \int \mathcal{D}\phi(\boldsymbol{r}) \exp\left(-\int \mathrm{d}^d r \left\{ b(\nabla\phi(\boldsymbol{r}))^2 + t\phi(\boldsymbol{r})^2 + u\phi(\boldsymbol{r})^4 \right\}\right). \tag{5.23}$$

This is identical to the ϕ^4-field theory already considered in Section 4.2.1. For space dimensions d close to the upper critical dimension (in this case $d_{\mathrm{uc}} = 4$, see Sections 2.10 and 4.2.1) the higher-order terms become irrelevant in the renormalization group sense at the Gaussian fixed point (Exercise 4.2) and thus can be dropped.

EXERCISE 5.1 The Hubbard–Stratonovich transformation is essentially a Gaussian integration. Consider the general model Hamiltonian on a hypercubic lattice

$$H = -\sum_{ij} J_{ij}\boldsymbol{S}_i \cdot \boldsymbol{S}_j, \tag{5.24}$$

where $\boldsymbol{S}_i$ has n components and is normalized as $|\boldsymbol{S}_i|^2 = 1$. The Ising model is a particular case of this Hamiltonian with $n = 1$. Assume that the system is translationally invariant and apply the same analysis as in the text to this n-component system. In particular, show that the effective Hamiltonian, taken to the quartic term in fields, is expressed as

$$\tilde{H} = \int \mathrm{d}^d r \left\{ b\sum_{j=1}^{n} (\nabla\phi_j(\boldsymbol{r}))^2 + t\sum_{j=1}^{n} \phi_j^2(\boldsymbol{r}) + u\left(\sum_{j=1}^{n} \phi_j^2(\boldsymbol{r})\right)^2 \right\}. \tag{5.25}$$

5.4 Integrating out degrees of freedom: Coarse graining

We cannot always apply the above Hubbard–Stratonovich methodology to arbitrary models to find the appropriate field theory. In general, there is no systematic standard procedure to find the action for any given lattice model with discrete variables. Let us assume now that we have already identified the relevant microscopic degrees of freedom and proceed to apply a coarse-graining, averaging, procedure to determine an effective field theory. Again, for simplicity, assume that the microscopic degrees of freedom are spins, not necessarily Ising spins, and define a *block variable* as

$$S_\zeta(\boldsymbol{r}) = \frac{1}{N_\zeta(\boldsymbol{r})} \sum_{i \in \text{block}_{\boldsymbol{r}}} S_i, \qquad (5.26)$$

where $\boldsymbol{r}$ represents the center of the block of linear dimension ζ, $a \ll \zeta \lesssim \xi$, with $N_\zeta(\boldsymbol{r})$ the number of spins in the block. Clearly, if $N_\zeta(\boldsymbol{r}) = 1$ then $S_\zeta(\boldsymbol{r}) = S_i$ with $ia = \boldsymbol{r}$. In this way, we construct new, approximately N/N_ζ (if all N_ζ are equal), coarse-grained variables S_ζ in terms of the original N spins S_i. Notice that the mapping of eqn (5.26) is not invertible, and therefore the procedure we will describe is not a mathematical isomorphism. One expects on physical and mathematical grounds that $S_\zeta(\boldsymbol{r})$ varies smoothly on a microscopic scale a, which means to lower the space resolution of the description, and only variations at a scale ζ will be appreciable, in the sense that it can vary from block to block.

Given the Hamiltonian $H(\{S_i\})$, written in terms of the original degrees of freedom $\{S_i\}$, we would like to determine the corresponding Hamiltonian for the coarse-grained variables $S_\zeta(\boldsymbol{r})$. Formally, one can perform the following mapping

$$\exp\left(-\bar{H}(\{S_\zeta(\boldsymbol{r})\})\right) = \text{Tr}\left\{ \exp\left(-H(\{S_i\})\right) \prod_{\boldsymbol{r}} \delta\left(S_\zeta(\boldsymbol{r}) - \frac{1}{N_\zeta(\boldsymbol{r})} \sum_{i \in \text{block}_{\boldsymbol{r}}} S_i\right)\right\}, (5.27)$$

where the trace operation $\text{Tr} = \sum_{\{S_i\}}$ represents the sum over the original spins $\{S_i\}$, with the constraint that only spin configurations that have a certain value $S_\zeta(\boldsymbol{r})$ are kept. The inverse temperature β has been incorporated into the Hamiltonian. Then, the fundamental idea consists in summing over the degrees of freedom corresponding to the shortest scales, thus generating effective models that describe the same long-distance physics by elimination of the short-distance, microscopic, structure. Moreover, in an ideal situation, one would like to keep the partition function of the original system invariant,

$$Z = \text{Tr} \exp\left(-H(\{S_i\})\right) = \sum_{\{S_\zeta(\boldsymbol{r})\}} \exp\left(-\bar{H}(\{S_\zeta(\boldsymbol{r})\})\right). \qquad (5.28)$$

One can continue with the construction of a new Hamiltonian by defining new block-spin variables from the coarse-grained $\{S_\zeta(\boldsymbol{r})\}$ ones. This coarse graining, or cell, procedure is simply the block-spin transformation of Chapter 4 and is the basis of the real-space renormalization group method. The main point, though, is that eventually the coarse-grained variables $\{S_\zeta(\boldsymbol{r})\}$ become a field $\phi(\boldsymbol{r})$, or set of fields $\{\phi_i(\boldsymbol{r})\}$, of a continuous variable $\boldsymbol{r}$ and the partition function can be expressed as a functional integral

$$Z = \int \mathcal{D}\phi(\boldsymbol{r}) \, \exp\left(-\tilde{H}(\phi(\boldsymbol{r}))\right) = \int \mathcal{D}\phi(\boldsymbol{r}) \, \exp\left(-S[\phi(\boldsymbol{r})]\right), \qquad (5.29)$$

the measure $\mathcal{D}\phi(\boldsymbol{r})$ meaning integration over the allowed configurations of the field. Strictly speaking, $\tilde{H}$ is not a Hamiltonian since the coarse-grained variables are not necessarily related by Poisson bracket relations; it simply determines the weight of the configurations. Because of its resemblance to the action of Euclidean field theories, we will also call it an action $S[\phi(\boldsymbol{r})]$. If two different, originally discrete, models have the same limiting fixed-point Hamiltonian after this systematic procedure, then both models belong to the same universality class.

Notice that there is nothing special about the real-space blocking procedure. One can alternatively write the original microscopic $H(\{S_i\})$ in terms of the Fourier-transformed, momentum $\boldsymbol{q}$, spin variables

$$S_{\boldsymbol{q}} = \sum_j \mathrm{e}^{\mathrm{i}\boldsymbol{q}\cdot\boldsymbol{r}_j}\, S_j \ , \ S_j = \frac{1}{N}\sum_{\boldsymbol{q}} \mathrm{e}^{-\mathrm{i}\boldsymbol{q}\cdot\boldsymbol{r}_j}\, S_{\boldsymbol{q}}, \tag{5.30}$$

and define a coarse-graining Hamiltonian in the momentum space through the averaging process

$$\exp\big(-\bar{H}(\{S_{\boldsymbol{q}}, q < \Lambda\})\big) = \sum_{\{S_{\boldsymbol{q}}, q>\Lambda\}} \exp\big(-H(\{S_{\boldsymbol{q}}\})\big). \tag{5.31}$$

Here, we achieve a lower spatial resolution by integrating out higher Fourier components $(q > \Lambda)$, Λ being the momentum cutoff, and $2\pi/\Lambda$ should be identified with the block size ζ.

We would like to make two remarks. The first is that, although this procedure can be made rigorous in the case of classical systems, it is impractical since, except for trivial cases, such as the one-dimensional Ising model of Section 3.6, the method generates a large number of additional interactions. The second remark is that an equivalent coarse-graining procedure for microscopic quantum-mechanical models is subject to additional mathematical subtleties.

5.5 Phenomenological Landau–Ginzburg approach

An alternative program to generate effective field theories is the *Landau–Ginzburg approach*. One constructs the effective Hamiltonian in a phenomenological way in terms of collective degrees of freedom expressed as the field of order parameter $\Phi(\boldsymbol{r}) = \{\phi_1(\boldsymbol{r}), \cdots, \phi_n(\boldsymbol{r})\}$, with the symmetry of the microscopic Hamiltonian taken into account. For example, a one-component scalar field $(n = 1)$ may describe the standard liquid–gas transition or the uniaxial Ising ferromagnet. A two-component, or complex, field describes a transition to a superfluid or superconducting phase, and a three-component, vector, field describes a classical magnetic transition. More esoteric order parameters include second-rank tensorial quantities that describe the transition to nematic or smectic liquid-crystal phases.

This process results in an effective field theory with couplings that are functions not only of the original microscopic couplings but also of external control parameters such as the temperature. Strictly speaking, the effective Hamiltonian $\tilde{H}$ is neither a proper Hamiltonian nor a free energy since, for example, couplings in $\tilde{H}$ depend on the temperature and fields. Also, the variables and their derivatives are not connected by canonical dynamical relations.

One should take into account several constraints in constructing the effective Hamiltonian. The first observation is that, if the original degrees of freedom are defined in a real space with local, short-range, interactions, one would expect that the resulting continuous fields define a *local field theory*,

$$\tilde{H}(\Phi(\boldsymbol{r})) = \int \mathrm{d}^d\boldsymbol{r}\, \tilde{\mathcal{H}}, \tag{5.32}$$

where $\tilde{\mathcal{H}}$ is a Hamiltonian *density*. Next, the effective Hamiltonian is a functional of the field and its derivatives. It also includes the explicit dependence on the coordinate $\boldsymbol{r}$ when the system is not uniform due, for example, to randomly distributed defects or impurities. The effective Hamiltonian density is therefore written generically as

$$\tilde{\mathcal{H}} = \tilde{\mathcal{H}}[\Phi(\boldsymbol{r}), \partial_\mu \Phi(\boldsymbol{r}), \partial_\mu \partial_\nu \Phi(\boldsymbol{r}), \cdots, \boldsymbol{r}], \tag{5.33}$$

where $\partial_\mu = \partial/\partial r^\mu$ and r^μ $(\mu = 1, \cdots, d)$ represent the components of the coordinate. The existence of derivatives in the argument of the effective Hamiltonian density reflects short-range interactions between nearby degrees of freedom in the microscopic Hamiltonian.

One of the most important elements to consider is the symmetry of the system. For instance, consider again the classical Ising Hamiltonian in the presence of an external magnetic field h and nearest-neighbor couplings J_{ij},

$$H = -\sum_{\langle ij \rangle} J_{ij} S_i S_j - h \sum_i S_i. \tag{5.34}$$

It is clear that, if we perform the transformation $S_i \to -S_i$, $\forall i$, the Hamiltonian remains invariant under such reflections as long as we also change the sign of the magnetic field $h \to -h$. This global symmetry transformation (involving all the spins in the system) forms the group $\mathbb{Z}_2$. The order parameter in this case, a scalar field $\phi(\boldsymbol{r})$ $(n = 1)$, is the magnetization and any effective Hamiltonian written in terms of (a functional of) $\phi(\boldsymbol{r})$ should satisfy this symmetry constraint

$$\tilde{H}(\phi(\boldsymbol{r}), h) = \tilde{H}(-\phi(\boldsymbol{r}), -h). \tag{5.35}$$

Moreover, if the system is translationally invariant, this symmetry should also be preserved in the effective Hamiltonian density as

$$\tilde{\mathcal{H}} = \tilde{\mathcal{H}}[\phi(\boldsymbol{r}), \partial_\mu \phi(\boldsymbol{r}), \partial_\mu \partial_\nu \phi(\boldsymbol{r}), \cdots], \tag{5.36}$$

without explicit dependence on the space coordinate $\boldsymbol{r}$. Another example is furnished by the classical ferromagnetic Heisenberg model without external field, where each spin has three components, $\boldsymbol{S}_i = (S_i^x, S_i^y, S_i^z)$. Then, the order parameter is represented by the vector magnetization $(n = 3)$. In the absence of an external magnetic field, all three directions of the spin are equivalent. Correspondingly, the effective Hamiltonian should be *rotationally invariant* with a group of symmetry called $SO(3)$.[2] Usually, the Hamiltonian is also translationally invariant.

Finally, the physical constraints of *boundedness* and *stability* should be taken into account in the design of a sensible field theory. It is necessary to keep in mind that the probability of a field configuration should remain finite. This implies well-defined mathematical constraints on the sign and magnitude of the coefficients in the

---

[2] Orthogonal transformation in three dimensions with determinant 1.

analytic expansion of the effective Hamiltonian density. Moreover, these coefficients are required to be analytic functions of the external parameters, such as the temperature or pressure.

All these physical constraints lead us to the following standard form of the Landau–Ginzburg (or Landau–Ginzburg–Wilson in the context of renormalization group theory) effective Hamiltonian density for a translationally and rotationally invariant system,[3]

$$\tilde{\mathcal{H}} = K \sum_{i=1}^{n} \sum_{\mu=1}^{d} \partial_\mu \phi_i(\boldsymbol{r}) \partial^\mu \phi_i(\boldsymbol{r}) + A \sum_{i=1}^{n} \phi_i^2(\boldsymbol{r})$$

$$+ B \left(\sum_{i=1}^{n} \phi_i^2(\boldsymbol{r}) \right)^2 - \sum_{i=1}^{n} h_i \, \phi_i(\boldsymbol{r}). \tag{5.37}$$

We have included only a few significant terms. The quartic invariant $\sum_{i=1}^{n} \phi_i^4(\boldsymbol{r})$ and higher-order terms have been omitted. We also added a source term, external field h_i, linearly coupled to the order parameter field. Since space variations of the order parameter should be penalized, the coefficient K is positive and represents the *stiffness*, favoring uniformity of the order parameter. Similarly, the highest-order power in the expansion of eqn (5.37) should have a positive coefficient, $B > 0$, because of stability reasons. We are thus left with the coefficient A. It is clear that the latter coefficient should change sign at the transition point and is the one that drives the transition. Therefore, $A > 0$ at high temperatures, favoring disorder, and $A < 0$ at low temperatures, favoring order.

The Landau mean-field theory of Chapter 2 is recovered as the saddle-point approximation of this effective field theory: It is obtained as the largest single contribution that maximizes the integrand of the functional integral

$$Z = \int \mathcal{D}\Phi(\boldsymbol{r}) \exp\left(-\tilde{H}(\Phi(\boldsymbol{r}))\right). \tag{5.38}$$

Let us illustrate the idea for the scalar field ($n = 1$).

Consider the partition function of eqn (5.38) with a Landau–Ginzburg Hamiltonian

$$\tilde{H}(\phi(\boldsymbol{r})) = \int d^d \boldsymbol{r} \left\{ (\nabla \phi(\boldsymbol{r}))^2 + t\phi(\boldsymbol{r})^2 + u\phi(\boldsymbol{r})^4 - h(\boldsymbol{r})\phi(\boldsymbol{r}) \right\}$$

$$\equiv \tilde{H}_0(\phi(\boldsymbol{r})) - \int d^d \boldsymbol{r} h(\boldsymbol{r})\phi(\boldsymbol{r}). \tag{5.39}$$

We apply the saddle-point approximation of Appendix A.1, which amounts to approximating the integral by its maximum value that corresponds to the most probable configuration $\phi_0(\boldsymbol{r})$. This is determined from

$$\left. \frac{\delta \tilde{H}_0(\phi(\boldsymbol{r}))}{\delta \phi(\boldsymbol{r})} \right|_{\phi_0(\boldsymbol{r})} = h(\boldsymbol{r}). \tag{5.40}$$

---

[3] It is not essential in this chapter to distinguish covariant and contravariant derivatives, ∂_μ and ∂^μ. See Appendix A.6.

Then, the partition function is approximated as

$$Z[h] = e^{-\tilde{H}(\phi_0(\boldsymbol{r}))}. \tag{5.41}$$

The magnetization field satisfies $\phi_m(\boldsymbol{r}) = \phi_0(\boldsymbol{r})$. The saddle-point free energy per unit volume V is minimized for the case of a uniform order parameter field $\phi_m(\boldsymbol{r}) = m$, with the result[4]

$$f = -\frac{T}{V} \log Z[h] = \tilde{t}m^2 + \tilde{u}m^4 - \tilde{h}m, \tag{5.42}$$

where $\tilde{t} = Tt$, $\tilde{u} = Tu$, and $\tilde{h} = Th$. This is the Landau theory of Section 2.3.

It is possible to consider the leading-order correction to the saddle-point approximation by the expansion of $\tilde{H}_0(\phi(\boldsymbol{r}))$ to second order in $\Delta = \phi(\boldsymbol{r}) - \phi_0(\boldsymbol{r})$ and a Gaussian integration. This is known as the *loop expansion*.

EXERCISE 5.2 Consider a system described by the following Landau–Ginzburg effective Hamiltonian (with dimension of energy)

$$T\tilde{H}(\phi(\boldsymbol{r})) = \int d^d\boldsymbol{r} \Big\{ c(\nabla\phi(\boldsymbol{r}))^2 + D(\nabla^2\phi(\boldsymbol{r}))^2 + t\phi(\boldsymbol{r})^2 + u\phi(\boldsymbol{r})^4 \Big\}, \tag{5.43}$$

with coefficients $D, u > 0$, and where $\phi(\boldsymbol{r})$ is a real scalar field. Assume that the other two coefficients may change as

$$c = c_0(\Delta - \Delta_c) \quad , \quad t = t_0(T - T_c) \tag{5.44}$$

as a function of some external parameter Δ and temperature T. Establish the phase diagram in the (Δ, T) plane near (Δ_c, T_c). Show that there are three distinct phases: A paramagnetic, disordered, phase with vanishing order parameter, a spatially homogeneous ordered phase, and an inhomogeneous, i.e. spatially modulated, ordered phase. The three phase boundaries meet at a critical point (Δ_c, T_c) known as the *Lifshitz point*. Determine the order of the phase transitions of the three phase boundaries.

Hint: Use a periodically modulated $\phi(\boldsymbol{r}) = \phi_0 \cos(\boldsymbol{q} \cdot \boldsymbol{r})$, with ϕ_0 a real constant, as the ansatz saddle-point solution.

5.6 Symmetry and its breakdown

Symmetry is one of the key concepts to characterize phase transitions, or physical phenomena at large. For example, it is well known that the invariance of a Hamiltonian or Lagrangian with respect to time translation leads to the conservation of energy. Similarly, invariance with respect to spatial translation is at the origin of momentum conservation. *Noether's theorem* as proved in Appendix A.6 is a general statement about a conservation law resulting from a continuous symmetry of the system.

The existence of symmetries in a physical system is formally expressed by the invariance of the Hamiltonian under the operation of elements of a group that specifies

---

[4] Recall that the temperature T is included in the Landau–Ginzburg effective Hamiltonian.

the symmetry. A simple example is the Ising model without external field, in which the overall (global) change of sign, $S_i \to -S_i \, (\forall i)$, keeps the Hamiltonian invariant. This operation is an element of the group $\mathbb{Z}_2$ consisting of two elements $\{1, -1\}$. The change of sign corresponds to the element -1, whereas the trivial element 1 changes nothing and represents the identity element. Another example of a global symmetry is the Heisenberg model, again in the absence of external field. The rotation of all spins by the same angle, $\boldsymbol{S}_i \to R \cdot \boldsymbol{S}_i \, (\forall i)$, where R is a matrix of rotation in the three-dimensional spin space, leaves Hamiltonian invariant because the interaction $\boldsymbol{S}_i \cdot \boldsymbol{S}_j$ is an inner product. The corresponding group is called $SO(3)$ and the above-mentioned rotation matrix R is a *representation* of this group. The language of group theory is often very useful to describe symmetry properties of a physical system. We give a brief introduction to group theory in Appendix A.7 for the reader's convenience. In particular, the concept and language of Lie group and Lie algebra are of central importance in the following chapter of conformal field theory. Although the text of this book is written as readably as possible without detailed knowledge of group theory, it is nevertheless useful to go through the appendix to better understand the background.

The symmetry of a Hamiltonian should be distinguished from the symmetry of a state of the same system. For instance, the Hamiltonian of the Ising model without external field has the $\mathbb{Z}_2$ symmetry, but a state, for example the all-up state, $S_i = 1 \, (\forall i)$, clearly changes into a different state, $S_i = -1 \, (\forall i)$, by the global reversal of the sign. This observation implies that the symmetry possessed by the Hamiltonian may be broken down in the states that are actually realized in the physical world. This is the phenomenon of *spontaneous symmetry breaking* and has been discussed already several times in this book. It should be noted that a global symmetry can be broken spontaneously but a local symmetry, the invariance of a Hamiltonian by an operation involving only a finite number of local degrees of freedom, is never broken spontaneously. This important comment will be further clarified in Section 7.7.

One of the central remarks in this section is that a symmetry can be broken spontaneously only in the thermodynamic limit. Let us again take the example of the Ising model. A quantitative measure of symmetry breaking is the order parameter, the spontaneous magnetization in the present example. The magnetization per site as a function of the external field, $m_N(h)$, for system size N, is an analytic function of h as long as N is finite. The reason is that the partition function $Z_N(h)$ is a sum of finite number of Boltzmann factors $\mathrm{e}^{-\beta H}$ and thus $m_N(h)$, the logarithmic derivative of $Z_N(h)$ with respect to h, is not a singular function of h. Thus, $m_N(h)$, an odd function of h, has a well-defined limit

$$\lim_{h \to +0} m_N(h) = \lim_{h \to -0} m_N(h) = 0. \tag{5.45}$$

The situation can change if we take the thermodynamic limit first,

$$\lim_{h \to \pm 0} \lim_{N \to \infty} m_N(h) \equiv m_0(\pm 0) \neq 0, \tag{5.46}$$

which defines the spontaneous magnetization m_0. A singularity at $h = 0$ may emerge due to the limiting process $N \to \infty$.

One can illustrate the mechanism leading to a state of spontaneously broken symmetry with a trivial example. Consider again the infinite-range model of Section 2.5, which is $\mathbb{Z}_2$ symmetric, in the presence of an external magnetic field h

$$H = -\frac{J}{2N}\left(\sum_i S_i\right)^2 - h\sum_i S_i. \tag{5.47}$$

The partition function is given by

$$Z_N(h) = \sqrt{\frac{KN}{2\pi}}\int_{-\infty}^{\infty}\mathrm{d}x\,\mathrm{e}^{-NKx^2/2}(2\cosh(Kx + \beta h))^N$$

$$= \sqrt{\frac{KN}{2\pi}}\int_0^{\infty}\mathrm{d}x\,\mathrm{e}^{-NKx^2/2}\left((2\cosh(Kx + \beta h))^N + (2\cosh(Kx - \beta h))^N\right), \tag{5.48}$$

where $K = \beta J$. This equation clearly indicates that $Z_N(h)$ is an even function of h, $Z_N(h) = Z_N(-h)$. The evaluation of the magnetization proceeds as

$$m_N(h) = \frac{1}{N}\frac{\partial}{\partial(\beta h)}\log Z_N(h)$$

$$= \frac{\displaystyle\int_{-\infty}^{\infty}\mathrm{d}x\,\mathrm{e}^{-NKx^2/2}\left(2\cosh(Kx + \beta h)\right)^N\tanh(Kx + \beta h)}{\displaystyle\int_{-\infty}^{\infty}\mathrm{d}x\,\mathrm{e}^{-NKx^2/2}\left(2\cosh(Kx + \beta h)\right)^N}$$

$$= -m_N(-h), \tag{5.49}$$

and $m_N(h)$ is an odd function of h, as expected. It is clear from the equation above that

$$\lim_{h\to\pm0}m_N(h) = 0, \tag{5.50}$$

and obviously

$$\lim_{N\to\infty}\lim_{h\to\pm0}m_N(h) = 0. \tag{5.51}$$

On the other hand, if one takes the thermodynamic limit first, keeping h finite,

$$\lim_{N\to\infty}m_N(h) = \tanh(Km_0(h) + \beta h), \tag{5.52}$$

where we have used the saddle-point method to compute the integrals in eqn (5.49). The quantity m_0 is determined from the maximization of the function

$$g(x) = -\frac{Kx^2}{2} + \log\left(2\cosh(Kx + \beta h)\right), \tag{5.53}$$

which appears in the integrands as $\mathrm{e}^{Ng(x)}$, in the limit of large N. Not surprisingly (see Section 2.5), the value of x that maximizes $g(x)$, i.e. m_0, satisfies the self-consistent mean-field equation $m_0(h) = \tanh(Km_0(h) + \beta h)$. Therefore, for temperatures

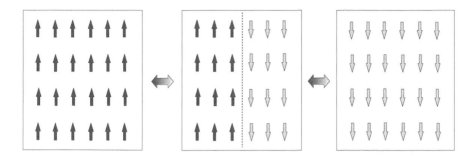

Fig. 5.2 In order for a ferromagnetic Ising system to go from a state with positive magnetization to another state with negative magnetization, an intermediate state with domain walls must be realized. For an infinitely large system, the creation of domain wall(s), indicated in a dotted line, costs an infinitely large energy.

T below the critical temperature $T_c = J$, spontaneous symmetry breaking (or spontaneous magnetization) manifests itself as the singular limit

$$\lim_{h \to \pm 0} \lim_{N \to \infty} m_N(h) = m_0(\pm 0) \neq 0. \tag{5.54}$$

Likewise, for $T > T_c$, $\lim_{h \to \pm 0} \lim_{N \to \infty} m_N(h) = 0$.

A related concept is *ergodicity breaking*. If we take the thermodynamic limit first, it becomes impossible that the system explores the whole phase space when symmetry breaking occurs for discrete symmetry. The system becomes trapped in a part of the phase space. Again, in the simple example of the ferromagnetic Ising model without external field, only the subspace of states satisfying $\sum_i S_i > 0$ can be accessed by the system when the symmetry is broken such that $\lim_{h \to +0} \lim_{N \to \infty} m_N(h) > 0$. The other subspace with $\sum_i S_i < 0$ is out of reach because the system has to go through the barrier of (infinitely) high (free) energy separating these two subspaces, see Fig. 5.2. Thus, ergodicity, which means that the system reaches all possible states, is broken.

The state of broken symmetry is characterized by *long-range order*, which is a very similar notion to spontaneous symmetry breaking but not exactly the same. Long-range order is defined by the existence of a finite (non-vanishing) limit of the two-point correlation function,

$$\lim_{h \to 0} \lim_{|\boldsymbol{r} - \boldsymbol{r}'| \to \infty} \lim_{N \to \infty} \langle O(\boldsymbol{r}) O(\boldsymbol{r}') \rangle \neq 0, \tag{5.55}$$

with $O(\boldsymbol{r})$ being the local order parameter, such as $\phi(\boldsymbol{r})$ for the ϕ^4 model and S_i for the Ising model. It is assumed that $O(\boldsymbol{r})$ is chosen such that the simple average $\langle O(\boldsymbol{r}) \rangle$ vanishes in the zero-field limit $h \to 0$ for a finite-size system.[5] Intuitively, if a symmetry is broken, the system is ordered in a global scale and the value of $O(\boldsymbol{r})$ at $\boldsymbol{r}$ is strongly correlated with the same quantity at a far position $\boldsymbol{r}'$. Thus, long-range order follows. To prove rigorously the equivalence of the existence of long-range order

---

[5] The Potts spin degree of freedom in its simple form does not satisfy this criterion. See Exercise 8.5 for more details.

and spontaneous symmetry breaking is a non-trivial mathematical problem, although physically quite plausible.

Also to be noted is the property known as *clustering*, in which the limit of eqn (5.55) reduces to the product of order parameters,

$$\lim_{h\to\pm0}\lim_{|r-r'|\to\infty}\lim_{N\to\infty}\langle O(r)O(r')\rangle = \lim_{h\to\pm0}\lim_{N\to\infty}\langle O(r)\rangle \cdot \lim_{h\to\pm0}\lim_{N\to\infty}\langle O(r')\rangle. \quad (5.56)$$

The sign of h should coincide in these limiting procedures.

5.7   Nambu–Goldstone modes

Attempts to change the particular broken-symmetry state to other possible broken-symmetry configurations cost energy along certain directions in the order-parameter space, as illustrated in Fig. 5.2 for the breaking of a discrete symmetry $\mathbb{Z}_2$. This may be interpreted as the system displaying a generalized rigidity or stiffness. If a continuous symmetry is spontaneously broken, on the other hand, the spectrum of the Hamiltonian generically has gapless collective excitations or soft modes. A soft mode means that changes along particular directions in the space of order parameters require no energy. The mexican-hat potential shown in Fig. 5.3 is a typical example, in which the system can move freely along the bottom of the potential.

These *emergent excitations* characterizing the ordered phase are known as the *Nambu–Goldstone modes*. Examples of these low-energy excitations include spin waves in the XY model, as explained in Chapter 7 (because of the spontaneous breaking of the spin rotational symmetry) and acoustic phonons in crystalline solids (because of the breaking of the space translational symmetry). The existence of these low-energy excitations is the essence of the *Goldstone theorem*. This theorem states essentially that, whenever a continuous symmetry is spontaneously broken in a system with short-range interactions, there exist modes with zero excitation energy and a continuous spectrum above it. As will be exemplified later, this fact manifests itself

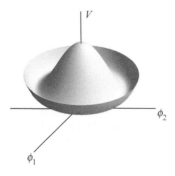

Fig. 5.3 A continuum of minima exists for the mexican-hat-type potential V. The system can continuously change its lowest-energy state from one of the continuous minima of the potential to another one without energy cost.

in the behavior of a correlation function in the Fourier space where that correlation function has a pole at zero wave number.

Let us present an example to understand the physics behind this theorem. A more formal manipulation to show the zero-energy mode is given in Appendix A.6.5. Consider eqn (5.37) with $n = 2$, a two-component system,

$$\tilde{\mathcal{H}}_0 = \big(\nabla\phi_1(\boldsymbol{r})\big)^2 + \big(\nabla\phi_2(\boldsymbol{r})\big)^2 + t\big(\phi_1(\boldsymbol{r})^2 + \phi_2(\boldsymbol{r})^2\big) + u\big(\phi_1(\boldsymbol{r})^2 + \phi_2(\boldsymbol{r})^2\big)^2. \quad (5.57)$$

We have set $K = 1, A = t, B = u$ and $h_i = 0$ in eqn (5.37). This system has the mexican-hat potential

$$V = t\big(\phi_1(\boldsymbol{r})^2 + \phi_2(\boldsymbol{r})^2\big) + u\big(\phi_1(\boldsymbol{r})^2 + \phi_2(\boldsymbol{r})^2\big)^2, \quad (5.58)$$

as a function of (ϕ_1, ϕ_2), as shown in Fig. 5.3 for $t < 0$. The Hamiltonian is

$$\tilde{H}_0(\Phi(\boldsymbol{r})) = \int \mathrm{d}^d\boldsymbol{r}\ \tilde{\mathcal{H}}_0, \quad (5.59)$$

with $\Phi(\boldsymbol{r}) = (\phi_1(\boldsymbol{r}), \phi_2(\boldsymbol{r}))$. This Hamiltonian displays an $SO(2)$ symmetry that involves transformations of the form

$$\begin{pmatrix} \phi'_1(\boldsymbol{r}) \\ \phi'_2(\boldsymbol{r}) \end{pmatrix} = \begin{pmatrix} \cos\theta & \sin\theta \\ -\sin\theta & \cos\theta \end{pmatrix} \begin{pmatrix} \phi_1(\boldsymbol{r}) \\ \phi_2(\boldsymbol{r}) \end{pmatrix} \quad (5.60)$$

leaving $\tilde{H}_0(\Phi(\boldsymbol{r}))$ invariant. As in the scalar case $(n = 1)$, when $t > 0$, the system is in the disordered phase with $\Phi(\boldsymbol{r}) = 0$. On the other hand, if $t < 0$, the potential V displays a set of minima at

$$\Phi(\boldsymbol{r})^2 = \phi_1(\boldsymbol{r})^2 + \phi_2(\boldsymbol{r})^2 = -\frac{t}{2u} > 0, \quad (5.61)$$

signaling a broken-symmetry, i.e. an ordered phase. This solution is infinitely degenerate: Any state in the circle labeled by the angle $\theta \in [0, 2\pi)$ is a possible and legitimate physical solution. A way to select one of those possible solutions is to apply an external field $\boldsymbol{h} = (h_1, h_2)$

$$\tilde{H}(\Phi(\boldsymbol{r})) = \tilde{H}_0(\Phi(\boldsymbol{r})) - \int \mathrm{d}^d\boldsymbol{r}\ \big(h_1\phi_1(\boldsymbol{r}) + h_2\phi_2(\boldsymbol{r})\big), \quad (5.62)$$

and choose, for instance, $\boldsymbol{h} = (h_1, 0)$ without loss of generality. This choice amounts to selecting a corresponding direction in the order parameter space

$$\tilde{\Phi}(\boldsymbol{r}) = (\tilde{\phi}_1(\boldsymbol{r}), \tilde{\phi}_2(\boldsymbol{r})) = \left(\sqrt{\frac{|t|}{2u}}, 0\right) \equiv (a, 0) \quad (5.63)$$

among the possible solutions of eqn (5.61) for $t < 0$.

We are interested in the stability of the state of broken symmetry of eqn (5.63). To this end, we write

$$\phi_1(\boldsymbol{r}) = a + \delta\phi_1(\boldsymbol{r}), \quad \phi_2(\boldsymbol{r}) = \delta\phi_2(\boldsymbol{r}) \quad (5.64)$$

and expand $\tilde{\mathcal{H}}_0$ to second order in $\delta\phi_1$ and $\delta\phi_2$. The former $\delta\phi_1$ is used to see the longitudinal stability because it is parallel to the vector order parameter $\tilde{\Phi} = (a, 0)$, and the latter $\delta\phi_2$ is a transverse perturbation to $\tilde{\Phi}$. The result is

$$\tilde{\mathcal{H}}_0 = \left(\nabla\delta\phi_1(\boldsymbol{r})\right)^2 + \left(\nabla\delta\phi_2(\boldsymbol{r})\right)^2 + t\left((a + \delta\phi_1(\boldsymbol{r}))^2 + (\delta\phi_2(\boldsymbol{r}))^2\right)$$

$$+ u\left((a + \delta\phi_1(\boldsymbol{r}))^2 + (\delta\phi_2(\boldsymbol{r}))^2\right)^2$$

$$= ta^2 + ua^4 + \left(\nabla\delta\phi_1(\boldsymbol{r})\right)^2 + (t + 6ua^2)\left(\delta\phi_1(\boldsymbol{r})\right)^2$$

$$+ \left(\nabla\delta\phi_2(\boldsymbol{r})\right)^2 + (t + 2ua^2)\left(\delta\phi_2(\boldsymbol{r})\right)^2$$

$$= -\frac{t^2}{4u} + \left(\nabla\delta\phi_1(\boldsymbol{r})\right)^2 + 2|t|\left(\delta\phi_1(\boldsymbol{r})\right)^2 + \left(\nabla\delta\phi_2(\boldsymbol{r})\right)^2. \tag{5.65}$$

A first observation is that $\delta\phi_1$ and $\delta\phi_2$ are decoupled (no cross-term $\delta\phi_1\delta\phi_2$) and hence the longitudinal and transverse modes can be analyzed separately. Secondly, a comparison with eqn (2.74) reveals that both fluctuations $\delta\phi_1$ and $\delta\phi_2$ are described by the Gaussian model. Thus, the present Hamiltonian reads in the Fourier representation, according to eqn (2.80),

$$\tilde{H} = \int \frac{d^d q}{(2\pi)^d} \left((2|t| + q^2)\, \delta\tilde{\phi}_1(\boldsymbol{q})\delta\tilde{\phi}_1(-\boldsymbol{q}) + q^2 \delta\tilde{\phi}_2(\boldsymbol{q})\delta\tilde{\phi}_2(-\boldsymbol{q})\right), \tag{5.66}$$

where we ignored the trivial constant term in eqn (5.65). If we interpret the coefficient of $\delta\tilde{\phi}_i(\boldsymbol{q})\delta\tilde{\phi}_i(-\boldsymbol{q})$ as the excitation energy of the mode i for the wave number (momentum) $\boldsymbol{q}$, the longitudinal mode ($i = 1$) has a positive lowest excitation energy $2|t|$ for $\boldsymbol{q} = 0$. In contrast, the transverse mode has a zero excitation energy for $\boldsymbol{q} = 0$ and a continuum spectrum q^2 above this zero-energy mode. Physically, the transverse mode $\delta\phi_2$ changes the broken-symmetry state $\tilde{\Phi} = (a, 0)$ to another ground state at a slightly different position along the bottom of the mexican-hat potential. It is clear that this change of the position costs no energy. The longitudinal mode, on the other hand, has a finite excitation energy because the system should leave the bottom of the potential if it tries to change the magnitude of the order parameter from $\tilde{\Phi} = (a, 0)$ along the first axis (i.e. keeping $\delta\phi_2 = 0$).

These statements may be re-expressed in terms of the correlation functions

$$\tilde{G}_{11}(\boldsymbol{q}) = \langle\delta\tilde{\phi}_1(\boldsymbol{q})\,\delta\tilde{\phi}_1(-\boldsymbol{q})\rangle = \frac{(2\pi)^d}{4|t| + 2q^2} \tag{5.67}$$

$$\tilde{G}_{22}(\boldsymbol{q}) = \langle\delta\tilde{\phi}_2(\boldsymbol{q})\,\delta\tilde{\phi}_2(-\boldsymbol{q})\rangle = \frac{(2\pi)^d}{2q^2} \tag{5.68}$$

$$\tilde{G}_{12}(\boldsymbol{q}) = \tilde{G}_{21}(\boldsymbol{q}) = \langle\delta\tilde{\phi}_1(\boldsymbol{q})\,\delta\tilde{\phi}_2(-\boldsymbol{q})\rangle = 0, \tag{5.69}$$

which have been derived using eqn (2.84) with $T = 1$. Equation (5.68) indicates that the transverse correlation function in the Fourier representation has a pole at the origin, which is an important characteristic of the Nambu–Goldstone mode, as stated at the beginning of this section.

If the interactions are not short ranged, the Hamiltonian is not expressed just by the squared gradients of fields, and the above arguments do not apply directly.

5.8 Topological defects

Another consequence of the spontaneous breaking of a symmetry is the emergence of defect structures such as vortices in superfluids, domain walls in ferromagnets, dislocations in periodic solids, or disclinations in nematic liquid crystals. These *topological defects* are responsible for determining important properties of real materials, such as *strain hardening*, which is the strengthening of a metal when subjected to plastic deformation. Although the energy of a macroscopic system is minimized when the symmetry is broken uniformly throughout the system, it turns out that the symmetry may be broken differently in different parts of the sample due to a variety of reasons. Under those circumstances, defects will appear, for instance, in the boundary separating those spatial regions characterized by states or configurations with different values of the order parameter Φ. For example, in a ferromagnet a domain wall may separate regions with different values or orientations of the macroscopic magnetization as shown in Fig. 5.2.

In this section we will show how concepts borrowed from topology provide the necessary tools to characterize, classify, and combine elementary defects. Let us start from the concept of an *order-parameter space* $\mathcal{U}$. Loosely speaking, this is the set of all possible values of the order parameter $\Phi(\boldsymbol{r})$ in a d-dimensional space, $\boldsymbol{r} \in \mathbb{R}^d$. A simple example is the XY model in two spatial dimensions. Suppose that low-temperature spin configurations in a subspace of $\mathbb{R}^2$, sometimes referred to as the *ordered medium*, are mapped to a set of points in the space $\mathbb{S}^1$ (the unit circle) by the rule

$$\frac{\Phi(\boldsymbol{r})}{|\Phi(\boldsymbol{r})|} = \big(\cos \phi(\boldsymbol{r}), \sin \phi(\boldsymbol{r}) \big) \in \mathbb{S}^1, \ \forall \boldsymbol{r} \in \Gamma \in \mathbb{R}^2, \tag{5.70}$$

where Γ is a closed loop in $\mathbb{R}^2$. See Fig. 5.4. In this case, $\mathbb{S}^1$ is the order-parameter space. Another familiar example is the Heisenberg model, in which the spin orientation at $\boldsymbol{r} \in \mathbb{R}^d$ is specified by a three-dimensional unit vector. The order-parameter space is then $\mathcal{U} = \mathbb{S}^2$, the surface of the unit sphere. Notice that the magnitude of the local (spin) variable is ignored and only its direction is considered in the analysis of topological defects. Examples of order-parameter spaces, $\mathbb{S}^1$, $\mathbb{S}^2$ and $\mathbb{P}^2$, are illustrated in Fig. 5.5.

One of the reasons to introduce the order-parameter space is its advantage in the classification of topological defects and their stabilities. Consider again the XY model in two dimensions. As shown in Fig. 5.4, the existence of a vortex, a typical topological defect, significantly influences the image in the order-parameter space. In panels (a) and (b), there is no vortex surrounded by the loop Γ, and the images in $\mathbb{S}^1$, drawn bold, are essentially the same in the sense that we can continuously deform the image in (b) into a single point as in (a). In contrast, the images in (c) and (d) are equivalent to each other but cannot be continuously reduced to a point as in (a) since those in (c) and (d) wind the circumference of $\mathbb{S}^1$. We say in the latter case that the *winding number* is $k = 1$. Also, in the real space $\mathbb{R}^2$, we can continuously change the spin configuration of (b) into (a) and also (c) changes into (d), the latter by rotating each

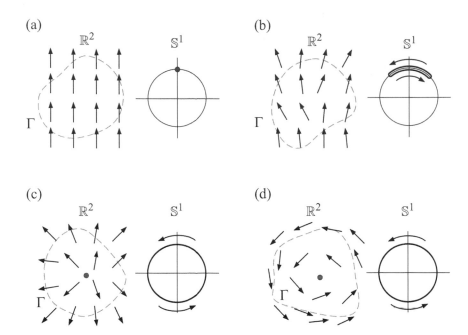

Fig. 5.4 The spin configurations in the real space $\mathbb{R}^2$ and the corresponding images in the order parameter spaces $\mathbb{S}^1$ for the two-dimensional XY model. The angle of each arrow along a loop Γ, shown dashed, in $\mathbb{R}^2$ is mapped to the corresponding point on $\mathbb{S}^1$. In (a) and (b) there is no vortex (topological defect) and the set of mapped points in (b) (drawn in bold) can be continuously deformed to a single point in (a). In (c) and (d), the loop Γ encircles a vortex and the corresponding images wind the circumference of $\mathbb{S}^1$. The arrows in the order-parameter space indicate the sense of motion of the images as one circles around the loop Γ in the real space.

spin roughly by 90°. Such a continuous rotation never succeeds in changing (a) to (c). The two configurations (a) and (b) are said to belong to the same *homotopy class*. Similarly, the homotopy class of (c) and (d) is the same. In general, two configurations are *equivalent* and belong to the same homotopy class if one of them can be deformed continuously to the other.

The winding number, which is formally defined as

$$k = \frac{1}{2\pi} \oint_\Gamma \nabla\phi(\boldsymbol{r}) \cdot \mathrm{d}\boldsymbol{r}, \qquad (5.71)$$

quantitatively characterizes the homotopy class and is an example of a *topological invariant*. The latter name comes from the stability of the winding number under continuous deformation. Remember that we refer to the topological stability, not the thermodynamics stability. Nevertheless, the former often leads to the latter, as will be studied in detail in Chapter 7.

It is possible to create a vortex in the XY model with winding numbers other than 0 (no vortex) or 1 by the field configuration $\phi(\boldsymbol{r}) = k\theta + \mathrm{const}$, where θ is the polar angle

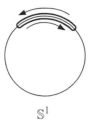
$\mathbb{S}^1$

$\mathbb{S}^2$

$\mathbb{P}^2$

Fig. 5.5 The order parameter spaces $\mathbb{S}^1$, $\mathbb{S}^2$ and $\mathbb{P}^2$ and images of closed loops in the real space. The left and center panels represent the spaces for the XY and Heisenberg models, respectively. The right panel corresponds to the nematic liquid crystal in $d = 3$, in which rod-like molecules are oriented as in the Heisenberg model but without the sense of the arrows. Thus, the up and down orientations are identified, and consequently only the upper half of a sphere constitutes the order parameter space known as $\mathbb{P}^2$. Correspondingly, the two points marked in black dots are identical and the curve drawn around the half sphere is a closed loop.

of the position vector $\boldsymbol{r}$ and k is an integer, $k \in \mathbb{Z}$. The image of spin configurations in $\mathbb{S}^1$ winds the circumference k times in this case. This fact is written formally as $\pi_1(\mathbb{S}^1) = \mathbb{Z}$. The subscript 1 of π_1 means that the loop Γ, a one-dimensional object, is used to map the configuration in $\mathbb{R}^2$ to $\mathbb{S}^1$. Thus, the homotopy class is classified by the group $\mathbb{Z}$, in which the usual rule of addition represents the multiplication as a group. The addition (or the aggregation) of two vortices of winding numbers k_1 and k_2 realizes a single vortex with winding number $k = k_1 + k_2$. This is one of the simplest examples of the *homotopy group*.

These discussions can be generalized to arbitrary types of topological defects, not just in spin systems but also in solids and liquid crystals. We, however, restrict ourselves to spin systems for simplicity of presentation and consider next the XY model in three spatial dimensions $\mathbb{R}^3$. The basic topological defect is a line of vortices created by stacking vortices, each on a two-dimensional cross-section. We write d_{d} for the dimensionality of the defect, and $d_{\mathsf{d}} = 1$ for a vortex line in $\mathbb{R}^3$. The previous case of two dimensions has $d_{\mathsf{d}} = 0$ since a vortex is an isolated point in $\mathbb{R}^2$. In the case of $\mathbb{R}^3$, the order parameter space is again $\mathbb{S}^1$ because the orientation of a spin can be specified by a point on $\mathbb{S}^1$. The homotopy group is also the same, $\pi_1(\mathbb{S}^1) = \mathbb{Z}$, each element being specified by the winding number counted along a loop surrounding a vortex line.

The Heisenberg model has three components and its order parameter space is $\mathbb{S}^2$ as mentioned already. The basic topological defect is a hedgehog structure, and in its most basic form, all spins point outward on the surface of $\mathbb{S}^2$. The topological invariant is the *wrapping number*. The above-mentioned simplest hedgehog has the wrapping number 1. The relevant homotopy group is $\pi_2(\mathbb{S}^2) = \mathbb{Z}$. The subscript 2 of π_2 is meant for the two-dimensional sphere that wraps the topological defect. If we choose a loop, a one-dimensional object, to encircle a topological defect of the Heisenberg model, the loop slips on the surface of the order parameter space $\mathbb{S}^2$ to eventually shrink to a point as depicted in Fig. 5.6. This fact is written as $\pi_1(\mathbb{S}^2) = 0$,

Fig. 5.6 A loop around a two-dimensional sphere slips on the surface by a continuous deformation and shrinks to a point.

Suppose in general that we surround a topological defect by an m-dimensional object Γ^m $(m < d)$ and consider an m-dimensional spherical closed surface $\mathbb{S}^m$. To surround a d_{d}-dimensional defect, one needs a sphere of dimension

$$m = d - d_{\mathsf{d}} - 1. \tag{5.72}$$

In the previous example of a vortex with $d_{\mathsf{d}} = 0$ in a $d = 2$ ordered medium, the surrounding object was a loop of dimension $m = 1$. If $d = 3$, for a loop to continue being the relevant surrounding object, the vortex should constitute a line defect, i.e. $d_{\mathsf{d}} = 1$. Similarly, a $d_{\mathsf{d}} = 0$ hedgehog defect in a three-dimensional ordered medium requires a surrounding object of dimension $m = 2$. In general, for a defect of dimensionality d_{d} to be topologically stable in a d-dimensional ordered medium, the mth homotopy group should not be trivial, $\pi_m(\mathcal{U}) \neq 0$. The *fundamental group*, which means $\pi_1(\mathcal{U})$, may be non-Abelian, but $\pi_m(\mathcal{U})$ with $m > 1$ is known to be always Abelian. A little more formal introduction to homotopy theory is found in Appendix A.8. Table 5.1 summarizes the discussion above.

When these ideas are applied to an ordered medium with an n-component order parameter $\Phi = (\phi_1, \cdots, \phi_n)$, it is easy to imagine that $\mathcal{U} = \mathbb{S}^{n-1}$ and we find

$$\pi_m(\mathbb{S}^m) = \mathbb{Z}, \quad \pi_l(\mathbb{S}^m) = 0 \ (l < m). \tag{5.73}$$

The latter relation for $l < m$ implies that an l-dimensional closed object on $\mathbb{S}^m$ with $l < m$ can always be shrunk to a point, which is a generalization of $\pi_1(\mathbb{S}^2) = 0$. We

Table 5.1 Types of defects, e.g. point defects, that may appear in a physical system of a given spatial dimensionality d. Textures refer to smooth d-dimensional configurations with fixed constant boundary conditions.

$d - d_{\mathsf{d}} - 1$	0	1	2	3
$d = 1$	point	texture		
$d = 2$	line	point	texture	
$d = 3$	surface	line	point	texture

conclude that, in order to have a topologically stable defect of dimension d_d, the relation

$$d_d = d - n \tag{5.74}$$

must be satisfied as eqn (5.72) with $m = n - 1$ suggests. Table 5.1 shows the types of defects.

In the case of an ordered medium with a discrete symmetry group, such as the $\mathbb{Z}_2$ symmetry of the Ising model, topological defects have the dimensionality $d_d = d - 1$. They are always generalized *domain walls* separating regions with different values of the order parameter Φ.

6
Conformal field theory

We have seen that a statistical system is scale invariant at criticality. Physical properties remain the same if we change the length scale by a constant factor. It is then natural to generalize the hypothesis of global scale invariance to an invariance under a coordinate-dependent local scaling factor. It turns out that this approach is enormously successful in two dimensions and produces a number of remarkable results. The present chapter is an introductory account of the basic concepts and important consequences of conformal symmetry, i.e. the invariance under local scale transformations, in field theories characterizing critical behavior. The goal is to catalog universality classes as a list of possible values of critical exponents and to find restrictions on the functional forms of correlation functions. From a mathematics standpoint, conformal symmetry applies to continuum theories, and therefore its obvious application to critical phenomena is formulated in the language of field theory. In this chapter, we do not discuss the microscopic model that gave origin to a particular continuum field theory, or in other words, that belongs to the same universality class. We will assume that such a statistical field theory exists and will study the physical and mathematical consequences of that theory being conformally invariant in the critical regime.

6.1   From scale invariance to conformal symmetry

In Chapters 1 and 3 we discussed the remarkable fact that scale invariance emerges close to a critical point. A scale transformation is mathematically represented as a *dilation*, i.e. a coordinate transformation $r \rightarrow r' = b^{-1}r$ with b a positive number. The hypothesis of scale invariance leads to many conclusions, for instance, that all critical exponents can be expressed in terms of a few scaling parameters, typically y_t and y_h. Then, it seems natural to think that at criticality (a point of self-similarity) more symmetries could emerge: One may wonder whether in the critical region a coordinate-dependent scale invariance is possible, i.e. $r \rightarrow r' = b(r)^{-1} \, r$, which would certainly have further implications. This is the extension of scale invariance to *conformal invariance*. The hypothesis of conformal invariance appears to be quite generally true in critical equilibrium systems as a result of the essential locality of the underlying statistical field theory. We will assume conformal invariance of critical field theories throughout this chapter.

   The predictive power of the use of conformal symmetry depends on the dimensionality d of the system under study. For $d \geq 3$, conformal symmetry fixes possible functional forms of some of the correlation functions. In two dimensions, much stronger results can be derived because the set of conformal transformations includes analytic functions of complex numbers and there are infinitely many of them. Conformal field theory then becomes an exercise in the theory of complex variables and analytic functions. As a consequence, a number of remarkable results emerge such as a list of possible critical exponents, admissible forms of correlation functions at criticality, and constraints on finite-size effects. Our discussions in the present chapter will therefore be focused on the two-dimensional case.

6.2 Conformal transformation

Conformal transformations are special coordinate mappings. A *conformal transformation* (*conformal mapping*) is an invertible map $\boldsymbol{r} \to \boldsymbol{r}'$ of the space $\mathbb{R}^d$ (or part of it) into itself (or part of it) that preserves angles between any two vectors but not necessarily their length scales.[1]

For $d \geq 3$ the conformal group, the set of conformal transformations, is composed of a finite number of elements, i.e. the following transformations,

$$\text{Translation by a constant vector } \boldsymbol{a} : \boldsymbol{r} \to \boldsymbol{r} + \boldsymbol{a} \tag{6.1}$$

$$\text{Rotation by a matrix } R : \boldsymbol{r} \to R\boldsymbol{r} \tag{6.2}$$

$$\textit{Dilation (dilatation) by a scale factor } b : \boldsymbol{r} \to b^{-1}\boldsymbol{r} \tag{6.3}$$

$$\text{Special conformal transformation} : \boldsymbol{r} \to \frac{\boldsymbol{r} + a\boldsymbol{r}^2}{1 + 2\boldsymbol{a} \cdot \boldsymbol{r} + a^2 r^2}. \tag{6.4}$$

The special conformal transformation is a combination of space inversion, translation, and another inversion,

$$\boldsymbol{r}' = \frac{\dfrac{\boldsymbol{r}}{r^2} + \boldsymbol{a}}{\left(\dfrac{\boldsymbol{r}}{r^2} + \boldsymbol{a}\right)^2}. \tag{6.5}$$

Equations (6.1) to (6.4) constitute *global conformal transformations*, which means mapping the whole space onto itself. In Appendix A.9 it is shown that these global mappings exhaust the possible conformal transformations for $d \geq 3$.

The two-dimensional case $d = 2$ is special in that an additional set of transformations are conformal because all analytic (holomorphic) functions of complex variables preserve local angles, as is known in complex analysis. Let us therefore introduce a complex coordinate,

$$z = r^1 + \mathrm{i}r^2 \ , \ \bar{z} = r^1 - \mathrm{i}r^2, \tag{6.6}$$

where $\boldsymbol{r} = (r^1, r^2)$ is the Cartesian coordinate. The two complex numbers z and $\bar{z}$ are considered independent variables because the degree of freedom has originally two,

---

[1] If we write $g_{\mu\nu}(\boldsymbol{r})$ for the metric tensor of the d-dimensional space under consideration, a conformal transformation is formally defined as a mapping that leaves the metric tensor invariant up to a scale, $g'_{\mu\nu}(\boldsymbol{r}') = \Omega(\boldsymbol{r})g_{\mu\nu}(\boldsymbol{r})$. Thus conformal transformations change the actual geometry of space.

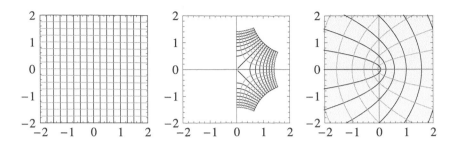

Fig. 6.1 Examples of conformal mappings in two dimensions. These maps correspond to the transformations $f(z) = z, \sqrt{z}$ and z^2 from left to right, respectively $((r^1, r^2) \in [-2, 2])$.

r^1 and r^2 components. Then, any analytic function $f(z)$, such that $f'(0) \neq 0$, defines holomorphic and antiholomorphic transformations

$$z \to f(z) , \quad \bar{z} \to \bar{f}(\bar{z}), \tag{6.7}$$

which are conformal around the origin $z = 0$. Similarly to z and $\bar{z}$, f and $\bar{f}$ are independent functions. For example, $f(z) = z^4$ is analytic, while $f(z) = |z| z^4$ is not and thus does not represent a conformal map. See Fig. 6.1 for other examples. Dilations and rotations are given by $z \to b^{-1} z$, $\bar{z} \to b^{-1} \bar{z}$, and $z \to e^{i\theta} z$, $\bar{z} \to e^{-i\theta} \bar{z}$, respectively.

In two dimensions, for flat space, the angle-preserving condition is well known to be rewritten as the Cauchy–Riemann equations

$$\frac{\partial r'^2}{\partial r^1} = -\frac{\partial r'^1}{\partial r^2}, \qquad \frac{\partial r'^1}{\partial r^1} = \frac{\partial r'^2}{\partial r^2}, \tag{6.8}$$

where we have written $f = r'^1 + i r'^2$, and the equivalent equations for antiholomorphic functions. In terms of complex coordinates, the Cauchy–Riemann equations (6.8) become

$$\partial_{\bar{z}} f(z, \bar{z}) = 0, \tag{6.9}$$

where $\partial_{\bar{z}} = \partial/\partial \bar{z} = (\partial_{r^1} + i \partial_{r^2})/2$ and $\partial_z = \partial/\partial z = (\partial_{r^1} - i \partial_{r^2})/2$. This is the mathematical statement that the conformal mapping must be holomorphic since $f(z, \bar{z}) = f(z)$ without $\bar{z}$ dependence. Similarly, we find $\partial_z \bar{f}(z, \bar{z}) = 0$, which implies $\bar{f}(z, \bar{z}) = \bar{f}(\bar{z})$.

Holomorphic and antiholomorphic functions define *local conformal transformations* in the sense they are not analytic in the whole complex plane except for the trivial case of a constant, as is well known in complex analysis.

In two dimensions the global conformal transformation of eqns (6.1) to (6.4) is summarized in a compact form

$$f(z) = \frac{az + b}{cz + d} \quad (ad - bc = 1, \ a, b, c, d \in \mathbb{C}), \tag{6.10}$$

which reproduces any one of eqns (6.1) to (6.4) by an appropriate choice of the coefficients a, b, c and d. The condition $ad - bc = 1$ comes from invertibility of the map, $ad - bc \neq 0$, which can be reduced to $ad - bc = 1$ by an appropriate normalization of the constants. Equation (6.10) for a global conformal mapping is called a *projective mapping* or a *Möbius mapping*.

EXERCISE 6.1 Confirm that eqn (6.10) can be reduced to any one of eqns (6.1) to (6.4).

Since the number of holomorphic/antiholomorphic functions is not limited, there exist an infinite number of local conformal transformations in two dimensions. This fact makes conformal symmetry a very powerful tool to analyze and characterize critical behavior in two dimensions: More symmetries impose more constraints. It ultimately provides a classification of all possible critical theories, as will be seen later in this chapter.

The decoupling between holomorphic and antiholomorphic coordinates is characteristic of conformal field theories in two dimensions. Let us consider infinitesimal conformal transformations

$$z \to z' = f(z) = z + \epsilon(z) \ , \ \bar{z} \to \bar{z}' = \bar{f}(\bar{z}) = \bar{z} + \bar{\epsilon}(\bar{z}), \tag{6.11}$$

where $\epsilon(z)$ and $\bar{\epsilon}(\bar{z})$ are holomorphic and antiholomorphic functions. We think of these functions as infinitesimal, although we have to keep in mind that it is not possible for them to be globally small unless they are constant. We will write them as Laurent expansions around $z = 0$

$$\epsilon(z) = \sum_{n=-\infty}^{\infty} \epsilon_n z^{n+1} \ , \ \bar{\epsilon}(\bar{z}) = \sum_{n=-\infty}^{\infty} \bar{\epsilon}_n \bar{z}^{n+1}. \tag{6.12}$$

It is illuminating to associate an algebra to such an infinitesimal transformation by noticing that

$$f(z + \epsilon(z)) - f(z) \approx f(z) + \epsilon(z)\partial_z f(z) - f(z) \equiv -\sum_{n=-\infty}^{\infty} \epsilon_n \ell_n f(z), \tag{6.13}$$

and similarly for the antiholomorphic part, using the generators of local conformal mappings

$$\ell_n = -z^{n+1}\partial_z \ , \ \bar{\ell}_n = -\bar{z}^{n+1}\partial_{\bar{z}}. \tag{6.14}$$

These generators form an infinite-dimensional Lie algebra called the *loop algebra*,[2]

$$[\ell_m, \ell_n] = (m-n)\ell_{m+n} \ , \ [\bar{\ell}_m, \bar{\ell}_n] = (m-n)\bar{\ell}_{m+n} \ , \ [\ell_m, \bar{\ell}_n] = 0, \tag{6.15}$$

as can be checked from eqn (6.14). This is infinite dimensional because n and m run from $-\infty$ to ∞. The holomorphic and antiholomorphic parts decouple due to the last commutation relation of eqn (6.15).

The subalgebra generated by the subset with $m, n = 0, \pm 1$ is closed, as one can verify by inserting $n, m = -1, 0, 1$ in eqn (6.15). This subalgebra is a set of generators of the global conformal mappings (6.1) to (6.4). The reason is that eqn (6.14) is

---

[2] See Appendix A.7 for the concept of Lie algebra.

non-singular at $z = 0$ if and only if $n \geq -1$ and, additionally, non-singular at $z \to \infty$ if and only if $n \leq 1$, as seen from the inversion $z \to 1/z$. The same holds for the antiholomorphic part. This implies that a global mapping is possible if and only if $-1 \leq n \leq 1$. Indeed it will be shown in the next section that ℓ_{-1} and $\bar{\ell}_{-1}$ generate translation, ℓ_0 and $\bar{\ell}_0$ are for rotation and dilation, and ℓ_1 and $\bar{\ell}_1$ generate the special conformal transformation.

6.3   Primary and quasi-primary operators

As already mentioned, the two-dimensional case is special from the standpoint of conformal symmetry since the local conformal group is isomorphic to the group of analytic transformations on the complex plane and thus it contains an infinite number of generators $\{\cdots, \ell_{-1}, \bar{\ell}_{-1}, \ell_0, \bar{\ell}_0, \cdots\}$. A conformally invariant theory satisfies an infinite number of constraints called the *conformal Ward identities*. These identities are written as differential equations for correlation functions. It is thus sometimes convenient to characterize the behavior of a theory by the transformation properties of its correlation functions rather than by the specific action S (which corresponds to the Hamiltonian in classical statistical mechanics) of the system. In the following, we will illustrate and demonstrate how these fundamental facts lead to very strong consequences.

The local field operators that transform under a global conformal mapping $z \to f(z), \bar{z} \to \bar{f}(\bar{z})$ as

$$\phi_j(z, \bar{z}) \to \phi_j'(f, \bar{f}) = (\partial_z f)^{-h_j} (\partial_{\bar{z}} \bar{f})^{-\bar{h}_j} \phi_j(z, \bar{z}) \qquad (6.16)$$

are called *quasi-primary* operators or quasi-primary fields.[3] The real numbers $h_j, \bar{h}_j$ are named the *conformal weights* of the operator ϕ_j. Correlation functions of quasi-primary operators thus satisfy

$$\langle \phi_1(z_1, \bar{z}_1) \phi_2(z_2, \bar{z}_2) \cdots \rangle = \left(\prod_i (\partial_{z_i} f_i)^{h_i} (\partial_{\bar{z}_i} \bar{f}_i)^{\bar{h}_i} \right) \langle \phi_1'(f_1, \bar{f}_1) \phi_2'(f_2, \bar{f}_2) \cdots \rangle, \quad (6.17)$$

where the expectation value is defined by the weight e^{-S},

$$\langle\langle \cdots \rangle\rangle = \frac{\int \mathcal{D}\phi \, (\cdots) e^{-S}}{\int \mathcal{D}\phi \, e^{-S}}, \qquad (6.18)$$

with S being the action or the effective Hamiltonian. As an example, a simple rescaling $f(z) = b^{-1} z, \bar{f}(\bar{z}) = b^{-1} \bar{z}$ gives

$$\phi_j'(f, \bar{f}) = b^{h_j + \bar{h}_j} \phi_j(z, \bar{z}), \qquad (6.19)$$

whereas a rotation $z \to e^{i\theta} z, \bar{z} \to e^{-i\theta} \bar{z}$ leads to

---

[3] Notice that the prime on ϕ_j is not a derivative.

$$\phi'_j(f, \bar{f}) = e^{-i(h_j - \bar{h}_j)\theta} \phi_j(z, \bar{z}). \tag{6.20}$$

Equations (6.19) and (6.20) make it possible to identify the sum and difference of the conformal weights with the scaling dimension x_j and the *spin* s_j of the operator ϕ_j, respectively,

$$x_j = h_j + \bar{h}_j \ , \ s_j = h_j - \bar{h}_j. \tag{6.21}$$

If ϕ_j transforms like eqn (6.16) for both global and local mappings, it is called a *primary operator*. Thus, a primary operator is quasi-primary but the converse may not necessarily be the case. Local scaling fields (operators) that do not transform as a primary are called *secondary operators*. Secondary operators may or may not be quasi-primary. The origin of these names, primary, quasi-primary, secondary operators, will become clearer in Section 6.7.

Let us next study the consequences of invariance of correlation functions under global conformal mappings. The infinitesimal transformations of eqn (6.11) induce a change in quasi-primary fields. According to eqn (6.16),

$$\delta_{\epsilon\bar{\epsilon}}\phi_j(z, \bar{z}) \equiv \phi_j(f, \bar{f}) - \phi_j(z, \bar{z})$$
$$= -(\epsilon(z)\partial + \bar{\epsilon}(\bar{z})\bar{\partial})\phi_j(z, \bar{z}) - \phi_j(z, \bar{z})(h_j\partial\epsilon(z) + \bar{h}_j\bar{\partial}\bar{\epsilon}(\bar{z})), \tag{6.22}$$

where $\partial = \partial_z$ and $\bar{\partial} = \partial_{\bar{z}}$. If we now impose the condition of conformal invariance by global mappings on the n-point correlation function of quasi-primary operators,

$$\delta_{\epsilon\bar{\epsilon}}\langle\phi_1(z_1, \bar{z}_1)\phi_2(z_2, \bar{z}_2)\cdots\phi_n(z_n, \bar{z}_n)\rangle = 0, \tag{6.23}$$

a differential equation results,

$$\sum_{i=1}^{n} (\epsilon_i\partial_i + h_i\partial_i\epsilon_i + \bar{\epsilon}_i\bar{\partial}_i + \bar{h}_i\bar{\partial}_i\bar{\epsilon}_i)\langle\phi_1(z_1, \bar{z}_1)\phi_2(z_2, \bar{z}_2)\cdots\phi_n(z_n, \bar{z}_n)\rangle = 0, \tag{6.24}$$

where $\epsilon_i = \epsilon(z_i)$, $\partial_i = \partial_{z_i}$, and similarly for the antiholomorphic parts.

For global (projective) conformal mappings, the Lie algebra (the loop algebra of eqn (6.15) in the present case) is generated by $\{\ell_{-1}, \ell_0, \ell_1\}$ and its antiholomorphic counterpart, as was described in the final paragraph of the previous section. The corresponding infinitesimal transformations are identified as $\epsilon = \bar{\epsilon} = 1$ for translation generated by $\ell_{-1}, \bar{\ell}_{-1}$, $\epsilon = z, \bar{\epsilon} = \bar{z}$ for rotation and dilation $(\ell_0, \bar{\ell}_0)$, and $\epsilon = z^2, \bar{\epsilon} = \bar{z}^2$ for special conformal transformation $(\ell_1, \bar{\ell}_1)$ as can be verified from eqns (6.1) to (6.4). For example, a rotation is $z \to z + i\delta \cdot z = (1 + i\delta)z \approx e^{i\delta}z$ (and similarly for $\bar{z}$) with δ a small real number.

EXERCISE 6.2 Show that $\epsilon = \bar{\epsilon} = 1$ represents translation, $\epsilon = z, \bar{\epsilon} = \bar{z}$ is for rotation and dilation, and $\epsilon = z^2, \bar{\epsilon} = \bar{z}^2$ corresponds to the special conformal transformation.

We then have a set of *projective Ward identities* for n-point correlation functions of quasi-primary operators, using eqn (6.24) for $\bar{\epsilon} = 0$,

$$\sum_{i=1}^{n} \partial_i \langle \phi_1(z_1, \bar{z}_1) \phi_2(z_2, \bar{z}_2) \cdots \phi_n(z_n, \bar{z}_n) \rangle = 0, \tag{6.25}$$

$$\sum_{i=1}^{n} (z_i \partial_i + h_i) \langle \phi_1(z_1, \bar{z}_1) \phi_2(z_2, \bar{z}_2) \cdots \phi_n(z_n, \bar{z}_n) \rangle = 0, \tag{6.26}$$

$$\sum_{i=1}^{n} (z_i^2 \partial_i + 2h_i z_i) \langle \phi_1(z_1, \bar{z}_1) \phi_2(z_2, \bar{z}_2) \cdots \phi_n(z_n, \bar{z}_n) \rangle = 0, \tag{6.27}$$

and a similar set of equations for the antiholomorphic variables, since both are independent.

Equations (6.25) to (6.27) fix the functional form of two-point correlation functions. Equation (6.25) reads

$$\left(\partial_1 + \partial_2 \right) \langle \phi_1 \phi_2 \rangle = 0, \tag{6.28}$$

where $\langle \phi_1 \phi_2 \rangle = \langle \phi_1(z_1, \bar{z}_1) \phi_2(z_2, \bar{z}_2) \rangle$. According to eqn (6.28), $\langle \phi_1 \phi_2 \rangle$ should depend on z_1 and z_2 only as their difference $z_1 - z_2 (\equiv z_{12})$ (and similarly for $\bar{z}_1$ and $\bar{z}_2$). This is simply a consequence of translation invariance as implied in $\epsilon = \bar{\epsilon} = 1$. Equation (6.26)

$$\left(z_1 \partial_1 + z_2 \partial_2 + h_1 + h_2 \right) \langle \phi_1 \phi_2 \rangle = 0 \tag{6.29}$$

and its antiholomorphic counterpart fix the correlation function as

$$\langle \phi_1 \phi_2 \rangle = \frac{1}{(z_{12})^{h_1 + h_2} (\bar{z}_{12})^{\bar{h}_1 + \bar{h}_2}}, \tag{6.30}$$

as can be verified by insertion of eqn (6.30) into eqn (6.29). We have fixed the normalization of ϕ_1 and ϕ_2 to reduce the numerator of the right-hand side to unity.[4] It is left as an exercise to show that eqn (6.27) for $\langle \phi_1 \phi_2 \rangle$ demands $h_1 = h_2 (= h)$ and $\bar{h}_1 = \bar{h}_2 (= \bar{h})$.

EXERCISE 6.3 Show that eqn (6.27) for a two-point correlation function requires that the conformal weight h_1 be equal to h_2 (and similarly for the antiholomorphic counterparts) if the correlation function is not to vanish.

We therefore have the explicit form of the two-point correlation function at criticality as

$$\langle \phi_1 \phi_2 \rangle = \frac{1}{(z_{12})^{2h} (\bar{z}_{12})^{2\bar{h}}}. \tag{6.31}$$

This equation suggests a relation between the conformal weights and critical exponents. For example, if ϕ_1 and ϕ_2 are both usual spin operators and $h = \bar{h}$, then eqn (6.31) will be

$$\langle S(\mathbf{r}_1) S(\mathbf{r}_2) \rangle = \frac{1}{(r_{12})^{4h}}, \tag{6.32}$$

---

[4] It is possible to leave the normalization arbitrary and write the numerator as C, but it does not change the essence of the theory.

because $z_{12}\bar{z}_{12} = r_{12}^2$. This relation is consistent with the identification of the scaling dimension x with $2h$ as in eqn (6.21). Equation (6.32) is to be compared with the generic critical form of the critical spin–spin correlation function

$$\langle S(\boldsymbol{r}_1)S(\boldsymbol{r}_2)\rangle \propto r_{12}^{-d+2-\eta}. \tag{6.33}$$

In two dimensions, we therefore have $4h = \eta$. In particular, the two-dimensional ferromagnetic Ising model is known to have $\eta = 1/4$, from which we infer $h = 1/16$.

Similar arguments lead to constraints on n-point ($n \geq 3$) correlation functions. The functional form of the three-point correlation function is fixed by conformal invariance. It is possible to write explicitly the four-point correlation function only in two dimensions.

6.4 Energy–momentum tensor and the Ward identity

An infinitesimal transformation $z \to z + \epsilon(z)$ cannot be holomorphic everywhere on the complex plane unless it is a trivial constant. Suppose that $\epsilon(z)$ is holomorphic inside a region D that includes the origin but not necessarily holomorphic outside D. Let us consider an n-point correlation function of primary operators,

$$\langle X_n \rangle = \langle \phi_1(z_1, \bar{z}_1) \cdots \phi_n(z_n, \bar{z}_n)\rangle, \tag{6.34}$$

with all its arguments $z_1, \bar{z}_1, \cdots, z_n, \bar{z}_n$ inside D. The change in this correlation function $\delta_{\epsilon\bar{\epsilon}}\langle X_n\rangle$ involves an infinitesimal change in the action δS because, for ϵ not necessarily conformal, $\delta_{\epsilon\bar{\epsilon}}$ applied to $\langle X_n\rangle$ affects not only X_n but also e^{-S}. The *energy–momentum tensor* $T_{\mu\nu}$, also known as the *stress tensor*, is defined as the rate of change in the action,

$$\delta S = -\frac{1}{2\pi}\int \mathrm{d}^2\boldsymbol{r}\, \partial^\mu \epsilon^\nu T_{\mu\nu}(\boldsymbol{r}), \tag{6.35}$$

where the integral is over the region outside D.[5] The indices μ and ν are for the two-dimensional Cartesian coordinates. A summation is implicit in the repeated indices, that is, the integrand is summed over μ and ν both from 1 to 2. In Appendix A.10, it is shown, using eqns (6.22) and (6.35), that invariance of the correlation function under the change $z \to z + \epsilon(z)$ and $\bar{z} \to \bar{z} + \bar{\epsilon}(\bar{z})$ leads to the following equation,

$$-\sum_{i=1}^{n}\left(\epsilon_i\partial_i + h_i\partial_i\epsilon_i + \bar{\epsilon}_i\bar{\partial}_i + \bar{h}_i\bar{\partial}_i\bar{\epsilon}_i\right)\langle X_n\rangle$$

$$= \frac{1}{2\pi\mathrm{i}}\oint_C \mathrm{d}w\, \epsilon(w)\,\langle T(w)X_n\rangle - \frac{1}{2\pi\mathrm{i}}\oint_C \mathrm{d}\bar{w}\, \bar{\epsilon}(\bar{w})\,\langle \bar{T}(\bar{w})X_n\rangle, \tag{6.36}$$

where C is the boundary of D, which means that the points $z_1, \cdots, z_n$ lie inside C and $\epsilon(z)$ is holomorphic inside C. The operator $T(w)$ and the antiholomorphic $\bar{T}(\bar{w})$ are defined in terms of the Cartesian components of $T_{\mu\nu}$,

---

[5] The prefactor $1/2\pi$ is for simplicity of later equations and is not essential.

$$T(w) = \frac{1}{4}\left(T_{11}(w) - T_{22}(w) - 2iT_{12}(w)\right) \tag{6.37}$$

$$\bar{T}(\bar{w}) = \frac{1}{4}\left(\bar{T}_{11}(\bar{w}) - \bar{T}_{22}(\bar{w}) + 2i\bar{T}_{12}(\bar{w})\right). \tag{6.38}$$

Equation (6.36) indicates that the energy–momentum tensor is the generator of an infinitesimal transformation of operators, which may be written symbolically as

$$\delta_{\epsilon\bar{\epsilon}}X_n = \frac{1}{2\pi i}\oint_C dw\,\epsilon(w)\,T(w)X_n - \frac{1}{2\pi i}\oint_C d\bar{w}\,\bar{\epsilon}(\bar{w})\,\bar{T}(\bar{w})X_n. \tag{6.39}$$

Notice that eqn (6.36) is a generic equation satisfied by $\langle X_n \rangle$, whereas eqn (6.24) is a consequence of the projective mapping $\epsilon(z)$ for which the integral over C on the right-hand side of eqn (6.36) vanishes because C can be chosen to be infinitely far away. Hereafter, we will often write only the holomorphic parts of the theory explicitly for simplicity, as long as no confusion is expected. The antiholomorphic parts will have the same expressions.

For $\bar{\epsilon} = 0$, eqn (6.36) is equivalent to

$$\langle T(w)X_n \rangle = \sum_{i=1}^{n}\left(\frac{1}{w - z_i}\partial_i + \frac{h_i}{(w - z_i)^2}\right)\langle X_n \rangle, \tag{6.40}$$

as one can verify by multiplying both sides of eqn (6.40) by $\epsilon(w)$ and integrating along C to reproduce eqn (6.36) for $\bar{\epsilon} = 0$. This equation suggests that the product $T(w)X_n$ is a meromorphic function[6] of w with singularities at $z_1, z_2, \cdots, z_n$. Equations (6.36) and (6.40) are called the *conformal Ward identities*.

Since eqns (6.36) and (6.40) are valid for arbitrary product of primary operators X_n and arbitrary $\epsilon(z)$ which is holomorphic inside C, we may write symbolically, for w close to z,

$$T(w)\phi_i(z, \bar{z}) = \frac{h_i}{(w - z)^2}\phi_i(z, \bar{z}) + \frac{1}{w - z}\partial\phi_i(z, \bar{z}) + \mathsf{regular}, \tag{6.41}$$

where the symbol 'regular' stands for terms analytic in w that do not contribute to the integral of eqn (6.36). This is an example of the *operator product expansion* (OPE), which expresses the product of operators as a series expansion in terms of a complete set of operators.

The energy–momentum tensor is expanded in a Laurent series, known as the *mode expansion*,

$$T(z) = \sum_{n=-\infty}^{\infty}\frac{L_n}{z^{n+2}}, \tag{6.42}$$

using the operators L_n, called *conformal generators*, acting on the space of all fields. This expression can be formally inverted to give

$$L_n = \frac{1}{2\pi i}\oint_C dz\,z^{n+1}T(z), \tag{6.43}$$

---

[6] A function that is holomorphic in a region, except at isolated poles.

where C surrounds the origin. It is useful to define here $\hat{T}$ as the operator acting on X_n on the right-hand side of eqn (6.36) for $\bar{\epsilon} = 0$,

$$\hat{T} = \frac{1}{2\pi i} \oint_C dw\, \epsilon(w) T(w) = \sum_{m=-\infty}^{\infty} \epsilon_m L_m, \tag{6.44}$$

where we used eqns (6.12) and (6.42). Then, one can rewrite the right-hand side of eqn (6.36) for $\bar{\epsilon} = 0$ as

$$\langle \hat{T} X_n \rangle = \sum_{m=-\infty}^{\infty} \epsilon_m \langle L_m X_n \rangle, \tag{6.45}$$

with the end result that the conformal generators must act as

$$\langle L_m X_n \rangle = -\sum_{i=1}^{n} \left(z_i^{m+1} \partial_i + h_i (m+1) z_i^m \right) \langle X_n \rangle. \tag{6.46}$$

Since the energy–momentum tensor T plays a crucial role in the following developments, it is useful to formulate the OPE of T with itself. If T were a primary operator, the OPE of T with itself would have an expression as eqn (6.41) with $\phi_i(z)$ replaced by $T(z)$. Equation (6.40), however, suggests that the OPE of $T(w)T(z)$ should have an additional term proportional to $(w-z)^{-4}$. To see this fact, it helps to replace X_n in eqn (6.40) with $T(z)X_n$ and expand $\langle T(z)X_n \rangle$, appearing in place of $\langle X_n \rangle$ on the right-hand side, once again using the same eqn (6.40). It is therefore reasonable to assume that the following expression is the correct OPE,[7]

$$T(w)T(z) = \frac{c}{2(w-z)^4} + \frac{2}{(w-z)^2} T(z) + \frac{1}{w-z} \partial T(z) + \text{regular}. \tag{6.47}$$

The first term on the right-hand side with a coefficient c, known as the *central charge*, is absent in eqn (6.41) for a primary operator and is often referred to as the *conformal anomaly*. The value of the central charge depends upon the particular conformal field theory under study and is an important number to classify critical field theories. Critical systems with real Boltzmann factors have real-valued central charges, as will be exemplified later.

The OPE of eqn (6.47) implies that, under an infinitesimal conformal transformation, the energy–momentum tensor $T(z)$ transforms as

$$\delta_\epsilon T(z) = \frac{1}{2\pi i} \oint dw\, \epsilon(w)\, T(w)T(z)$$

$$= \epsilon(z)\partial T(z) + 2\big(\partial \epsilon(z)\big) T(z) + \frac{c}{12} \partial^3 \epsilon(z), \tag{6.48}$$

---

[7] Remember that eqn (6.40) is valid for X_n that is the product of primary operators. If X_n here is replaced by $T(z)X_n$, there is no guarantee that the same equation holds since $T(z)X_n$ may not necessarily be primary. Thus, the argument in the text is heuristic at best. The justification of the OPE of eqn (6.47) is better expressed by its consistency with the non-infinitesimal transformation developed later in this section.

which is to be compared with eqn (6.39) with $\bar{\epsilon} = 0$ for primary operators. The existence of the third term of conformal anomaly is specific to the energy–momentum tensor, which is not primary, and vanishes for an infinitesimal projective conformal mapping represented as a second-order polynomial of $\epsilon(z)$. This observation suggests that the energy–momentum tensor is quasi-primary but not primary.

Justification of the OPE of eqn (6.47) lies in its consistency with the non-infinitesimal version of the transformation. It is useful to rewrite the infinitesimal transformation (6.48) as a finite map,

$$T(z) \to (\partial_z f)^2 T(f) + \frac{c}{12}\{f, z\}, \tag{6.49}$$

where

$$\{f, z\} = \frac{\partial_z^3 f}{\partial_z f} - \frac{3}{2}\left(\frac{\partial_z^2 f}{\partial_z f}\right)^2 \tag{6.50}$$

is called the *Schwarz derivative* or *Schwarzian*. It is not difficult to check that eqn (6.49) reduces to eqn (6.48) for an infinitesimal transformation.

> **EXERCISE 6.4** Show that eqn (6.49) reduces to eqn (6.48) for an infinitesimal transformation.

In solving Exercise 6.4, one notices that the coefficient 2 of $(w - z)^{-2}$ on the right-hand side of eqn (6.47) comes from the square $(\partial_z f)^2$ on the right-hand side of eqn (6.49). This square, in turn, reflects the fact that the energy–momentum tensor has two coordinate indices $\mu\nu$ and therefore should be multiplied by a factor of $(\partial_z f)^2$ under the map $z \to f(z)$. This explains the conformal weight $h = 2$ of the energy–momentum tensor. It is also instructive to notice that eqn (6.50) satisfies the consistency condition that two successive transformations $z \to f(z) \to u(f)$ are equivalent to a single transformation $z \to u(z)$.

> **EXERCISE 6.5** Confirm that a transformation of T by $z \to u(z)$ is reproduced by two successive transformations $z \to f(z) \to u(f)$ using eqn (6.49). For this purpose, first show that the following relation is satisfied by the Schwarzian,
>
> $$\{u, z\} = \{u, f\}\left(\frac{\partial f}{\partial z}\right)^2 + \{f, z\}. \tag{6.51}$$
>
> Then, verify that eqn (6.49) applied to two successive transformations $z \to f(z) \to u(f)$ reproduces the relation for the single transformation $z \to u(z)$.

These arguments make it clear that eqn (6.49) is the legitimate finite extension of the infinitesimal transformation (6.48). In other words, eqn (6.49) is equivalent to eqn (6.48) and hence to eqn (6.47). This justifies the OPE of eqn (6.47).

We notice that the Schwarz derivative of a global conformal mapping (6.10) vanishes. Hence, we again confirm that the energy–momentum is a quasi-primary operator, though it is not primary.

EXERCISE 6.6 Show that the Schwarz derivative of a global map vanishes.

6.5 Virasoro algebra

The algebra of the operators $\{L_n, \bar{L}_n\}_n$, called the *Virasoro algebra*, is of central importance in conformal field theory. To find the commutation relation of these operators, we first define the integral representation of the product of two L_ns as

$$L_m L_n = \oint_{|z|>w} \frac{dz}{2\pi i}\, z^{m+1} \oint_C \frac{dw}{2\pi i}\, w^{n+1}\, T(z)T(w), \tag{6.52}$$

where C surrounds the origin and the integral over z runs on a contour surrounding w. Then, the commutator is

$$L_m L_n - L_n L_m = \oint_C \frac{dw}{2\pi i}\, w^{n+1} \oint_w \frac{dz}{2\pi i}\, z^{m+1} T(z)T(w), \tag{6.53}$$

as can be seen from Fig. 6.2.

We therefore have

$$
\begin{aligned}
[L_m, L_n] &= \oint_C \frac{dw}{2\pi i} \oint_w \frac{dz}{2\pi i}\, w^{n+1} z^{m+1} \left(\frac{c}{2(z-w)^4} + \frac{2}{(z-w)^2} T(w) + \frac{1}{z-w}\partial T(w) \right) \\
&= \oint_C \frac{dw}{2\pi i}\, w^{n+1}\left(\frac{c}{12}(m+1)m(m-1)w^{m-2} + 2(m+1)w^m T(w) + w^{m+1}\partial T(w) \right) \\
&= \frac{c}{12}m(m^2-1)\delta_{m+n,0} + 2(m+1)L_{m+n} - \oint_C \frac{dw}{2\pi i}\,(m+n+2)w^{m+n+1}T(w) \\
&= \frac{c}{12}m(m^2-1)\delta_{m+n,0} + (m-n)L_{m+n}. \tag{6.54}
\end{aligned}
$$

The same relation holds for the antiholomorphic part. The Virasoro algebra is thus summarized as

$$[L_m, L_n] = (m-n)L_{m+n} + \frac{c}{12}m(m^2-1)\delta_{m+n,0} , \tag{6.55}$$

$$[\bar{L}_m, \bar{L}_n] = (m-n)\bar{L}_{m+n} + \frac{\bar{c}}{12}m(m^2-1)\delta_{m+n,0} , \tag{6.56}$$

$$[L_m, \bar{L}_n] = 0. \tag{6.57}$$

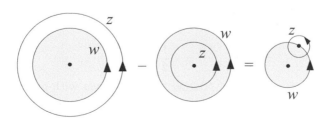

Fig. 6.2 The difference of the two integral contours for $L_m L_n$ and $L_n L_m$.

This is called the *central extension* of the loop algebra of eqn (6.15). For $n = 0, \pm 1$, the final terms involving c and $\bar{c}$ vanish and the Virasoro algebra reduces to a simple loop algebra independent of the central charges c and $\bar{c}$.

Any conformal field theory is characterized by c and $\bar{c}$, which are not necessarily equal. It is an important problem to identify the value of the central charge for a given field theory.

6.6 Gaussian theory

The abstract formalism developed so far may be better digested through a simple example. The Gaussian theory is best suited for this purpose since everything can be worked out explicitly.

The action of the critical $d = 2$ Gaussian theory is

$$S = \int d^2 \boldsymbol{r} \left((\nabla \phi)^2 + a \phi^2 \right) \quad (a \to 0), \tag{6.58}$$

which is the same as eqn (2.74). This is the same quantity as the generalized Landau free energy F in eqn (2.74) with $b = 1$. Also, the inverse temperature β in Section 2.9 is to be replaced by 1 because the Boltzmann factor is e^{-S} here, whereas it was $e^{-\beta F}$ in Section 2.9. The condition $a = 0$ ensures criticality and justifies the application of the conformal field theory methodology.

The two-point correlation function has been evaluated in Section 2.9. According to eqn (2.85) and Exercise 2.12, its explicit form for $a > 0$ is

$$\langle \phi(\boldsymbol{r}_2) \phi(\boldsymbol{r}_2) \rangle = \frac{1}{2} \int d\boldsymbol{q}\, e^{i\boldsymbol{q}\cdot(\boldsymbol{r}_1 - \boldsymbol{r}_2)} \frac{1}{a + q^2} = \pi K_0(\sqrt{a}\, r) \quad (r = |\boldsymbol{r}_1 - \boldsymbol{r}_2|), \tag{6.59}$$

where $K_0(x)$ is the modified Bessel function of the second kind. The asymptotic form of $K_0(\sqrt{a}\, r)$ for small a gives

$$\langle \phi(\boldsymbol{r}_1) \phi(\boldsymbol{r}_2) \rangle = -\pi \log(\sqrt{a}\, r) \quad (\sqrt{a} \ll 1). \tag{6.60}$$

We drop $\log \sqrt{a}$ as it does not play a role in the analysis of the r dependence of the correlation function. This is an elementary instance of the process of regularization, in which one subtracts a diverging constant. In order to simplify some of the expressions appearing below, we multiply the field ϕ by an appropriate factor to yield

$$\langle \phi(\boldsymbol{r}_1) \phi(\boldsymbol{r}_2) \rangle = -\log r. \tag{6.61}$$

This rescaling of ϕ is allowed because it does not affect the criticality condition $a = 0$ and hence all the universal properties are kept intact. In terms of complex variables, eqn (6.61) reads

$$\langle \phi(z, \bar{z}) \phi(w, \bar{w}) \rangle = -\frac{1}{2} \log(z - w)(\bar{z} - \bar{w}), \tag{6.62}$$

where we have chosen $r = |z - w|$. Taking the derivatives with respect to z and w, we have

$$\langle \partial \phi(z, \bar{z}) \partial \phi(w, \bar{w}) \rangle = -\frac{1}{2} \frac{1}{(z - w)^2}. \tag{6.63}$$

Comparison of this equation with eqn (6.30) suggests that $\partial\phi(z,\bar{z})$ is a primary operator with the conformal weights $h = 1, \bar{h} = 0$.[8] The field $\phi(z,\bar{z})$ itself is not primary as the correlation (6.62) is not of the form of eqn (6.30). Similarly, the antiholomorphic $\bar{\partial}\phi(z,\bar{z})$ is primary with $h = 0, \bar{h} = 1$. We hereafter suppress the antiholomorphic variables $\bar{z}, \bar{w}$ to simplify the notation.

In Appendix A.11 it is shown that the holomorphic part of the energy–momentum tensor of the Gaussian theory can be chosen as $-(\partial\phi)^2$. Since this expression involves a divergence, as one sees in eqn (6.63) in the limit $z \to w$, we subtract the trivial divergent constant, which does not affect the essence of the theory[9] and is another case of regularization. We therefore define

$$T(z) = -\lim_{w \to z}\left(\partial\phi(z)\partial\phi(w) - \langle\partial\phi(z)\partial\phi(w)\rangle\right) \equiv -:\partial\phi(z)\partial\phi(z):. \qquad (6.64)$$

The last expression with colons on both ends is named the *normal order*.

Since the primary operator $\partial\phi$ has $h = 1$ as mentioned above, the OPE of T and $\partial\phi$ is, according to eqn (6.41),

$$T(w)\partial\phi(z) = \frac{1}{(w-z)^2}\partial\phi(z) + \frac{1}{w-z}\partial^2\phi(z) + \text{regular}. \qquad (6.65)$$

To determine the central charge, we take the expectation value of eqn (6.47),

$$\langle T(z)T(w)\rangle = \frac{c}{2(w-z)^4}, \qquad (6.66)$$

where we have used $\langle T(z)\rangle = 0$. See the footnote of page 152 and Exercise 6.7 for this relation. We therefore evaluate $\langle\partial\phi(z)\partial\phi(z)\partial\phi(w)\partial\phi(w)\rangle$. This expectation value of the product of four operators is decomposed into the product of two-point correlation functions in the Gaussian theory, since all higher-order cumulants vanish, other than the second order one, as discussed in Appendix A.4. To express the four-point correlation function in terms of products of two-point functions, there are two ways to combine the first z in the product $\partial\phi(z)\partial\phi(z)\partial\phi(w)\partial\phi(w)$ with w: The first z can be combined with the third w as well as with the fourth w, giving two identical contributions.[10]

This is a part of the general result under the name of *Wick's theorem*. The reader may feel convinced by the following simple example of a Gaussian integral,

$$\langle f(x,y)\rangle = \frac{1}{2\pi}\int_{-\infty}^{\infty}\mathrm{d}x\mathrm{d}y\, f(x,y)\mathrm{e}^{-(5x^2-6xy+5y^2)/8}. \qquad (6.67)$$

One easily verifies the relation numerically

$$\langle xxyy\rangle = \langle x^2\rangle\langle y^2\rangle + 2\langle xy\rangle^2 \qquad (6.68)$$

---

[8] The normalization is achieved if we multiply $\partial\phi$ by $i\sqrt{2}$.

[9] Such as the conservation law of the energy–momentum tensor, eqn (A.169), and the tracelessness, eqn (A.171).

[10] If we combine the first z with the third w, the second z automatically combines with the fourth w. The combination of the first z with the second z drops by regularization.

by inserting the results of direct calculations, $\langle x^2 y^2 \rangle = 43/16$, $\langle x^2 \rangle = \langle y^2 \rangle = 5/4$ and $\langle xy \rangle = 3/4$. The term corresponding to $\langle x^2 \rangle \langle y^2 \rangle$ drops in our Gaussian theory problem due to regularization.

Then, we find by using eqn (6.63)

$$\langle T(z)T(w) \rangle = 2\langle \partial\phi(z)\partial\phi(w) \rangle^2 = \frac{1}{2(z-w)^4}. \tag{6.69}$$

This implies $c = 1$.

6.7   Operator formalism

We now rewrite the formulation developed so far in terms of operators. This is necessary to develop a representation theory of the Virasoro algebra in the next section, from which strong constraints can be placed on the possible values of the central charge. In the operator formalism, a primary field is replaced by a state or a vector in a vector space and the conformal generator L_n acts on those states.[11] Notice that we write the formulas only using the holomorphic part of the variables. The complete expressions should include the antiholomorphic parts.

6.7.1   State and operator

Let us start with the vacuum state $|0\rangle$, which corresponds to the identity field at the origin,

$$|0\rangle \longleftrightarrow \mathbf{1}(z = 0). \tag{6.70}$$

This means that the quantity on the right-hand side is rewritten as in the left-hand side, though they represent essentially the same content. According to eqns (6.43) and (6.41), $L_n \, (n \geq -1)$ acts on $\mathbf{1}(z = 0)$ as

$$L_n \mathbf{1}(z = 0) = \frac{1}{2\pi i} \oint_C dw \, w^{n+1} T(w) \mathbf{1}(0) = 0 \tag{6.71}$$

because the conformal weight of $\mathbf{1}(z = 0)$ is 0: The identity does not change under conformal transformations. This relation is translated into the operator formalism as

$$L_n |0\rangle = 0 \quad (n \geq -1). \tag{6.72}$$

Equivalently, the requirement of regularity at $z = 0$ of the following expression

$$T(z)|0\rangle = \sum_{n=-\infty}^{\infty} \frac{L_n}{z^{n+2}} |0\rangle \tag{6.73}$$

---

[11] The function $\phi_j(z)$, primary or not, has so far often been called an operator. This 'operator' has actually been a classical quantity and will be called a 'field' in this section. The role of such a field is played by a state (or a vector) in the operator formalism, as will now be seen. An operator in the present section refers to, typically, $T(z)$ and L_n, which are considered to act on the states.

leads also to eqn (6.72). Actually, $L_1|0\rangle = L_2|0\rangle = 0$ is sufficient for $L_n|0\rangle = 0$ with $n \geq 1$ as one is convinced by induction: Using the Virasoro algebra (6.55), we find

$$L_3|0\rangle = [L_2, L_1]|0\rangle = 0 \tag{6.74}$$

if $L_1|0\rangle = L_2|0\rangle = 0$ and similarly for $L_n|0\rangle$ with $n \geq 4$.

The state vector $|h_j\rangle$ is defined as a quantity corresponding to the primary $\phi_j(z)$ with the conformal weight h_j in the limit $z \to 0$,

$$|h_j\rangle \longleftrightarrow \lim_{z \to 0} \phi_j(z), \tag{6.75}$$

or equivalently

$$|h_j\rangle = \lim_{z \to 0} \phi_j(z)|0\rangle. \tag{6.76}$$

An important property of this state is

$$L_0|h_j\rangle = h_j|h_j\rangle, \quad L_n|h_j\rangle = 0 \quad (n \geq 1). \tag{6.77}$$

It is straightforward to prove these relations from eqns (6.43) and (6.41), for $n \geq 0$,

$$
\begin{aligned}
L_n|h_j\rangle \longleftrightarrow L_n\phi_j(0) &= \frac{1}{2\pi i} \oint_C dw\, w^{n+1} T(w)\phi_j(0) \\
&= \frac{1}{2\pi i} \oint_C dw\, w^{n+1} \left(\frac{h_j}{w^2} \phi_j(0) + \frac{1}{w} \partial\phi_j \right) \\
&= h_j\phi_j(0)\delta_{n,0} \\
&\longleftrightarrow h_j|h_j\rangle\delta_{n,0}.
\end{aligned}
\tag{6.78}
$$

Similarly to the case of $L_n|0\rangle$, the relations $L_1|h_j\rangle = L_2|h_j\rangle = 0$ automatically guarantees $L_n|h_j\rangle = 0$ for $n \geq 3$. We note in passing that L_{-1} can be identified with the differential operator,

$$
\begin{aligned}
L_{-1}\phi_j(0) &= \frac{1}{2\pi i} \oint_C dw\, T(w)\phi_j(0) \\
&= \frac{1}{2\pi i} \oint_C dw \left(\frac{h_j}{w^2} \phi_j(0) + \frac{1}{w} \partial\phi_j(0) \right) \\
&= \partial\phi_j(0).
\end{aligned}
\tag{6.79}
$$

6.7.2 Conformal family

We next define a set of states generated from $|h_j\rangle$ by repeated operations of $L_{-1}, L_{-2}, \cdots,$[12]

$$|m_1 m_2 \cdots m_n; h_j\rangle \equiv L_{-1}^{m_1} L_{-2}^{m_2} \cdots L_{-n}^{m_n}|h_j\rangle. \tag{6.80}$$

[12] Remember that the action of L_n with $n \geq 1$ onto $|h_j\rangle$ gives a trivial vanishing result.

This state is called a *descendant* of $|h_j\rangle$ or a *secondary state* derived from the primary $|h_j\rangle$. This is an eigenstate of the operator L_0,

$$L_0|m_1 m_2 \cdots m_n; h_j\rangle = \left(h_j + \sum_{k=1}^{n} k m_k\right)|m_1 m_2 \cdots m_n; h_j\rangle. \qquad (6.81)$$

This equation is a consequence of the Virasoro algebra $[L_0, L_{-k}] = kL_{-k}$ $(k \geq 1)$. To see it, we first note that the Virasoro algebra leads to

$$L_0 L_{-k}^{m_k} = L_{-k}^{m_k} L_0 + k m_k L_{-k}^{m_k}, \qquad (6.82)$$

which can be shown by induction, for example,

$$L_0 L_{-k} = L_{-k} L_0 + k L_{-k} \qquad (6.83)$$

$$\begin{aligned}
L_0 L_{-k}^2 &= (L_{-k} L_0 + k L_{-k}) L_{-k} \\
&= L_{-k}(L_{-k} L_0 + k L_{-k}) + k L_{-k}^2 \\
&= L_{-k}^2 L_0 + 2k L_{-k}^2,
\end{aligned} \qquad (6.84)$$

and similarly for higher powers of L_{-k}. Equation (6.82) makes it possible to push L_0 to the right of the product of $L_{-k}^{m_k}$ in eqn (6.80) multiplied by L_0 from the left, yielding finally eqn (6.81).

The set of secondary states (or secondary fields) derived from a primary is called the *conformal family* or the *conformal tower*. Also, the name *Verma module* is used. The conformal family includes the derivatives of the primary, as L_{-1} is the derivative, as well as fields not expressed as derivatives of the primary.

The conformal family in its state-vector representation is somewhat analogous to the bases of the representation of angular momentum. The primary $|h_j\rangle$ corresponds to the state $|\ell\rangle$ with the largest value of the z-component of the angular momentum operator. The raising operator ℓ_+ annihilates $|\ell\rangle$, whereas the lowering operator ℓ_- generates other states with smaller values of the z-component $|\ell - 1\rangle, |\ell - 2\rangle, \cdots$. The operation $\ell_+|\ell\rangle = 0$ corresponds to $L_n|h_j\rangle = 0$ for $n \geq 1$. The lowering operator ℓ_- corresponds to L_{-n} with $n \geq 1$. The state $|h_j\rangle$ is called the *highest weight state* from this analogy. An important difference is that the conformal family is infinite dimensional as it includes infinitely many states, whereas the angular momentum is represented in a finite-dimensional vector (Hilbert) space.

A few additional remarks will be useful for later developments of the representation theory of the Virasoro algebra. The descendants of a primary $|h_j\rangle$ are classified by their *levels*. The level $\mathcal{N}$ of the state of eqn (6.80) is defined as $\sum_{k=1}^{n} k m_k$. For instance, descendants at levels 1, 2 and 3 are

$$\text{Level 1}: \quad L_{-1}|h_j\rangle \qquad (6.85)$$

$$\text{Level 2}: \quad L_{-2}|h_j\rangle, \ L_{-1}^2|h_j\rangle \qquad (6.86)$$

$$\text{Level 3}: \quad L_{-3}|h_j\rangle, \ L_{-1}L_{-2}|h_j\rangle, \ L_{-1}^3|h_j\rangle. \qquad (6.87)$$

Clearly, the number of states at level $\mathcal{N}$ is the number of ways $P(\mathcal{N})$ to divide $\mathcal{N}$ into a sum of natural numbers. For example, $P(3) = 3$ as 3 can be divided in three ways as $3, 1 + 2, 1 + 1 + 1$, corresponding to the three states in eqn (6.87). The $P(\mathcal{N})$ states at level $\mathcal{N}$ are not necessarily all independent. This important fact will be elucidated further in the next section.

The descendants of the vacuum have special properties:

$$\text{Level 1}: \quad L_{-1}|0\rangle = 0 \tag{6.88}$$

$$\text{Level 2}: \quad L_{-2}|0\rangle = T(0)|0\rangle, \quad L_{-1}^2|0\rangle = 0 \tag{6.89}$$

$$\text{Level } n: \quad L_{-n}|0\rangle = \partial^{n-2}T(0)|0\rangle, \cdots (n \geq 3). \tag{6.90}$$

The first relation is due to $\partial \mathbf{1} = 0$. The relations for $L_{-2}|0\rangle$ and $L_{-n}|0\rangle$ come from eqn (6.43). It is seen from the second relation that the energy–momentum tensor is not primary but a descendant of the vacuum.

6.7.3 Conjugate state

We have to introduce conjugate states to define the inner product. For a primary field $\phi_j(z, \bar{z})$, we define the conjugate field $\phi_j^\dagger(z, \bar{z})$ by way of the global mapping $w = -1/z, \bar{w} = -1/\bar{z}$,

$$\phi_j^\dagger(z, \bar{z}) = w^{2h_j} \bar{w}^{2\bar{h}_j} \phi_j(w, \bar{w}). \tag{6.91}$$

Correspondingly, the conjugate state is defined as

$$\langle h_j| = \lim_{z, \bar{z} \to 0} \langle 0|\phi_j^\dagger(z, \bar{z}) = \lim_{w, \bar{w} \to \infty} \langle 0|\phi_j(w, \bar{w}) w^{2h_j} \bar{w}^{2\bar{h}_j}. \tag{6.92}$$

Then, the inner product has the desirable property,

$$\langle h_j | h_k \rangle = \lim_{w, \bar{w} \to \infty} \lim_{z, \bar{z} \to 0} \langle 0|\phi_j(w, \bar{w}) w^{2h_j} \bar{w}^{2\bar{h}_j} \phi_k(z, \bar{z})|0\rangle$$

$$= \lim_{w, \bar{w} \to \infty} w^{2h_j} \bar{w}^{2\bar{h}_j} w^{-2h_j} \bar{w}^{-2\bar{h}_j} \delta_{h_j, h_k} = \delta_{h_j, h_k}, \tag{6.93}$$

where we used eqn (6.30) and the result of Exercise 6.3.

The conjugate operator of L_n is defined as

$$L_n^\dagger = L_{-n}. \tag{6.94}$$

Then, the operation of $L_n^\dagger$ to the left is derived by conjugation. For a primary $|h_j\rangle$, we have from $L_0|h_j\rangle = h_j|h_j\rangle$ and $L_n|h_j\rangle = 0 \ (n \geq 1)$,

$$\langle h_j|L_0^\dagger = \langle h_j|h_j, \quad \langle h_j|L_{-n}^\dagger = 0 \ (n \leq -1). \tag{6.95}$$

Using the definition of the conjugate state, we can show that the vacuum expectation value of the energy–momentum tensor vanishes, $\langle 0|T(z)|0\rangle = 0$. Since $L_n|0\rangle = 0$

for $n \geq -1$, we find

$$\langle 0|T(z)|0\rangle = \sum_{n=-\infty}^{\infty} z^{-n-2}\langle 0|L_n|0\rangle = \sum_{n=-\infty}^{-2} z^{-n-2}\langle 0|L_n|0\rangle = \sum_{n=2}^{\infty} z^{n-2}\langle 0|L_{-n}|0\rangle.$$
(6.96)

According to the Virasoro algebra (6.55), $L_0 L_{-n} - L_{-n} L_0 = n L_{-n}$ $(n \geq 2)$, and thus

$$\langle 0|T(z)|0\rangle = \sum_{n=2}^{\infty} z^{n-2} \frac{1}{n} \langle 0|L_0 L_{-n} - L_{-n} L_0|0\rangle = 0,$$
(6.97)

because of $L_0|0\rangle = 0$ and $\langle 0|L_0 = \langle 0|L_0^\dagger = 0$.

EXERCISE 6.7 Use the Virasoro algebra to prove

$$\langle 0|T(w)T(z)|0\rangle = \frac{c}{2(w-z)^4}.$$
(6.98)

This provides a way to compute the central charge c.

EXERCISE 6.8 Use the Virasoro algebra to show

$$\langle 0|L_2 L_{-2}|0\rangle = \frac{c}{2}, \quad \langle h_j|L_1 L_{-1}|h_j\rangle = 2h_j.$$
(6.99)

These equations imply $c \geq 0$ and $h_j \geq 0$ if the norms are to be non-negative.

6.7.4   Correlations of secondary fields

An interesting consequence of the classification of fields in terms of primary and secondary fields is that the correlation functions of secondary fields are completely specified by those of primary fields. To illustrate this important statement, consider the correlation function

$$\langle (L_{-m}\phi_j)(z)X_n\rangle = \oint_{C_z} \frac{dw}{2\pi i} \frac{1}{(w-z)^{m-1}} \langle T(w)\phi_j(z)X_n\rangle \quad (m \geq 1),$$
(6.100)

where X_n is the product of primary fields $\phi_1(w_1)\phi_2(w_2)\cdots\phi_n(w_n)$, each with the weight h_i. The contour C_z closely encircles z such that the points $w_1, w_2, \cdots, w_n$ are not included. Using eqn (6.40), we rewrite $T(w)X_n$ and change the integral contour to encircle $w_1, w_2, \cdots, w_n$ in the opposite direction, as in Fig. 6.3,

$$\langle (L_{-m}\phi_j)(z)X_n\rangle$$

$$= -\sum_{i=1}^{n} \oint_{C_{w_i}} \frac{dw}{2\pi i} \frac{1}{(w-z)^{m-1}} \langle \phi_j(z) \Big(\frac{h_i}{(w-w_i)^2}\phi_i(w_i) + \frac{1}{w-w_i}\partial\phi_i(w_i) \Big) \rangle$$

$$= \mathcal{L}_{-m}\langle \phi_j(z)X_n\rangle,$$
(6.101)

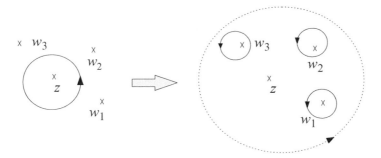

Fig. 6.3 The contour of the integral is changed from a circle around z to those around $w_1, w_2, \cdots$ in the opposite direction. The large contour surrounding all points, shown as dotted, does not contribute as the integrand decays sufficiently rapidly.

with the differential operator

$$\mathcal{L}_{-m} = \sum_{i=1}^{n} \left(\frac{(m-1)h_i}{(w_i - z)^m} - \frac{1}{(w_i - z)^{m-1}} \partial_{w_i} \right). \tag{6.102}$$

6.8   Unitary representation of the Virasoro algebra

The unitary representation of the Virasoro algebra is constructed in the vector space defined in the previous section. The requirement of unitarity of the representation leads to strong constraints on the values of the central charge and conformal weight. A part of such constraints has already been discussed in Exercise 6.8, in which the conditions $c \geq 0$ and $h_j \geq 0$ were derived from positive semi-definiteness of the norms of $L_{-2}|0\rangle$ and $L_{-1}|h_j\rangle$. The result $h_j \geq 0$ is natural also from the physics point of view that the two-point correlation function should not increase as the distance between two points increases, see eqn (6.30).

It is known that we can catalog the possible values of the central charge and conformal weight by studying the condition for the existence of states with vanishing norm, called the *null states*, in the Verma module. Since the proof is highly involved and beyond the modest scope of this book, we just present the results and refer the interested reader to more advanced books listed at the end of the volume.

Let us start with the trivial case of the vacuum $|0\rangle$ as a primary. The descendant of the vacuum at level 1 is $L_{-1}|0\rangle$. This state vanishes because L_{-1} is a differential operator. The state $L_{-1}|0\rangle = 0$ is thus a null state, and all its descendants vanish.

A non-trivial example starts from level 2 of a non-vacuum primary $|h_j\rangle$. There are two independent basis vectors at level 2, $L_2|h_j\rangle$ and $L_{-1}^2|h_j\rangle$, from which we may construct a null state as

$$L_{-2}|h_j\rangle + \alpha L_{-1}^2|h_j\rangle = 0 \tag{6.103}$$

by appropriately choosing the coefficient α. The requirement of the existence of a null state leads to constraints on c and h_j as follows. Let us apply L_1 to eqn (6.103) from the left. Using eqn (6.77) with $n = 1$ and the Virasoro algebra (6.55), we have

$$L_1 L_{-2}|h_j\rangle + \alpha L_1 L_{-1}^2|h_j\rangle = [L_1, L_{-2}]|h_j\rangle + \alpha[L_1, L_{-1}^2]|h_j\rangle$$
$$= (3 + 2\alpha(2h_j + 1))L_{-1}|h_j\rangle = 0. \qquad (6.104)$$

Since $L_{-1}|h_j\rangle \neq 0$ except for the trivial case of the vacuum ($h_j = 0$), we conclude $\alpha = -3/(4h_j + 2)$. Similarly, the application of L_2 to eqn (6.103) from the left yields

$$[L_2, L_{-2}]|h_j\rangle + \alpha[L_2, L_{-1}^2]|h_j\rangle = \left(4h_j + \frac{c}{2} + 6\alpha h_j\right)|h_j\rangle = 0. \qquad (6.105)$$

Hence, the central charge satisfies

$$c = -4h_j(3\alpha + 2) = \frac{2h_j(5 - 8h_j)}{2h_j + 1}. \qquad (6.106)$$

Therefore, the conformal weight satisfies, given the central charge c,

$$16h_j^2 + 2(c - 5)h_j + c = 0 \qquad (6.107)$$

if we require the existence of a null vector at level 2. This equation relates the central charge with the conformal weight. As will be mentioned later, the two-dimensional Ising model has a central charge $c = 1/2$, and consequently the conformal weights are $h_j = 1/16$ and $1/2$.

For general level $\mathcal{N}$, the problem is reduced to the analysis of a determinant in $P(\mathcal{N})$ dimensions. The conclusion of detailed studies is that the Virasoro algebra has a non-trivial, irreducible, unitary representation with a finite number of primary fields if the central charge is expressed as

$$c = 1 - \frac{6}{m(m+1)} \quad m = 3, 4, 5, \cdots. \qquad (6.108)$$

The corresponding primary fields have conformal weights characterized by two integers p and q,

$$h_{p,q} = h_{m-p,m+1-q} = \frac{((m+1)p - mq)^2 - 1}{4m(m+1)} \quad (1 \le p \le m-1,\ 1 \le q \le m). \qquad (6.109)$$

The field theories having these values of c and h are called *minimal models*. Also, the case $c > 1, h \ge 0$ has a unitary representation but with an infinite number of primary fields.

Minimal models correspond to common statistical models. The Ising model is known to have $c = 1/2, m = 3$. Other examples include the tricritical Ising model $m = 4$ ($c = 7/10$) and the three-state Potts model $m = 5$ ($c = 4/5$).

We have concentrated on the representations of the holomorphic part of the Virasoro algebra. The representations of the antiholomorphic components are built in the same fashion. The overall representations of the algebra are the tensor products of the holomorphic and antiholomorphic representations.

The value of c alone does not define a unique model. In general, there may be more than one model for a given $c < 1$, each with different (p, q) and different physical significance of the field $\phi_{p,q}$. Most importantly, unitary theories with $c < 1$ can be completely classified, which means that one can catalog all fixed points and thus determine all possible universality classes. This goal has been achieved not by solving all models explicitly but by symmetry arguments coming from conformal invariance.

6.9 Ising model

The simplest non-trivial minimal model has $m = 3$ $(c = 1/2)$, which is considered to correspond to the Ising model at criticality for the following reason. The primary fields of the theory are described by the conformal weights $(h_j, \bar{h}_j)$. According to eqn (6.109), the possible conformal weights of primary fields for $m = 3$ are

$$h_{1,1} = 0 \; , \; h_{1,2} = \frac{1}{16} \; , \; h_{2,1} = \frac{1}{2}. \tag{6.110}$$

All other $h_{p,q}$ are equal to one of these three cases due to the symmetry $h_{p,q} = h_{m-p,m+1-q}$. The associated primary fields are

$$\phi_{1,1} = 1 \; , \; \phi_{1,2} \; , \; \phi_{2,1}, \tag{6.111}$$

and the states are written as

$$|0\rangle \; , \; \left|\frac{1}{16}\right\rangle = \phi_{1,2} |0\rangle \; , \; \left|\frac{1}{2}\right\rangle = \phi_{2,1} |0\rangle. \tag{6.112}$$

In the full theory with holomorphic and antiholomorphic parts, the primary fields are of the form $\Phi_{p,q}(z, \bar{z}) = \phi_{p,q}(z)\bar{\phi}_{p,q}(\bar{z})$. These results are summarized in the following table.

Operator	$h_j (= \bar{h}_j)$	x_j
$\phi_{1,1}$	0	0
$\phi_{1,2}$	$\frac{1}{16}$	$\frac{1}{8}$
$\phi_{2,1}$	$\frac{1}{2}$	1

We need next to identify the field operators $\phi_{p,q}$ with the scaling operators of the critical Ising model. It is known from the exact solution of the two-dimensional Ising model that the spin–spin and energy–energy correlation functions at the critical point behave as

$$\langle S_0 S_r \rangle \propto \frac{1}{r^\eta} \; , \; \langle E_{nn}(0) E_{nn}(r) \rangle \propto \frac{1}{r^{4-2/\nu}}, \tag{6.113}$$

with critical exponents $\eta = 1/4$ and $\nu = 1$, where $E_{nn}(r) = S_r S_{r+1}$. See eqns (3.80) and (3.82). The energy–energy correlation function is a four-point correlation in terms of spins. On the other hand, from the expression for the two-point correlation function of primary fields, eqn (6.31),

$$\langle \phi_{1,2}(z,\bar{z})\phi_{1,2}(0,0)\rangle = \frac{1}{z^{1/8}\bar{z}^{1/8}} = \frac{1}{r^{1/4}},$$

$$\langle \phi_{2,1}(z,\bar{z})\phi_{2,1}(0,0)\rangle = \frac{1}{z^{1}\bar{z}^{1}} = \frac{1}{r^{2}}. \tag{6.114}$$

These equations justify the identification of $\phi_{1,2}$ with the Ising spin field S and $\phi_{2,1}$ with the energy density field E_{nn}. In this way, the conformal field theory with $m = 3$ ($c = 1/2$) is considered to represent the Ising model at the critical point.

6.10   Finite-size effects

One of the most remarkable applications of the conformal field theory to critical phenomena is found in the analysis of finite-size effects. One can extract information on the scaling dimension and conformal charge from finite-size computations. This method provides a very useful practical tool to identify the universality class of a model system out of numerical data of finite-size systems.

The idea starts from the conformal transformation

$$z \longrightarrow f(z) = \frac{L}{2\pi} \log z, \tag{6.115}$$

which maps the whole complex plane to an infinitely long cylinder of circumference L. If we write $z = re^{i\theta}$, then the new coordinate on the cylinder is

$$f \equiv u + iv = \frac{L}{2\pi} \log r + i\frac{L\theta}{2\pi}. \tag{6.116}$$

The range $\theta : 0 \to 2\pi$ is mapped to $v : 0 \to L$ with v being understood to be periodic with period L. We apply this mapping to the two-point correlation function. If we write $h, \bar{h}$ for the conformal weight of a primary operator ϕ,

$$\langle \phi(z_1,\bar{z}_1)\phi(z_2,\bar{z}_2)\rangle = \Big((\partial_z f)(z_1)(\partial_z f)(z_2)\Big)^{h} \Big((\partial_{\bar{z}}\bar{f})(\bar{z}_1)(\partial_{\bar{z}}\bar{f})(\bar{z}_2)\Big)^{\bar{h}} \langle \phi(f_1,\bar{f}_1)\phi(f_2,\bar{f}_2)\rangle. \tag{6.117}$$

Using the explicit form of the mapping of eqn (6.115) and also eqn (6.31) for the left-hand side, the above relation is rewritten as

$$\langle \phi(f_1,\bar{f}_1)\phi(f_2,\bar{f}_2)\rangle$$
$$= \left(\frac{2\pi}{L}\right)^{2h+2\bar{h}} \frac{(z_1 z_2)^h (\bar{z}_1 \bar{z}_2)^{\bar{h}}}{(z_1 - z_2)^{2h}(\bar{z}_1 - \bar{z}_2)^{2\bar{h}}}$$
$$= \left(\frac{\pi}{L}\right)^{2h+2\bar{h}} \left(\sinh \frac{\pi}{L}(f_1 - f_2)\right)^{-2h} \left(\sinh \frac{\pi}{L}(\bar{f}_1 - \bar{f}_2)\right)^{-2\bar{h}}. \tag{6.118}$$

For a large separation of z_1 and z_2 along the infinitely long direction of the cylinder, $u_1 - u_2 \gg L$, the above form of the correlation function reduces to

$$\langle \phi(u_1, v_1)\phi(u_2, v_2)\rangle \approx \left(\frac{2\pi}{L}\right)^{2x} \exp\left(-\frac{2\pi x}{L}(u_1 - u_2)\right), \qquad (6.119)$$

where we have written $x = h + \bar{h}$ for the scaling dimension of ϕ and have assumed that the spin vanishes, $s = h - \bar{h} = 0$. It is natural that the correlation function decays exponentially for a finite-size cylinder as it is essentially a one-dimensional system. A remarkable point of eqn (6.119) is that the correlation length is related directly to the scaling dimension,

$$\xi = \frac{L}{2\pi x}. \qquad (6.120)$$

This formula allows us to estimate the scaling dimension x from the correlation length ξ for a finite-size system.

Another application concerns a finite-size correction to the free energy. We again use the transformation (6.115) and consider a small change $(u, v) \to (u + \epsilon u, v)$. Then, the free energy as a function of the linear size, $F(L)$, changes as

$$e^{-F(L+\delta L)} = \int \mathcal{D}\phi \, e^{-S-\delta S}, \qquad (6.121)$$

where $\delta L = \epsilon L$. Expanding both sides to first order in δ, we obtain

$$e^{-F(L)}\left(1 - \delta L \frac{\partial F}{\partial L}\right) = \int \mathcal{D}\phi \, e^{-S} - \int \mathcal{D}\phi \, \delta S \, e^{-S}, \qquad (6.122)$$

or

$$\frac{\partial F}{\partial L} = \frac{\langle \delta S\rangle}{\delta L}. \qquad (6.123)$$

We can express $\langle \delta S\rangle$ on the right-hand side of this equation as, using eqns (6.35), (6.37) and (6.38) and noticing $\partial^\mu \epsilon^\nu \neq 0$ only for $\mu = \nu = u$,

$$\langle \delta S\rangle = -\frac{\epsilon}{2\pi}\int_0^L dv \, \langle T_{uu}\rangle = -\frac{\epsilon}{2\pi}\int_0^L dv \, \langle T(f) + \bar{T}(f)\rangle. \qquad (6.124)$$

The expectation value of the energy–momentum tensor is

$$\langle T(f)\rangle = \langle \bar{T}(f)\rangle = -\frac{c}{24}\left(\frac{2\pi}{L}\right)^2, \qquad (6.125)$$

because, according to eqn (6.49),

$$(\partial_z f)^2 \langle T(f)\rangle + \frac{c}{12}\{f, z\} = \langle T(z)\rangle = 0, \qquad (6.126)$$

and the mapping (6.115) yields $\{f, z\} = 1/(2z^2)$.[13] We thus arrive at

$$\frac{\partial F}{\partial L} = \frac{\pi c}{6L^2},$$ (6.127)

which leads to

$$F = F_0 - \frac{\pi c}{6L}.$$ (6.128)

The integration constant F_0 actually corresponds to the bulk part of the free energy proportional to L. Equation (6.128) shows that the correction of the order of L^{-1} has a universal coefficient $-\pi c/6$, which can be used to evaluate c from finite-size numerical calculations.

---

[13] Remember that $\langle T(z) \rangle = 0$ due to eqn (6.40) with $X_n = 1$, which has $h_i = 0$. The expectation value of T for the restricted geometry $\langle T(f) \rangle$ does not vanish.

7

Kosterlitz–Thouless transition

As the spatial dimensionality d decreases, fluctuations become larger and the stability of the low-temperature ordered state deteriorates. Consequently, for instance, the Ising model in one dimension does not display long-range order at finite temperatures, i.e. does not have an ordered phase. If the basic variables and symmetries are continuous as in the XY and Heisenberg models, the (long-range) ordered state at any finite temperature disappears already in two dimensions. The XY model, nevertheless, undergoes an unusual phase transition without an onset of long-range order in two dimensions, which is known as the Kosterlitz–Thouless transition. We describe the theory of such interesting behavior in this chapter. Also elucidated is Elitzur's theorem for the absence of spontaneous symmetry breaking in lattice gauge theories.

7.1 Peierls argument

Mean-field theory correctly describes conventional critical phenomena above four dimensions (the upper critical dimension). As the spatial dimensionality d decreases, the effects of interactions between a spin and its neighbors become weaker due mainly to the decrease in the number of neighbors, and eventually long-range order disappears at finite temperatures below a certain dimension. This borderline dimensionality is the lower critical dimension d_{lc}. Systems with discrete degrees of freedom such as the Ising model have $d_{\mathrm{lc}} = 1$ and systems with continuous symmetries have typically $d_{\mathrm{lc}} = 2$. In the present section we introduce an argument that makes clear the difference, as far as long-range order at finite temperatures is concerned, between the one- and two-dimensional ferromagnetic Ising models. The argument can generically be applied to other systems such as the antiferromagnetic Ising and Potts models. Indeed, it constitutes a very useful tool to argue for the existence of long-range order at finite temperature in cases where no exact solution is available. The following sections will discuss the conditions for the existence and absence of long-range order in the XY model.

The first example is the one-dimensional Ising model, for which we develop a physical picture for the absence of long-range order not by solving the model explicitly (see Section 9.1) but by comparing the energy and entropy contributions to the free energy. Let us fix the left-most spin in the up (or +) state in the Ising model on a chain with length L. The right-most spin remains free. The ground state to be realized at $T = 0$ has all spins up because of our particular boundary condition. As the temperature increases from zero, excited states appear, in which some spins have the opposite direction (down or −) as in the left panel of Fig. 7.1. A parallel pair of

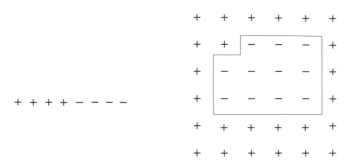

Fig. 7.1 Spin configurations where some spins are reversed from the ferromagnetic (all +) ground state in one dimension (left), and two dimensions (right).

spins (++ or −−) have energy −J and antiparallel pairs (+− or −+) have energy J, with $J > 0$. Thus, the energy to reverse the state of a pair from the lower (parallel) to higher (antiparallel) energy configuration is +2J. The number of locations where such a reversal may happen is $L − 1 \approx L$, for large L, which implies an entropy of $\log L$. It follows that the free-energy increase due to the spin reversal is

$$\Delta F = 2J - T \log L. \tag{7.1}$$

This formula shows that, for a fixed value of J, the free energy decreases $\Delta F < 0$ by the reversal of spin pairs for a sufficiently large system ($L \gg 1$) and arbitrary finite temperature ($T > 0$). The perfectly (long-range) ordered state with all spins up is therefore easily destroyed by any small but finite temperature. Thus, a simple argument consisting of a comparison between the energy and entropy contributions to the free energy in the Ising model reveals the absence of long-range order in one dimension at finite temperatures.

A two-dimensional version of this argument proceeds as follows. Consider a configuration of spins such as the one shown in the right panel of Fig. 7.1. Suppose that an island of down (−) spins with perimeter Γ emerged in the Ising model of N sites with a boundary condition consisting of all spins up (+), as in the right panel of Fig. 7.1. This boundary condition, which destroys the up/down symmetry of the model, i.e. $\mathbb{Z}_2$, favors a particular ground state (the one with all +), and mimics an infinitesimal magnetic field. The continuous line separating the + spins from the − spins in the island is called a *domain wall*. In general, the latter is not a closed polygon and it can be open, but the chosen boundary condition forces the domain walls to be closed polygons.

The energy to generate such an island is 2$J\Gamma$. The corresponding entropy is evaluated by counting the number of different ways to generate an island with perimeter Γ. This is the number of paths that return to the original position after Γ steps (the lattice spacing is the unit of length) with the constraint to pass through a single bond only once. On the square lattice, a single step to go from a site to the next has three possibilities because three bonds out of four are allowed to be chosen to avoid going back onto the same bond as in the preceding step. We therefore have roughly $3^\Gamma N$ possibilities for Γ steps. The factor N results from the fact that there are at most N sites to start from.[1] Then, the entropy is $\log(3^\Gamma N)$, and the free energy cost to generate an island of reversed spins is

$$\Delta F = 2J\Gamma - T\Gamma \log 3 = \Gamma(2J - T \log 3), \qquad (7.2)$$

if we choose an island such that $3^\Gamma \gg N$. For temperatures lower than $T_c \equiv 2J/\log 3 = 1.8J$, the island of reversed $(-)$ spins costs a positive free energy $(\Delta F > 0)$ and consequently is unlikely to happen, implying the stability of ferromagnetic long-range order. At high temperatures $T > T_c$, on the other hand, many islands can exist and the long-range order is destroyed. We therefore conclude that the two-dimensional Ising model undergoes a phase transition around $T_c = 1.8J$, which is fairly close to the exact solution $2.2J$. These ideas constitute the *Peierls argument* to show the existence of a phase transition in the two-dimensional Ising model. The Peierls argument is not a rigorous proof for the existence of long-range order. It, nevertheless, gives us a lucid physical picture that the existence of long-range order is determined by the balance between energy and entropy. A more formal mathematical proof of the existence of long-range order (ferromagnetic phase at finite temperature) can be developed based on the Peierls argument, see Appendix A.12.

The above argument applies to models with short-range interactions. If the system has long-range interactions, such as $J_{ij} = J/|i - j|^{1+\sigma}$, which decays in a power of the relative distance between two sites, a one-dimensional model can have a phase transition at a finite temperature.[2] We, nevertheless, concentrate ourselves on short-range interactions throughout this book unless otherwise stated explicitly (as in the infinite-range model of Section 2.5) because they represent typical situations.

7.2   Lower critical dimension of the XY model

Long-range order of systems with continuous degrees of freedom and symmetries, such as the XY model $(n = 2)$ and Heisenberg model $(n = 3)$, is vulnerable to instabilities due to thermal fluctuations and is actually absent at finite temperatures in two dimensions. Order is fragile in these systems and can be destroyed more easily than in systems with discrete degrees of freedom. Let us study in some detail how the lower critical dimension becomes two.

The following XY model on a hypercubic lattice as introduced in Section 1.5 will be discussed as a concrete example,

$$H = -J \sum_{\langle ij \rangle} \cos(\phi_i - \phi_j). \qquad (7.3)$$

This model has been used to study the critical behavior of the superfluid to normal phase transition in liquid ^{4}He and displays a global $U(1)$ symmetry, which amounts to a change $\phi_i \to \phi_i + \alpha$ on every site, with α a real number. Suppose that the temperature is very low and neighboring spins are aligned almost parallel to each other. Then, the

---

[1] This simple evaluation fails to take into account the avoidance condition of overlaps with more than a few steps before. A more accurate estimate leads to $a^\Gamma N$ possibilities with a slightly less than 3, which, however, does not affect our conclusion qualitatively.

[2] There are examples of classical systems with short-range interactions, such as Kittel's zipper model, that display true thermodynamic phase transitions in $d = 1$. Typically, these short-range models include hard-core interactions.

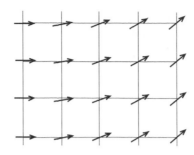

Fig. 7.2 The stable spin configuration with the left and right boundaries fixed.

argument $\phi_i - \phi_j$ of the cosine is very small compared to π, and it would be a good approximation to expand the cosine to second order. Since we are interested in the behavior of the system over a large spatial scale, we are allowed to ignore discreteness of the spatial coordinates of the lattice. We therefore construct an effective Hamiltonian, valid at low temperatures, by using a Fourier representation as

$$H \approx -J \sum_{\langle ij \rangle} \left(1 - \frac{1}{2}(\phi_i - \phi_j)^2 \right) \approx \frac{J}{2} \int d\boldsymbol{r} (\nabla \phi)^2 = \frac{J}{2} \int \frac{d\boldsymbol{q}}{(2\pi)^d} q^2 \tilde{\phi}(\boldsymbol{q}) \tilde{\phi}(-\boldsymbol{q}), \quad (7.4)$$

where we have dropped an additive constant that does not play a role, and considered the gradient along the spatial directions of the lattice $\phi_i - \phi_j \rightarrow (\boldsymbol{r}_i - \boldsymbol{r}_j) \cdot \nabla \phi((\boldsymbol{r}_i + \boldsymbol{r}_j)/2)$ with $\phi(\boldsymbol{r})$ a continuous function. This is the *spin-wave approximation* around an assumed ordered state, a quadratic Hamiltonian, which is expected to be valid at low temperatures. In this approximation no phase transition may occur, as seen in the Gaussian model with $t = 0$ in eqn (2.80). In general, effective Hamiltonians such as the one of eqn (7.4) with an analytic expansion, in terms of gradients of a slowly varying hydrodynamic (i.e. phenomenological) variable, characterize *generalized elasticity* problems. In the case of magnetic systems the parameter of *rigidity J* is known as the *spin-wave stiffness* or the *helicity modulus*, and it is proportional to the *superfluid density* in the case of superfluids, the latter being discussed in Section 7.4.

To understand what type of spin configurations are stable under the spin-wave approximation, we adopt the variational principle with respect to a local angle variable (or a scalar field) $\phi(\boldsymbol{r})$, $\delta H/\delta \phi(\boldsymbol{r}) = 0$, to find the following Laplace equation,[3]

$$\nabla^2 \phi = 0. \quad (7.5)$$

We solve this Laplace equation under the boundary condition that the left boundary $(x = 0)$ has $\phi = 0$ and the right $(x = L)$ has $\phi = \phi_0$, along the chosen x-direction. The solution is the uniformly rotated state, $\phi = x\phi_0/L$, as depicted in Fig. 7.2. Then, the energy of this configuration has the value

$$E = \frac{J}{2} L^{d-2} \phi_0^2, \quad (7.6)$$

---

[3] Integration by parts in the third expression of eqn (7.4) changes the integrand to $-\phi \nabla^2 \phi$. Functional variation of this expression yields the result.

as can be verified by inserting $\nabla \phi = (\phi_0/L, 0, 0, \cdots)$ to the intermediate expression of eqn (7.4), assuming that the volume of the system is L^d. This result shows that energy increases indefinitely as $L \to \infty$ for $d > 2$. A very large energy is needed to twist both sides of the system by a finite angle, suggesting that the system is robust against the change of boundary conditions. The information on the specific state that one end has propagates through the system to the other end. The system is thus considered to have rigid long-range order. If $d < 2$, the twist energy (7.6) does not increase with system size and the effects of the boundaries do not propagate deep into the system. Hence no long-range order exists. This simple argument therefore illustrates that the stability of the long-range order in the XY model changes when the dimension of the system is $d = 2$. The case $d = 2$ is marginal and needs more careful scrutiny.

So far, the theory was just developed for the susceptibility of the energy against a change of boundary conditions and did not include the effects of temperature. We therefore must evaluate the behavior of the system at finite temperatures to confirm the validity of the conclusion that $d = 2$ is the borderline dimensionality, i.e. the lower critical dimension. The goal is to calculate the fluctuation of the relative orientation of two spins as a function of their distance. The fluctuation of the relative orientation is written as, using the Fourier representation,

$$\langle(\phi(\boldsymbol{r}) - \phi(0))^2\rangle = \int \frac{\mathrm{d}\boldsymbol{q}_1 \mathrm{d}\boldsymbol{q}_2}{(2\pi)^{2d}} (e^{-i\boldsymbol{q}_1 \cdot \boldsymbol{r}} - 1)(e^{-i\boldsymbol{q}_2 \cdot \boldsymbol{r}} - 1)\langle\tilde{\phi}(\boldsymbol{q}_1)\tilde{\phi}(\boldsymbol{q}_2)\rangle. \tag{7.7}$$

The expectation value of Fourier-transformed angle variables in the integrand is to be calculated from the Hamiltonian (7.4). Since eqn (7.4) is a quadratic form of $\tilde{\phi}(\boldsymbol{q})$, i.e. a Gaussian theory, it is straightforward to apply the computations of Section 2.9 with $t = 0$ and $b = J/2$ to find

$$\langle\tilde{\phi}(\boldsymbol{q}_1)\tilde{\phi}(\boldsymbol{q}_2)\rangle = \frac{(2\pi)^d T}{J q_1^2}\delta(\boldsymbol{q}_1 + \boldsymbol{q}_2). \tag{7.8}$$

We then have

$$\langle(\phi(\boldsymbol{r}) - \phi(0))^2\rangle = \frac{T}{J} \int \frac{\mathrm{d}\boldsymbol{q}}{(2\pi)^d} \frac{|e^{-i\boldsymbol{q}\cdot\boldsymbol{r}} - 1|^2}{q^2} \propto \frac{T}{J}\int_{r^{-1}}^{a^{-1}} \mathrm{d}q\, q^{d-3}, \tag{7.9}$$

where the upper limit of the integral, the largest allowed wave number, has been replaced by the largest absolute value of the wave number, which is proportional to the inverse of the lattice constant a^{-1}. The lower limit of the integral is chosen to be the inverse of the distance, r^{-1}, because the integrand in the middle expression is very small for $q < r^{-1}$, i.e. $e^{-i\boldsymbol{q}\cdot\boldsymbol{r}} \approx 1$.[4] For $d > 2$, the last integral of eqn (7.9) is

$$\frac{T}{J(d-2)}(a^{2-d} - r^{2-d}). \tag{7.10}$$

---

[4] A more rigorous evaluation of the integral leads to the same conclusion that the lower critical dimension of the XY model is two.

Since $2 - d < 0$, this expression converges to a finite value for large r. This suggests that fluctuations of the angle difference stay finite and long-range order is not destroyed because two spins far apart share essentially the same angle. If the dimensionality is exactly two, the integral (7.9) yields a logarithmic term, $\log r$, which diverges as r tends to infinity. Fluctuations grow indefinitely as the distance increases. Therefore, the two angle variables $\phi(\boldsymbol{r})$ and $\phi(0)$ become uncorrelated in the limit $r \to \infty$ as long as the temperature is finite. We conclude that long-range order does not exist at finite temperatures in the two-dimensional XY model. The same is true for $d < 2$.

We have used several approximate estimates in the above discussion. It is possible to derive the same result rigorously by using Schwarz inequalities, as will be shown in the next section.

7.3 Mermin–Wagner theorem: Absence of spontaneous magnetization

The *Mermin–Wagner theorem* states that two-dimensional short-range interacting systems with continuous degrees of freedom and symmetries do not have spontaneous magnetization at finite temperatures. More generally, it is a statement relating the dimensionality of a system with continuous symmetry with the existence of the phenomenon of spontaneous symmetry breaking. We have made use of the spin-wave approximation in the previous section, and the same conclusion is derived rigorously in the present section. The original theorem was given in the context of quantum spin systems, but we explain here a classical version since it is slightly simpler and does not need the introduction of quantum spin operators. A common physical mechanism, related to symmetry and fluctuations, lies behind the formal proofs both for quantum and classical systems. The quantum version of the theorem is proved in Appendix A.13. Although illuminating, the reader who is not interested in the details of the proof can skip this section.

The Hamiltonian of the XY model in the presence of a finite external field h is

$$H = -J \sum_{\langle ij \rangle} \cos(\phi_i - \phi_j) - h \sum_i \cos \phi_i. \tag{7.11}$$

The sum in the first term on the right-hand side runs over nearest-neighbor pairs on the square lattice. The starting point of the proof is the following *Schwarz inequality* that holds under very general conditions,

$$\langle AA^* \rangle \geq \frac{|\langle AB^* \rangle|^2}{\langle BB^* \rangle}, \tag{7.12}$$

where A and B are functions of angle variables (ϕ_i, ϕ_j) and $\langle \cdots \rangle$ denotes the thermal average with respect to the canonical ensemble Boltzmann weight $\mathrm{e}^{-\beta H}$ with $\beta = 1/T$. The crux of the proof is to choose A and B as follows,

$$A = \frac{1}{N} \sum_j \mathrm{e}^{-i\boldsymbol{q} \cdot \boldsymbol{r}_j} \sin \phi_j, \quad B = \frac{1}{N} \sum_l \mathrm{e}^{-i\boldsymbol{q} \cdot \boldsymbol{r}_l} \frac{\partial H}{\partial \phi_l}. \tag{7.13}$$

Here, q is a wave vector and r_j and r_l are position vectors. We insert these definitions into the Schwarz inequality (7.12) and sum both sides over q for a finite-size system with periodic boundary conditions. The left-hand side is bounded as

$$\sum_q \langle AA^* \rangle = \frac{1}{N^2} \sum_{i,j} \sum_q e^{-i q \cdot (r_i - r_j)} \langle \sin \phi_i \sin \phi_j \rangle = \frac{1}{N} \sum_i \langle \sin^2 \phi_i \rangle \leq 1. \qquad (7.14)$$

The numerator of the right-hand side of eqn (7.12) is

$$\langle AB^* \rangle = \frac{Tm}{N}, \qquad (7.15)$$

where m is the magnetization per spin. To derive this identity, we first note that

$$\langle AB^* \rangle = \frac{1}{N^2} \sum_{j,l} e^{-i q \cdot (r_j - r_l)} \left\langle \sin \phi_j \frac{\partial H}{\partial \phi_l} \right\rangle, \qquad (7.16)$$

whose last expectation value is rewritten, after integration by parts, as

$$\frac{1}{Z} \int_0^{2\pi} \prod_i d\phi_i e^{-\beta H} \sin \phi_j \frac{\partial H}{\partial \phi_l} = \frac{T}{Z} \int_0^{2\pi} \prod_i d\phi_i e^{-\beta H} \cos \phi_j \delta_{lj}, \qquad (7.17)$$

where we used $e^{-\beta H} \partial H / \partial \phi_l = -\beta^{-1} \partial e^{-\beta H} / \partial \phi_l$. It then follows

$$\langle AB^* \rangle = \frac{T}{N^2} \sum_j \langle \cos \phi_j \rangle = \frac{Tm}{N}. \qquad (7.18)$$

The denominator of the right-hand side of eqn (7.12) is upper-bounded as ($q = |q|$)

$$\langle BB^* \rangle \leq T \left(\frac{Jq^2 + h}{N} \right). \qquad (7.19)$$

To understand this inequality it is useful first to insert the definition of eqn (7.13),

$$\langle BB^* \rangle = \frac{1}{N^2} \sum_{l,j} e^{-i q \cdot (r_l - r_j)} \left\langle \frac{\partial H}{\partial \phi_l} \frac{\partial H}{\partial \phi_j} \right\rangle. \qquad (7.20)$$

We next use $e^{-\beta H} \partial H / \partial \phi_l = -\beta^{-1} \partial e^{-\beta H} / \partial \phi_l$ to rewrite the expectation value on the right-hand side, after integration by parts, as

$$\frac{1}{Z} \int_0^{2\pi} \prod_i d\phi_i e^{-\beta H} \frac{\partial H}{\partial \phi_l} \frac{\partial H}{\partial \phi_j} = \frac{T}{Z} \int_0^{2\pi} \prod_i d\phi_i e^{-\beta H} \frac{\partial^2 H}{\partial \phi_l \partial \phi_j}. \qquad (7.21)$$

The second order derivative appearing here is evaluated according to the combination of indices.

- For $l = j$.

$$\frac{\partial^2 H}{\partial \phi_l \partial \phi_j} = \frac{\partial^2 H}{\partial \phi_l^2} = J \sum_\delta \cos(\phi_l - \phi_{l+\delta}) + h \cos \phi_l. \qquad (7.22)$$

Here, $\boldsymbol{\delta}$ is the vector to the nearest-neighbor site on the square lattice. There are four of them.

- For neighboring l and j.

$$\frac{\partial^2 H}{\partial \phi_l \partial \phi_j} = \frac{\partial}{\partial \phi_l}\left(J \sum_{\boldsymbol{\delta}} \sin(\phi_j - \phi_{j+\boldsymbol{\delta}}) + h \sin \phi_j\right) = -J \cos(\phi_j - \phi_l). \qquad (7.23)$$

- Otherwise, it is 0.

Combining these three cases, we have

$$\beta\langle BB^* \rangle$$

$$= \frac{1}{N^2} \sum_l \left(J \sum_{\boldsymbol{\delta}} \langle \cos(\phi_l - \phi_{l+\boldsymbol{\delta}}) \rangle + h\langle \cos \phi_l \rangle\right) - \frac{J}{N^2} \sum_{l,\boldsymbol{\delta}} e^{-i\boldsymbol{q}\cdot\boldsymbol{\delta}}\langle \cos(\phi_l - \phi_{l+\boldsymbol{\delta}}) \rangle$$

$$= \frac{J}{N^2} \sum_l (4 - \sum_{\boldsymbol{\delta}} e^{-i\boldsymbol{q}\cdot\boldsymbol{\delta}})\langle \cos(\phi_l - \phi_{l+\boldsymbol{\delta}}) \rangle + \frac{h}{N^2} \sum_l \langle \cos \phi_l \rangle, \qquad (7.24)$$

where $\langle \cos(\phi_l - \phi_{l+\boldsymbol{\delta}}) \rangle$ respects the symmetry of the square lattice in the sense that it is independent of $\boldsymbol{\delta}$. Now, note that the sum over vectors to neighboring sites gives[5]

$$\sum_{\boldsymbol{\delta}} e^{-i\boldsymbol{q}\cdot\boldsymbol{\delta}} = 2\cos q_x + 2\cos q_y. \qquad (7.25)$$

Using the trivial inequality $1 - \cos q \leq q^2/2$, which can be verified by graphical means, and another trivial relation $\langle \cos(\cdots) \rangle \leq 1$, we can rewrite the final expression of eqn (7.24) as

$$\beta\langle BB^* \rangle = \frac{J}{N^2} \sum_l (4 - 2\cos q_x - 2\cos q_y)\langle \cos(\phi_l - \phi_{l+\boldsymbol{\delta}}) \rangle + \frac{h}{N}m$$

$$\leq \frac{J}{N}(q_x^2 + q_y^2) + \frac{h}{N} = \left(\frac{Jq^2 + h}{N}\right). \qquad (7.26)$$

This is eqn (7.19).

Replacement of the relations (7.14), (7.15) and (7.19) into the corresponding expressions in the Schwarz inequality (7.12) gives

$$1 \geq \frac{T}{N}m^2 \sum_q \frac{1}{Jq^2 + h}. \qquad (7.27)$$

In the thermodynamic limit $N \to \infty$ the sum becomes an integral,

$$1 \geq Tm^2 \int \frac{d\boldsymbol{q}}{(2\pi)^2} \frac{1}{Jq^2 + h}. \qquad (7.28)$$

---

[5] We normalize the lattice constant to unity, i.e. $a = 1$.

In two dimensions, for any $T > 0$, this integral diverges as $h \to 0$ due to the singularity at the origin (*infrared divergence*). The inequality is satisfied only if $m \to 0$ as $h \to 0$, i.e. no spontaneous symmetry breaking can occur. This ends the proof that there is no spontaneous magnetization in the XY model at finite temperatures in two dimensions.[6]

The square of the wave number q^2 in the denominator of eqn (7.28) represents essentially the same long-range processes as in the evaluation of the energy of a spin wave in eqn (7.4). In the spin-wave approximation this q^2 leads to the power of $d - 3$ in eqn (7.9) for the fluctuation of relative angles, and as a result, the integral diverges in two dimensions. The same mechanism is seen to work to give a diverging integral as $h \to 0$ in eqn (7.28). We would like to mention that the general Mermin–Wagner theorem of this section and Appendix A.13 can be applied to other classical or quantum models with short-range interactions. Indeed, the original formulation was applied to the classical Heisenberg model to prove that ferromagnetism (or antiferromagnetism) cannot be present in $d \leq 2$. When applied to other models, the starting point is always the inequality of eqn (7.12) with appropriately chosen functions A and B depending on the model. Notice that the theorem does not exclude the possibility to have spontaneous symmetry breaking at $T = 0$. Indeed, at exactly $T = 0$ the two-dimensional classical XY model has long-range order.

> **EXERCISE 7.1** Generalize the proof of this section to an arbitrary dimension d and show that there is no spontaneous magnetization for $d < 2$ and that it is impossible to show the same result for $d > 2$.

7.4   Kosterlitz–Thouless transition

We have seen that the two-dimensional XY model has no spontaneous magnetization (long-range order) at finite temperatures and consequently has no ordinary (Landau-type) phase transition. This system is, nevertheless, known to have a special type of phase transition without long-range order. The low-temperature phase does not display long-range order but has clearly different correlation properties from the high-temperature paramagnetic phase. While correlation functions decay exponentially in the paramagnetic phase, they slowly decrease as a power law in the low-temperature phase (except at exactly $T = 0$ where there is long-range order). This power-law correlation is reminiscent of what happens at the critical point. An important difference from the usual critical point, though, is that this power-law behavior extends over a finite temperature range. Sometimes this phase is referred to as a low-temperature *critical phase*. Systems with power-law decay of (potential) order parameter correlation functions are said to have *quasi-long-range order*. The XY model is at the lower critical dimension in two dimensions, and this fact causes such singular behavior. This special transition from quasi-long-range order to disorder is

---

[6] Rigorously speaking, long-range order is not identical to spontaneous magnetization, the former being defined by the limiting value of a correlation function as discussed at the end of Section 5.6. We, however, often use these two names interchangeably in this book because they are physically equivalent.

known as the *Kosterlitz–Thouless (KT) transition*, and the critical phase is called the *Kosterlitz–Thouless (KT) phase*.[7]

Let us calculate the correlation function using the spin-wave approximation, which is valid at low temperatures, to verify its power-law behavior. The correlation function of the XY model is expressed as

$$\left\langle \cos\big(\phi(\boldsymbol{r}) - \phi(0)\big)\right\rangle = \left\langle e^{i(\phi(\boldsymbol{r})-\phi(0))}\right\rangle. \tag{7.29}$$

Note that the imaginary part of the right-hand side vanishes due to the symmetry of global reversal of angle variables, $\phi(\boldsymbol{r}) \to -\phi(\boldsymbol{r})$ $\forall \boldsymbol{r}$, under which the Hamiltonian remains invariant. The goal is to evaluate the expectation value on the right-hand side using the spin-wave Hamiltonian (7.4). Since this Hamiltonian is quadratic in the angle variables, the corresponding Gibbs–Boltzmann distribution is Gaussian. We therefore use the fact that the cumulants of order higher than the second order vanish for the Gaussian distribution (see Appendix A.4) to find

$$\left\langle \cos\big(\phi(\boldsymbol{r}) - \phi(0)\big)\right\rangle = \left\langle e^{i(\phi(\boldsymbol{r})-\phi(0))}\right\rangle = e^{-\frac{1}{2}\langle(\phi(\boldsymbol{r})-\phi(0))^2\rangle}. \tag{7.30}$$

The expression in the exponent can be evaluated as in the integral (7.9) for the case $d = 2$,

$$\left\langle (\phi(\boldsymbol{r}) - \phi(0))^2 \right\rangle \approx \frac{T}{J(2\pi)^2} \int_0^{2\pi} d\theta \int_{r^{-1}}^{a^{-1}} \frac{dq}{q} = \frac{T}{\pi J} \log\left(\frac{r}{a}\right). \tag{7.31}$$

Inserting this result into eqn (7.30), we have $(r = |\boldsymbol{r}|)$

$$\left\langle \cos\big(\phi(\boldsymbol{r}) - \phi(0)\big)\right\rangle = r^{-T/2\pi J}. \tag{7.32}$$

Usually, correlation functions do not decay with distance if long-range order exists and decay exponentially in the paramagnetic phase. Equation (7.32) shows a slow, power-law decay in between, which is characteristic of systems exactly at the critical point. However, the exponent $\eta = T/2\pi J$ is not universal since it explicitly depends on T and J. Remarkably, this result holds for any temperature as long as the spin-wave approximation is a justified assumption. We conclude that the system is critical for a finite temperature range. This may also be understood as if there exists a fixed line, i.e. a set of fixed points, under the renormalization group. Thus, this behavior is called quasi-long-range order. The relative angle of the spin variables does not have (long-range) order but changes slowly.

> **EXERCISE 7.2** Calculate the correlation function (7.29) for general $d(\neq 2)$ and confirm that $d = 2$ is the borderline dimension of stability, the lower critical dimension.

The spin-wave quadratic model is critical and it displays no phase transition. However, at sufficiently high temperatures we would expect the XY model to be in a paramagnetic phase with exponentially decaying relative angle correlations.

[7] J. M. Kosterlitz and D. J. Thouless, J. Phys. C **6** (1973) 1181. A similar idea was proposed by V. L. Berezinskii in Sov. Phys. JETP **34** (1972) 610.

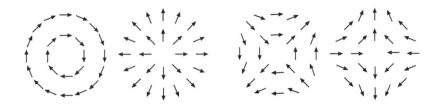

Fig. 7.3 Vortices with $n = 1, 1, -1, -1$ and $c = -\pi/2, 0, \pi/2, \pi$ from left to right, respectively, where n and c are defined by $\phi(\boldsymbol{r}) = n\theta + c$.

As the temperature increases from the low-temperature limit, it must happen that the angle variables change more drastically than those expressed in the spin-wave approximation. The Laplace equation $\nabla^2\phi = 0$ admits, besides the uniform solutions $\phi = $ const. studied above, non-uniform singular solutions. Topological vortex solutions (see Section 5.8) are inhomogeneous states that cannot be described by a continuous function such as the smoothly varying spin waves. In particular, vortex configurations, not taken into account in the spin-wave approximation of the XY model, gradually affect the state of the system, eventually destroying the quasi-long-range order.

To express configurations with vortices, we write the angle field variable $\phi(\boldsymbol{r})$ as a function of the angle θ measured from the x-axis as $\phi(\boldsymbol{r}) = n\theta + c$, with n an integer known as the winding number (a topological invariant) and c a constant, Fig. 7.3. Then, the derivative of the angle variable $\nabla\phi$ appearing in the spin-wave Hamiltonian (7.4) has as radial and azimuthal components as

$$(\nabla\phi)_r = \frac{\partial\phi}{\partial r} = 0, \quad (\nabla\phi)_\theta = \frac{1}{r}\frac{\partial\phi}{\partial\theta} = \frac{n}{r}. \tag{7.33}$$

An excitation with a positive winding number is a vortex and one with a negative value of n is also known as an *antivortex*. Then, the energy needed to create such a vortex configuration is, according to eqn (7.4),

$$E = \frac{J}{2}\int\frac{n^2}{r^2}r\mathrm{d}r\mathrm{d}\theta = n^2\pi J\log\frac{R}{r_0} + E_{\mathrm{C}}. \tag{7.34}$$

Here, R is the linear size (radius) of the system, r_0 is the radius of the vortex core (the lower bound of the integral and the short-distance cutoff), and E_{C} is the vortex core energy. We have assumed that the upper bound of the integral is R, the lower bound r_0, and the contribution below the lower bound $r < r_0$ gives a finite energy E_{C}. Thus, the total energy of a vortex has two contributions, the first term in eqn (7.34) represents the elastic energy, while the energy E_{C} is a microscopic contribution associated with the destruction of the uniform order at the core of the vortex. The entropy of a single vortex S is given by the logarithm of the number of ways to place the center of the vortex in the region having radius between r_0 and R. This number of ways is expected to be proportional to the relevant area of the region. We thus have

$$S = \log\left\{\left(\frac{R}{r_0}\right)^2 \cdot \mathrm{const}\right\}. \tag{7.35}$$

The combination of these estimates of energy and entropy leads to the following free energy needed to create a vortex with winding number $n = 1$,

$$\Delta F = (\pi J - 2T) \log R + \text{const.} \tag{7.36}$$

For the temperature range $T > T_{\mathrm{KT}} \equiv \pi J/2$, the creation of vortices causes a free-energy decrease. We therefore conclude that a phase transition takes place at $T_c = T_{\mathrm{KT}}$, above which a large number of vortices proliferate and the spin-wave approximation breaks down. For $T > T_{\mathrm{KT}}$, the angle variable around a vortex changes very quickly, invalidating the spin-wave approximation, and the relative angle between spins far apart are correlated only very weakly. Then, the quasi-long-range order is destroyed and the system becomes paramagnetic. This is the Kosterlitz–Thouless (KT) transition. The low-temperature region below T_{KT} with quasi long-range order is the KT phase.

In the KT phase the creation of a vortex increases the free energy, and a single vortex is not stable. Nevertheless, a pair of vortices with different signs of their winding number n may be stable if the distance between them is not too large. To show this result we assume that several vortices exist around the origin. Then, the angle field far from the origin is written as, instead of eqn (7.33),

$$(\nabla \phi)_r = \frac{\partial \phi}{\partial r} = 0, \quad (\nabla \phi)_\theta = \frac{1}{r} \frac{\partial \phi}{\partial \theta} = \frac{\sum_i n_i}{r}. \tag{7.37}$$

The energy corresponding to eqn (7.34) is

$$E = \left(\sum_i n_i \right)^2 \pi J \log \frac{R}{r_0} + E_{\mathrm{C}}, \tag{7.38}$$

where E_{C} is the total vortex core energy. We may therefore conclude that several vortices may exist even in the low-temperature KT phase as long as the condition of neutrality $\sum_i n_i = 0$ is satisfied. In particular, a pair of vortices having the same absolute value but opposite signs of their winding numbers, $\pm n$, are allowed to exist.

Since at low temperatures vortices can be bound in pairs, the Kosterlitz–Thouless transition is physically associated with the unbinding of vortices. The simple and heuristic energy–entropy argument developed in this section neglects interactions between vortices that will be studied in the next section. A more sophisticated renormalization group analysis, to be developed in a later section, shows that this qualitative picture is the correct description of the transition.

EXERCISE 7.3 Draw vortex configurations with $n = 2$ and $n = -2$ similarly to Fig. 7.3.

EXERCISE 7.4 Derive the condition for the term $\sum_i \cos(p\phi_i)$ with p a natural number to be relevant in the sense of renormalization group for the XY model. It will be useful first to estimate the scaling dimension x_p from the calculation of the corresponding correlation function $\langle \cos(p\phi(\boldsymbol{r}) - p\phi(0)) \rangle$ generalizing the discussions in the first half of the present section. The result will reveal the condition for the exponent y_p to be positive. In particular, show that the term $\sum_i \cos(p\phi_i)$ is relevant if p is larger than a threshold p_0 and the temperature is lower than some T_p. This implies that this term is irrelevant in the temperature range $T_p < T < T_{\mathrm{KT}}$ where the KT phase is realized. On the other hand, this term is relevant for $T < T_p$ and the system has the same properties as the *clock model* in which the angles assume only discrete values $\phi_i = 2\pi k/p \ (k = 0, 1, 2, \cdots, p - 1)$.

A few words are in order on the superfluid transition of liquid helium. The kinetic energy of a thin film of superfluid helium is written as

$$E = \frac{\rho_s}{2} \int v_s^2 \, \mathrm{d}x\mathrm{d}y = \frac{\rho_s}{2} \left(\frac{\hbar}{m}\right)^2 \int (\nabla\psi)^2 \, \mathrm{d}x\mathrm{d}y, \qquad (7.39)$$

where ρ_s is the superfluid density per area and v_s is the superfluid velocity, and we have used the Landau–Ginzburg relation $v_s = (\hbar/m)\nabla\psi$. Comparison with eqn (7.4) or eqn (7.48) in the next section reveals the relation $J = \rho_s \hbar^2 / m^2$. It follows that the ratio of ρ_s to T at the transition point $T_c = T_{\mathrm{KT}} (= \pi J/2)$ assumes a universal value independent of experimental details,

$$\frac{\rho_s}{T_{\mathrm{KT}}} = \frac{2m^2}{\pi\hbar^2}. \qquad (7.40)$$

Since $\rho_s = 0$ above the transition due to the absence of superfluidity, ρ_s/T jumps from the above finite value to zero at the transition. This is called the *universal jump* of the superfluid density (stiffness) and has been confirmed experimentally. Also, the problem of a roughening transition of equilibrium crystal surfaces displays a universal jump in the smoothness parameter and has also been confirmed experimentally.

Equation (7.40) represents a quantity proportional to the ratio of J and T. This turns out to lead to the fact that the critical exponent η assumes a specific number $\eta(T_{\mathrm{KT}}) = 1/4$ at the transition point. The reason is that eqn (7.32) implies that $\eta = T/2\pi J$, proportional to the ratio between T and J, has the value $1/4$ at the transition point because $T_{\mathrm{KT}} = \pi J/2$. The relation $\eta(T_{\mathrm{KT}}) = 1/4$ is often used to check if a transition belongs to the same universality class as the KT transition.

7.5 Interaction energy of vortex pairs

We have learned that a vortex pair can exist in a stable manner around the origin if the neutrality condition is satisfied. The physical properties of vortices are better understood when we study the energy of vortices in their general configurations. In the following we will establish a connection between the XY model and a neutral Coulomb gas in two dimensions with charges n_i, such that $\sum_i n_i = 0$.

The angle variable or field for a single vortex with winding number $n = 1$ located at the origin is written as

$$(\nabla\phi)_r = 0, \quad (\nabla\phi)_\theta = \frac{1}{r}. \qquad (7.41)$$

Let us integrate the field $v \equiv \nabla \phi$ around the vortex, along a closed circuit,

$$\oint v \cdot \mathrm{d}r = \int_0^{2\pi} \frac{1}{r} r \mathrm{d}\theta = 2\pi. \tag{7.42}$$

Then, the Stokes theorem implies that $(\mathrm{curl}\, v)_z = 2\pi\delta(r)$, where the z-direction is perpendicular to the two-dimensional xy plane. For a more general configuration with many vortices, we have

$$(\mathrm{curl}\, v)_z = 2\pi \sum_i n_i \delta(r - r_i) \equiv 2\pi N(r), \tag{7.43}$$

where r_i is the position of the vortex i, i.e. the location of its core.

Now, the Cartesian components of the vector field $v = \nabla\phi$ for a single vortex with $n = 1$ are, according to eqn (7.41),

$$v_x = \frac{\partial\phi}{\partial x} = -\frac{y}{r^2}, \quad v_y = \frac{\partial\phi}{\partial y} = \frac{x}{r^2}. \tag{7.44}$$

If we introduce a new scalar field $\psi = -\log(r/r_0)$, the above components are expressed as

$$v_x = \frac{\partial\psi}{\partial y}, \quad v_y = -\frac{\partial\psi}{\partial x}. \tag{7.45}$$

In a superfluid the real physical vortices can be described by these variables, in which case v is called the superfluid velocity, ϕ the velocity potential, and ψ the stream function. For a generic case with many vortices, we generalize $\psi = -\log(r/r_0)$ to

$$\psi(r) = -\sum_i n_i \log \frac{|r - r_i|}{r_0}. \tag{7.46}$$

It is instructive to verify the validity of this generalization. From eqns (7.45) and (7.46), we find

$$(\mathrm{curl}\, v)_z = \frac{\partial v_y}{\partial x} - \frac{\partial v_x}{\partial y} = -\frac{\partial^2\psi}{\partial x^2} - \frac{\partial^2\psi}{\partial y^2} = -\nabla^2\psi = 2\pi N(r), \tag{7.47}$$

which is consistent with eqn (7.43).

The total energy of a (neutral) system with many vortices is therefore

$$E = \frac{J}{2}\int(v_x^2 + v_y^2)\mathrm{d}x\mathrm{d}y = \frac{J}{2}\int(\nabla\psi)^2\mathrm{d}x\mathrm{d}y$$

$$= -\frac{J}{2}\int(\psi\nabla^2\psi)\mathrm{d}x\mathrm{d}y$$

$$= -\pi J \sum_{i \neq j} n_i n_j \log \frac{|r_i - r_j|}{r_0} + E_{\mathrm{C}}. \tag{7.48}$$

This equation can be interpreted as the total energy of a set of charged vortices (with charge n_i) interacting via a two-dimensional Coulomb potential having a logarithmic dependence on the relative distance. Notice that the energy of a pair of vortices with opposite charges, $n_i n_j < 0$, is minimized when they are close to each other, i.e. tightly bound. At low temperatures two vortices with different signs (such as $n = \pm 1$) are bound together, creating a gas of vortex–antivortex pairs or *molecules*, and the system may be regarded as a dielectric. Above the KT transition point those pairs are destroyed (melted) by thermal fluctuations and single vortices freely move around, forming a plasma-like state. Unbound vortices correspond to free or mobile charges. In this sense, the physics of the KT transition is equivalent to the statistical mechanics of a two-dimensional Coulomb gas. Equation (7.48) suggests that the coupling constant J is related to the dielectric constant ϵ by $\pi J = 1/\epsilon$. The effective interaction energy between vortices (7.48) will be rederived in Section 10.3.4 without the *ad hoc* introduction of vortex degrees of freedom.

7.6 Renormalization group analysis

An analysis of the KT transition by the renormalization group method is a prominent example in which the real-space renormalization works very successfully.

7.6.1 Renormalization group equation to describe the KT transition

Let us first identify the variables that determine the critical behavior of the present system. Physical intuition is useful to find the relevant variables, and we eventually write renormalization group equations for these variables. The temperature is clearly the most important variable. The corresponding scaling field x is chosen as $x = 2 - \pi K (= 2 - \pi J/T)$ such that it vanishes at the fixed point.[8] The variable x is actually not relevant but marginal. In conventional critical phenomena, the temperature variable is relevant and renormalizes toward zero if the initial value is below the critical point, as illustrated in Fig. 1.5. However, in the KT transition, there is no isolated fixed point and all temperatures below the critical point are attracted to corresponding points on a fixed line, which represents a set of fixed points. The KT transition point does not correspond to an unstable fixed point characteristic of a relevant scaling field. Nevertheless, the temperature is not irrelevant but is marginal.

Another important variable to take into consideration is the number of vortices. If there are few vortices, the spin-wave approximation describes the system qualitatively faithfully and the system is in the KT phase. As the number of vortices increases, the slow and smooth change of angles, as assumed in the spin-wave approximation, is not respected and the angles vary quickly near vortices, eventually leading to the KT transition into the paramagnetic phase. It is therefore reasonable to introduce as a relevant variable the chemical potential μ of vortices, which controls the number of vortices, or equivalently the *fugacity* obtained by exponentiating the chemical potential, $y_0 = \mathrm{e}^{-\beta \mu}$. For small y_0 (large chemical potential) the number of vortices

---

[8] Strictly speaking, a fixed point should be distinguished from a critical point. Thus, K appearing here is not the interaction constant before renormalization $K = J/T$ (*bare coupling*) but is the variable after many steps of renormalization (*renormalized coupling*). The difference between these two concepts will be explained in more detail later.

is small and the system is in the KT phase, whereas the paramagnetic phase has a large y_0 with very many vortices. This means that we have to see whether the fugacity increases or decreases under a renormalization group transformation.

A more quantitative analysis is facilitated by the energy

$$E(\boldsymbol{r}_1, \boldsymbol{r}_2) = 2\pi J \log \frac{|\boldsymbol{r}_1 - \boldsymbol{r}_2|}{r_0} + E_C, \tag{7.49}$$

which describes a pair of vortices with opposite winding numbers $n = \pm 1$, located at $\boldsymbol{r}_1$ and $\boldsymbol{r}_2$. This equation is derived from eqn (7.48) when we set $n_1 = -n_2 = 1$ or $n_1 = -n_2 = -1$ with all other n_is vanishing. The chemical potential for two vortices is 2μ, and the corresponding fugacity is y_0^2. Then, the probability for the above configuration of a pair of vortices to appear should be proportional to the following expression

$$y_0^2 e^{-\beta E(\boldsymbol{r}_1, \boldsymbol{r}_2)} = y_0^2 e^{-\beta E_C} \left| \frac{\boldsymbol{r}_1 - \boldsymbol{r}_2}{r_0} \right|^{-2\pi K}. \tag{7.50}$$

The correlation function of vortex variables $\langle |n(\boldsymbol{r}_1) n(\boldsymbol{r}_2)| \rangle$ is calculated from contributions of non-vanishing values of $|n(\boldsymbol{r}_1)|$ and $|n(\boldsymbol{r}_2)|$ and hence is proportional to the probability of eqn (7.50) when there are only a small number of vortices ($|n(\boldsymbol{r}_1)|, |n(\boldsymbol{r}_2)| = 0$ or 1). This observation leads to two interesting facts. First, the distance-independent part of the above equation, $y_0^2 e^{-\beta E_C}$, implies that the fugacity y_0 always appears as a product with $e^{-\beta E_C/2}$. We may thus adopt $y = y_0 e^{-\beta E_C/2}$ as a basic scaling field instead of y_0. Secondly, the scaling dimension of a vortex is $v_x = \pi K$.[9]

We are now ready to write the renormalization group equation for y, which controls the number of vortices, using the relation between the scaling dimension v_x and the renormalization group exponent v_y, $v_y = d - v_x$. The relation $y' = b^{v_y} y$ with the scaling factor b reduces in the limit of an infinitesimal scaling factor $b = 1 + \mathrm{d}l$ to

$$\frac{\mathrm{d}y}{\mathrm{d}l} = v_y \cdot y = (2 - v_x)y = (2 - \pi K)y = xy, \tag{7.51}$$

if we notice that $b^{v_y} \approx 1 + v_y \mathrm{d}l$. This is the differential renormalization group equation for y.

To derive the renormalization group equation for the scaling field x, we assume that the system is close to the KT transition point ($|x| \ll 1$). Moreover, we are interested in whether or not the number of vortices increases by a renormalization group transformation. These aspects justify keeping only the lowest-order term in the Taylor expansion of the right-hand side of the renormalization group equation (beta function) in powers of x and y. Since vortices show up as pairs in the KT phase, the

---

[9] The scaling fields of the two-dimensional XY model are often written as x and y for historical reasons. This notation may be confused with the scaling dimension or the exponent of the renormalization group eigenvalue. In this book we write v_x and v_y for the scaling dimension and exponent, respectively, of the vortex numbers. Do not confuse this notation with the Cartesian components of the velocity field $\boldsymbol{v}$ of the previous section.

second-order term is the lowest-order one in the variable y. As will be shown later, the renormalization group equation for x is readily written only in terms of this effect,[10]

$$\frac{\mathrm{d}x}{\mathrm{d}l} = a^2 y^2. \tag{7.52}$$

Since the presence of vortices disorders the system, x should increase by the effects of y, and the coefficient on the right-hand side is chosen to be positive, a^2. Let us confirm that no other additive terms including a constant and a low-order term in x appear. It is first clear that a constant term does not exist because $x^* = y^* = 0$ is a fixed point. A term proportional to x represents an instability of the fixed point $x^* = 0$ since a temperature lower than the critical one renormalizes to still lower temperatures, see Fig. 4.4, which is in conflict with the physical picture that the KT phase corresponds to a fixed line, not an isolated fixed point. Similarly, x^2 increases the temperature with $x < 0$ toward $x = 0$, and the KT transition point $x = 0$ becomes a stable, isolated fixed point, again incompatible with our physical intuition. Similar considerations exclude all terms written as functions of x on the right-hand side of eqn (7.52). We therefore conclude that eqns (7.52) and (7.51) are the right renormalization group equations to describe the KT transition. They are called the *Kosterlitz equations*.

7.6.2   Solving the Kosterlitz equations

To solve the Kosterlitz equations, it is useful to note that the scale variable l can be eliminated by taking the ratio between both sides of eqns (7.52) and (7.51),

$$\frac{\mathrm{d}x^2}{\mathrm{d}y^2} = a^2. \tag{7.53}$$

The result is an equation for x and y, and its analytic solution is a hyperbola

$$x^2 - a^2 y^2 = \mathsf{const.} \tag{7.54}$$

Since the KT transition point is at $x = y = 0$, the solution corresponding to the transition point has a vanishing constant on the right-hand side of eqn (7.54). This means that the line $y = \pm x/a$ should go through the transition point on the xy plane. The renormalization group flow is drawn in Fig. 7.4. Within the KT phase, the spin-wave approximation captures the essence of the relevant physics and vortices do not play an essential role, irrelevant in the renormalization group sense, and hence y is renormalized to 0. This is the situation to the left of the line $y = -x/a$ in Fig. 7.4. In the paramagnetic phase y renormalizes to larger and larger values, as in the region to the right of the line $y = -x/a$. These considerations lead to the renormalization group flow illustrated in Fig. 7.4. On the low-temperature side, the KT phase, y is renormalized to 0 and x is to a finite value that corresponds to the initial (bare) value, and the system is attracted to the fixed line $y^* = 0, x < 0$. On the high-temperature side, the paramagnetic phase, y increases indefinitely and more and more vortices are created as the renormalization group process goes on. The line $y = -x/a$ separates these two phases and is known as a *separatrix*.

---

[10] Do not confuse the constant a of this section with the lattice constant.

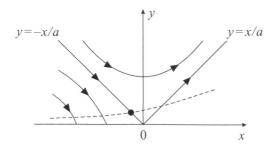

Fig. 7.4 Renormalization group flow of the Kosterlitz equations. The dashed line represents the bare XY model before renormalization. The small black dot on this line is the KT transition point written in terms of the bare coupling.

We have seen that the scaling field y is related to the fugacity y_0 and the chemical potential μ for vortices by the relation $y = y_0 e^{-\beta E_C/2} = e^{-\beta\mu - \beta E_C/2}$. These quantities, fugacity and chemical potential, have been introduced artificially to study the situation with a small number of vortices and do not exist in the original XY model. The original problem corresponds to $\mu = 0$ or $y_0 = 1$, and hence the bare couplings lie on the curve $y = e^{-\beta E_C/2}$ drawn as a dashed line in Fig. 7.4. This figure shows that the value of x increases toward a slightly larger value on the line $y = 0, x < 0$ after renormalization even within the KT phase (to the left of the line $y = -x/a$). Correspondingly, the temperature is also renormalized to a larger value. The fixed line is reached after many steps of the renormalization group transformation, and the dashed line of the original system itself is not invariant under renormalization. It should also be noticed that the fixed-point condition $x^* = 0$ ($T = T_{KT} = \pi J/2$) representing the KT transition point is to be described by the renormalized temperature. The transition temperature in terms of the original variable is at the crossing point of the line $y = -x/a$ and the dashed curve, the small black dot in Fig. 7.4. Since this crossing point has $x < 0$, the KT transition temperature in terms of the original variable is smaller than $\pi J/2$.

We next study the singularities of physical quantities near the transition point using the solution of the Kosterlitz equations. Since $x^2 - a^2y^2 = 0$ at the transition point, we will have $x^2 - a^2y^2 = -ct$ ($t = (T - T_{KT})/T_{KT}, c > 0$) slightly above the transition point. Then, the solution to the Kosterlitz equation

$$\frac{\mathrm{d}x}{\mathrm{d}l} = a^2 y^2 = x^2 + ct \tag{7.55}$$

is

$$l = l_0 + \frac{1}{\sqrt{ct}} \arctan \frac{x}{\sqrt{ct}}. \tag{7.56}$$

Equation (7.55) indicates that x increases with l ($x \neq 0$) independent of its sign, meaning that the renormalization flows to the right in Fig. 7.4.

EXERCISE 7.5 Confirm that the Kosterlitz equation near the transition point (7.55) has the solution (7.56).

Now, suppose that l (essentially the same as the scale b of renormalization) has reached the size of the system after many renormalization steps. If we write l_f for this l and set the value of arctan on the right-hand side of eqn (7.56) to its largest possible value $\pi/2$, we have

$$l_f = l_0 + \frac{k}{\sqrt{t}} \quad \left(k = \frac{\pi}{2\sqrt{c}}\right). \tag{7.57}$$

We will later use the fact that vortices cease to exist in pairs when l reaches l_f since almost all degrees of freedom have been traced out. In order to connect this l_f with the correlation length, we notice that the correlation length $\tilde{\xi}'$ measured from the standard of the renormalized system of scale b is related to the correlation length $\tilde{\xi}$ in the original scale as $\tilde{\xi}' = \tilde{\xi}/b$. Then, the renormalization group equation for the infinitesimal scaling $b = 1 + dl$ is

$$\frac{d\tilde{\xi}}{dl} = -\tilde{\xi}. \tag{7.58}$$

This equation is solved as $\tilde{\xi} \propto e^{-l}$. The value of $\tilde{\xi}$ obtained by integration of eqn (7.58) is the correlation length measured in the standard after renormalization. If this value is A, the correlation length measured in the unit of the unrenormalized system is $\xi = Ae^l$. We therefore conclude that the limit length for a vortex pair to exist, measured in the original scale (standard of length) is, from eqn (7.57),

$$\xi \approx \xi_0 e^{l_f} \approx \exp\left(\frac{k}{\sqrt{t}}\right). \tag{7.59}$$

The correlation length diverges exponentially, with a non-universal coefficient k, as the temperature approaches the transition point from above. An exponential divergence is very strong. For example, when $t = 10^{-2}$, eqn (7.59) gives $\xi \approx 2 \times 10^5$ if k is unity. It is therefore necessary to take sufficient care in numerical studies of the KT transition.

In order to check the singularity of the free energy, we note that the scaling law $f(t) = b^{-d}f(b^{y_t}t)$ can be rewritten as $f = b^{-d}g(b/\xi)$ because we have $f(b^{y_t}t) = f((b/t^{-\nu})^{y_t}) \equiv g(b/\xi)$ from $\xi = t^{-\nu} = t^{-1/y_t}$. This expression $f = b^{-d}g(b/\xi)$ for the scaling function is valid even when the correlation length diverges exponentially, not polynomially, because the argument is written as the ratio of b and ξ without reference to a power of t explicitly. Let us set $b = \xi$ in $f = b^{-d}g(b/\xi)$ and apply eqn (7.59) to find

$$f = \xi^{-d}g(1) \approx \exp\left(-\frac{2k}{\sqrt{t}}\right), \tag{7.60}$$

which expresses the essential singularity of the free energy. This equation shows that the free energy has a very weak singularity that is differentiable arbitrarily many times. Consequently, the same is true for the specific heat. The essential singularity in the specific heat at T_{KT} is very weak and unobservable experimentally and in numerical simulations. Indeed the specific heat has a broad non-universal peak slightly above the transition point and has no sign of singularity.

Those peculiar exponential singularities in the correlation function and specific heat reflect the lower critical dimension (two) of the XY model. As seen in

Section 3.6.3, similar exponential singularities are shared by the Ising model at its lower critical dimension $d = 1$ near the transition point $T = 0$.

7.7 Lattice gauge theory and Elitzur's theorem

In this section we digress from the main topic of KT transition and discuss the absence of spontaneous symmetry breaking in systems with local (gauge) symmetry. The theorem of Mermin and Wagner claims that continuous global symmetries do not break down spontaneously in two or lower dimensions. The same is true for discrete global symmetries in one dimension. We show in the present section that there exist no spontaneous symmetry breaking in any dimensions for local symmetries. This result contrasts the difference between global and local symmetries.

For this purpose we analyze the *lattice gauge theory*, which has been introduced to understand the mechanism of confinement of quarks. Although the physical motivation is different, some models in the lattice gauge theory show phase transitions, whose properties are controlled by the symmetry and dimensionality of the system, similarly to conventional spin systems.

7.7.1 Lattice gauge theory

Symmetries of a physical system can be classified into global or local (gauge) depending on the character of the transformation realizing the mathematical mapping.[11] For example, in the conventional Ising model, one needs to change the sign of *all* spins $(S_i \rightarrow -S_i, \forall i)$ to realize the global $\mathbb{Z}_2$ discrete symmetry of the model. The same happens in the XY model, where *all* angle variables need to be transformed by the same amount $(\phi_i \rightarrow \phi_i + \alpha, \forall i)$ to realize the global $U(1)$ continuous symmetry.[12] On the other hand, there are models where transforming only *some* degrees of freedom is enough to achieve invariance. A *gauge theory* is defined by a Hamiltonian or action, classical or quantum, that is invariant under a local or gauge transformation. It can be defined on a lattice or in the continuum, e.g. as a field theory. According to the *gauge principle* adopted widely in field theory, all fundamental physical interactions in nature arise from actions or Hamiltonians that are invariant under local transformations. The primary motivation to study lattice gauge theories is to provide a non-perturbative approach for the standard theory of strong interactions in high-energy physics, also known as quantum chromodynamics, and thus to attempt to explain the phenomenon of quark confinement. This is well beyond the scope of this book, and we will only concentrate on some of the aspects of critical phenomena in classical models of the lattice gauge theory.

An example of a classical model that displays discrete gauge symmetries is the $\mathbb{Z}_2$ lattice gauge theory, also known as the $\mathbb{Z}_2$ *gauge theory* or the *Ising lattice gauge theory*. Consider Ising spins $S_i = \pm 1$ that reside on the bonds i,[13] and not on the vertices (sites), of a three-dimensional cubic lattice. Then, the Hamiltonian of the $\mathbb{Z}_2$

---

[11] Sometimes, a gauge symmetry is referred to as a gauge structure instead of a symmetry since two states related by this gauge transformation are the same state but with a different label.

[12] The group $U(1)$ is composed of rotations on the complex plane.

[13] A bond is often called a *link* in gauge theories.

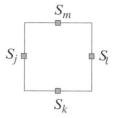

Fig. 7.5 The product of four spins on bonds around a plaquette constitutes the basic interaction $B_\square$ in the Z_2 gauge theory.

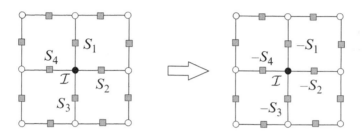

Fig. 7.6 The signs of spin variables on bonds emanating from a site $\mathcal{I}$ are changed. This is a local, gauge transformation and keeps the Hamiltonian invariant.

gauge theory is defined as

$$H = -J \sum_{\square} B_{\square}, \tag{7.61}$$

where the sum spans all possible square plaquettes $\square$ on the lattice each containing four spins and, as we will see, is invariant under the $\mathcal{G} = \mathbb{Z}_2$ gauge group. The interaction term comprises the product of these four spins, as depicted in Fig. 7.5,

$$B_{\square} = \prod_{i \in \square} S_i. \tag{7.62}$$

In addition to the global $\mathbb{Z}_2$ symmetry, $S_i \to -S_i$ ($\forall i$), this Hamiltonian has a $\mathbb{Z}_2$ gauge symmetry, which consists of the following transformation (see Fig. 7.6): Select any vertex $\mathcal{I}$ of the lattice shared by six bonds (four bonds for $d = 2$ as in Fig. 7.6), flip the sign of the spins on these six (four) bonds, $S_i \to -S_i$ if i emanates from $\mathcal{I}$. Since each plaquette connected with vertex $\mathcal{I}$ has two spins flipped, their product $B_{\square}$, and thus the overall Hamiltonian, remains invariant.

Notice that, while the lowest-energy state in the usual Ising model is two-fold degenerate, the ground-state degeneracy is much more enormous in the $\mathbb{Z}_2$ gauge theory. For example, if the configuration on the left panel of Fig. 7.6 is a ground state, the one on the right panel is also a ground state. This gauge transformation that

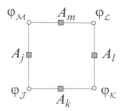

Fig. 7.7 Allocation of variables on a plaquette to generate a gauge transformation in the $U(1)$ gauge theory.

generates another state with the same energy is applicable to any vertex. Consequently, there exist a large number of degenerate states (exponential in the system size).

Another example of a classical model of gauge theory is the $U(1)$ *gauge theory* ($U(1)$ *lattice gauge theory*) with a continuous local symmetry

$$H = -J \sum_{\square} A_{\square}, \tag{7.63}$$

where the plaquette interaction $A_{\square}$ is defined in terms of elements of the group $U(1)$, i.e. complex fields $\phi_j = e^{iA_j}$ with gauge variables $-\pi < A_j \leq \pi$ defined on the bonds j, as

$$A_{\square} = \frac{1}{2} \left(\phi_j \phi_k \phi_l^* \phi_m^* + \phi_j^* \phi_k^* \phi_l \phi_m \right) = \cos\left(A_j + A_k - A_l - A_m \right), \tag{7.64}$$

with j, k, l, m bonds belonging to the plaquette $\square$ as in Fig. 7.7. We assume that those plaquette interaction terms are defined on a general hypercubic lattice.

The following $U(1)$ gauge transformation is a symmetry of the Hamiltonian of eqn (7.63),

$$A_j \to A_j + \varphi_{\mathcal{J}} - \varphi_{\mathcal{M}} \tag{7.65}$$

$$A_k \to A_k + \varphi_{\mathcal{K}} - \varphi_{\mathcal{J}} \tag{7.66}$$

$$A_l \to A_l + \varphi_{\mathcal{K}} - \varphi_{\mathcal{L}} \tag{7.67}$$

$$A_m \to A_m + \varphi_{\mathcal{L}} - \varphi_{\mathcal{M}}, \tag{7.68}$$

with arbitrary c-number functions $\varphi_{\mathcal{J}}$ defined on the vertices $\mathcal{J}$ of the lattice as indicated in Fig. 7.7. It is straightforward to check that eqns (7.65) to (7.68) keep eqn (7.64) invariant. This transformation is *Abelian* because the group $U(1)$ is Abelian; two successive changes of angles are equivalent to the changes in the other order and are thus commutable. There are several non-Abelian generalizations of these models.

7.7.2 Elitzur's theorem

The presence of local (gauge) invariance has important physical consequences. One of those consequences is *Elitzur's theorem*, which states that non-gauge-invariant (or gauge-variant) local physical quantities cannot exhibit spontaneous breaking of gauge

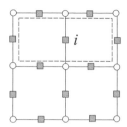

Fig. 7.8 An example of a sublattice $\mathcal{C}_i$, which includes bond i, is shown as a dashed contour.

symmetries, discrete or continuous, in any dimensions. This does not imply that a phase transition, signaled as a singularity in the free energy, cannot occur, as we will see in the example of the three-dimensional Z_2 gauge theory below. Therefore, the expectation value of a gauge-variant quantity cannot be used as a Landau-type local order parameter to describe such a phase transition. Since symmetry breaking of local quantities is precluded, differences in the behavior of correlation functions in different phases have to manifest themselves in non-local quantities written in terms of the original local degrees of freedom.

We now prove Elitzur's theorem. The essence of the proof is as follows. Consider the absolute value of the average of any local quantity $f(\phi_i)$ (involving only a finite number of fields or variables $\{\phi_i\}$ like S_i), which is bounded and non-invariant under a gauge symmetry group $\mathcal{G}$ of a Hamiltonian H (such as $\mathbb{Z}_2$ in the Z_2 gauge theory). This $|\langle f(\phi_i)\rangle|$ is shown to be bounded from above by the absolute mean value of the same quantity computed for a zero-dimensional Hamiltonian $\bar{H}$ (i.e. it involves a finite number of degrees of freedom) which is globally invariant under $\mathcal{G}$ and preserves the range of the interactions. This upper bound is shown to vanish in the zero-field limit after the thermodynamic limit due to the local character of the symmetry.

More explicitly, to determine if spontaneous symmetry breaking occurs, we evaluate

$$\langle f(\phi_i)\rangle = \lim_{h\to 0} \lim_{N\to\infty} \langle f(\phi_i)\rangle_{h,N}, \tag{7.69}$$

where $\langle f(\phi_i)\rangle_{h,N}$ is the average value of $f(\phi_i)$ calculated on a finite lattice of N sites and in the presence of a symmetry breaking field h. Simple examples of $f(\phi_i)$ are S_i for the Z_2 gauge theory and $\mathrm{e}^{i\phi_i}$ for the $U(1)$ gauge theory. Since the lattice Λ is formed out of the union of smaller finite sublattices, $\Lambda = \bigcup_l \mathcal{C}_l$, the bond i belongs at least to one subset (see Fig. 7.8).

It is convenient to rename the fields in the following way: $\phi_l = \psi_l$ if $l \notin \mathcal{C}_i$ and $\phi_i = \eta_i$ if $i \in \mathcal{C}_i$. Then, we can separate the variables to write $\langle f(\phi_i)\rangle_{h,N}$ as

$$\langle f(\phi_i)\rangle_{h,N} = \frac{\sum_{\{\phi_l\}} f(\phi_i)\mathrm{e}^{-\beta\left(H(\{\phi_l\})+h\sum_l \phi_l\right)}}{\sum_{\{\phi_l\}} \mathrm{e}^{-\beta\left(H(\{\phi_l\})+h\sum_l \phi_l\right)}} \tag{7.70}$$

$$= \frac{\sum_{\{\psi_l\}} z_{\{\psi\}}\mathrm{e}^{-\beta h\sum_{l\notin \mathcal{C}_i}\psi_l}\, g(\{\psi_l\})}{\sum_{\{\psi_l\}} z_{\{\psi\}}\mathrm{e}^{-\beta h\sum_{l\notin \mathcal{C}_i}\psi_l}}, \tag{7.71}$$

where

$$z_{\{\psi\}} = \sum_{\{\eta_i\}} e^{-\beta\left(H(\{\psi_l,\eta_i\})+h\sum_{i\in c_i}\eta_i\right)}, \tag{7.72}$$

and

$$g(\{\psi_l\}) = \frac{\sum_{\{\eta_i\}} f(\eta_i) e^{-\beta\left(H(\{\psi_l,\eta_i\})+h\sum_{i\in c_i}\eta_i\right)}}{z_{\{\psi\}}}. \tag{7.73}$$

From eqn (7.71), since $z_{\{\psi\}} e^{-\beta h\sum_{l\notin c_i}\psi_l}$ is positive definite, $\langle f(\phi_i)\rangle_{h,N}$ can be bounded as follows

$$|\langle f(\phi_i)\rangle_{h,N}| \leq |g(\{\bar{\psi}_l\})|, \tag{7.74}$$

where $\{\bar{\psi}_l\}$ is the particular configuration of fields ψ_l that maximizes $g(\{\psi_l\})$ in eqn (7.71). The quantity $H(\{\psi_l,\eta_i\})$ is a zero-dimensional Hamiltonian in that it involves only a finite number of bonds as far as the field variables η_i are concerned. This zero-dimensional Hamiltonian $H(\{\psi_l,\eta_i\})$ is invariant under the global symmetry group of transformations $\mathcal{G}$ over the fields η_i, e.g. $S_i \to -S_i$ ($\forall i$) in the Z_2 gauge theory.

Let us define $\bar{H}(\{\eta_i\}) \equiv H(\{\bar{\psi}_l,\eta_i\})$. The range of the interactions between the η-fields in $\bar{H}(\{\eta_i\})$ is clearly the same as the range of the interactions between the ϕ-fields in the original Hamiltonian $H(\{\phi_l\})$. Since $\bar{H}(\{\eta_i\})$ is a zero-dimensional Hamiltonian with only a finite number of variables, $g(\{\bar{\psi}_l\})$ is an analytic function of h for any N including the thermodynamic limit. The exponential in the numerator of eqn (7.73) is invariant under the global transformation $\mathcal{G}$ in the limit $h \to 0$ after $N \to \infty$ but the function $f(\eta_i)$ changes the sign in the Z_2 gauge theory, e.g. $f(\eta_i)$ may be S_i, which changes as $S_i \to -S_i$. In the case of the $U(1)$ gauge theory, the phase changes like $f(\eta_i) = e^{i\phi_i} \to e^{i(\phi_i+\phi)}$. Thus, $g(\{\bar{\psi}_l\}) = -g(\{\bar{\psi}_l\})$ for the Ising (Z_2) case and $g(\{\bar{\psi}_l\}) = e^{i\phi}g(\{\bar{\psi}_l\})$ for the $U(1)$ gauge theory, any one of which is satisfied only if $g(\{\bar{\psi}_l\}) = 0$. This completes the proof.

Notice that the frozen variables $\bar{\psi}_l$ act like external fields in $\bar{H}(\{\eta_i\})$, which do not break the global symmetry group of transformations $\mathcal{G}$. From a physics standpoint, a gauge symmetry involves a few degrees of freedom and it costs only a finite amount of energy to change a stable state to another one, which is in marked contrast to the case of global symmetry depicted in Fig. 5.2. This is the essence of the above proof.

7.7.3 Phase transitions in the lattice gauge theory

The three-dimensional Z_2 gauge theory of eqn (7.61) is dual (i.e. essentially equivalent) to the three-dimensional Ising model, as explained in Section 10.2. The latter has a phase transition at finite temperature. This means that the free energy of the Z_2 gauge theory shows the same singularity at the critical temperature T_c as the conventional Ising model. However, the phase transition in the Z_2 gauge theory does not manifest itself as a spontaneous symmetry breaking in the local spin variables due to Elitzur's theorem; the Z_2 gauge theory does not have a Landau-type local order parameter.

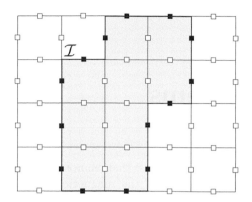

Fig. 7.9 A Wilson loop consists of the product of variables along a closed path Γ (thick lines). In this figure the variables on Γ are denoted in black squares. A gauge transformation of variables around a vertex $\mathcal{I}$ on Γ keeps W_Γ invariant.

To characterize the low- and high-temperature phases, one must use a correlation function that is gauge invariant. A well-known physical quantity used to characterize the phases of gauge models is the *Wilson loop*, constructed for the Z_2 gauge theory, for example, as

$$W_\Gamma = \left\langle \prod_{i \in \Gamma} S_i \right\rangle, \tag{7.75}$$

where i runs over the bonds forming a closed path or loop Γ as in Fig. 7.9. This quantity W_Γ is gauge invariant. For example, if one changes the signs of S_is connected to a vertex $\mathcal{I}$ located on Γ, two of the S_is on bonds emanating from $\mathcal{I}$ and on Γ change the sign and thus W_Γ remains invariant.

From the dependence of W_Γ on Γ in the limit of large loops, one can determine the nature of the phases of the model. In the high-temperature phase, the Wilson loop has an exponential decay controlled by the area of the loop, an *area law* $W_\Gamma \approx \mathrm{e}^{-c|A|}$ $(c > 0)$, where $|A|$ is the size of the area surrounded by Γ (shown in gray in Fig. 7.9). At low temperatures, it is controlled by the length of the loop, a *perimeter law* $W_\Gamma \approx \mathrm{e}^{-c|\Gamma|}$ $(c > 0)$, where $|\Gamma|$ is the length of Γ. The temperature at which there is a change in the asymptotic behavior of W_Γ defines the transition point T_c.

The two-dimensional version of the Z_2 gauge theory is trivially solvable by a high-temperature expansion, as elucidated in Section 10.2. It displays no finite-temperature phase transition, and the lower critical dimension of the Z_2 gauge theory is $d_{lc} = 2$.

8
Random systems

Real materials always contain randomness that cannot be expressed by idealized simple model systems. For example, some of the magnetic atoms carrying spins may be replaced by impurities without spins or the strength of interactions between spins may change from bond to bond due to irregularities in the crystal structure. In the present chapter we study the effects of randomness on phase transitions and critical phenomena. At the initial stage of studies on randomness, some people believed that randomness may obscure singular behavior such as the divergence of physical quantities at the critical temperature. It is now established that well-defined phase transitions continue to exist as long as randomness is not too strong, but the critical behavior may get modified with respect to the pure sample. We will show what phase transitions exist under the influence of randomness.

8.1 Random fields

Hamiltonians describing phase transitions and critical phenomena usually consist of interaction and field terms. These are competing relevant terms, in the sense of renormalization group, that determine the values of the exponents y_t and y_h. The most basic model does not have other relevant operators. It follows that the effects of randomness can be studied by its influences on those two terms. The present section treats the field term with randomness. The randomness in interactions will be discussed in the next section.

The Hamiltonian with randomness in the field term is written as

$$H = -J \sum_{\langle ij \rangle} S_i S_j - \sum_i h_i S_i. \tag{8.1}$$

For simplicity, we analyze only the Ising model in this chapter, unless stated otherwise. The value of the external field is assumed to depend on the site index h_i, reflecting randomness. Since in most cases it is virtually impossible to identify the values of randomness h_i at each site i from experiments, it is customary to adopt a model distribution function of random fields $\{h_i\}$. Typical examples are the following Gaussian and binary distributions,[1]

$$P(h_i) = \frac{1}{\sqrt{2\pi} h_0} \exp\left(-\frac{h_i^2}{2h_0^2}\right) \tag{8.2}$$

$$P(h_i) = \frac{1}{2}\delta(h_i - h_0) + \frac{1}{2}\delta(h_i + h_0). \tag{8.3}$$

---

[1] The standard deviation of the Gaussian distribution is usually denoted as σ but we write instead h_0 so that the notation is the same as the one used in the binary distribution.

We assume that randomness at different sites is not strongly correlated and thus accept independence of the distribution of randomness at different sites.

The *random-field Ising model* of eqn (8.1) may not directly represent realistic magnetic materials unless much milder distributions than those in eqns (8.2) and (8.3) are used, mild in the sense that the variance is small. It is usually considered impossible to realize site-dependent fields that change the sign from site to site. It, nevertheless, turns out that a randomly diluted antiferromagnet under uniform field and fluids in random, porous media in their lattice-gas representation are examples that are well described by the random-field Ising model.

8.1.1 Quenched system and self-averaging

The time scale for the change of randomness in fields is usually much longer than that of thermal fluctuations. For example, when randomness comes from the random mixture of magnetic and non-magnetic atoms, the positions of atoms do not change within the experimental time scale but the orientation of spins changes quickly. The corresponding theoretical framework is first to generate a set $\{h_i\}$ from the distribution function $P(h_i)$ and then apply the usual statistical-mechanical prescription to calculate the free energy and other physical quantities for the given fixed values of $\{h_i\}$. Randomness with this property is called *quenched randomness*, and the corresponding system is named a *quenched system*. In contrast, if the degrees of freedom of randomness change in a similar time scale as that of microscopic degrees of freedom, the system is called an *annealed system*.[2] We treat quenched systems that correspond to most experimental situations.

It is difficult to calculate the free energy explicitly as a function of the quenched randomness of $\{h_i\}$, i.e. N variables $h_1, \cdots, h_N$. Fortunately, it turns out, in the thermodynamic limit $N \to \infty$, that many physical quantities including the free energy do not depend upon the values of fields $\{h_i\}$ themselves but only on the distribution function $P(h_i)$ as eqns (8.2) and (8.3). This fact is called the *self-averaging property*.

To understand the self-averaging property let us divide a system into subsystems, as depicted in Fig. 8.1. The size of the whole system is L_0^d and that of a subsystem is L_1^d, where d is the spatial dimension. We assume $L_0 \gg L_1 \gg 1$, that is, the whole system and the subsystems are both very large and in addition the former is much larger than the latter. From $L_1 \gg 1$, the size of the interface between subsystems (the surface of a subsystem) L_1^{d-1} is much smaller than the size of the subsystem itself L_1^d, as we have $L_1^{d-1}/L_1^d = L_1^{-1} \ll 1$. Then, the sum of the free energy of each independent subsystem F_{sub} is very close to the free energy of the total system F_{tot},

$$F_{\mathrm{tot}} = \sum_{j=1}^{M} \left(F_{\mathrm{sub}}^{(j)} + \mathcal{O}(L_1^{d-1}) \right) = \sum_{j=1}^{M} F_{\mathrm{sub}}^{(j)} + \mathcal{O}(L_0^d L_1^{-1}). \tag{8.4}$$

---

[2] 'Quenching' means to quickly cool the system and freeze the degrees of freedom of randomness. 'Annealing' means to slowly cool the system down to a low temperature and thus the atoms (randomness) have time to reach their equilibrium locations.

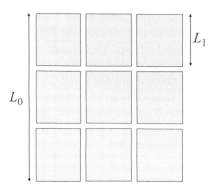

Fig. 8.1 It is useful to divide the system into subsystems to understand the self-averaging property.

Here, j is the index labeling the subsystem and M is the number of subsystems $(L_0/L_1)^d (\gg 1)$. The last term of $\mathcal{O}(\cdot)$ corresponds to the surface free energy related to the interface between subsystems. If we divide both sides by the total number of spins $N = L_0^d$ to find the free energy per spin $f_{\text{tot}} = F_{\text{tot}}/N$, we have

$$f_{\text{tot}} = \frac{1}{M} \sum_{j=1}^{M} \frac{1}{L_1^d} F_{\text{sub}}^{(j)} + \mathcal{O}(L_1^{-1}) \approx \frac{1}{M} \sum_{j=1}^{M} f_{\text{sub}}^{(j)}. \tag{8.5}$$

The right-hand side is the average of the free energy of many $(M \gg 1)$ independent systems, each with different values of random fields $\{h_i\}$ generated from the same distribution function $P(h_i)$. The left-hand side is, in contrast, the free energy of a single large system with given values of random fields $\{h_i\}$. Equation (8.5) suggests that the free energy as a function of a fixed set of random fields $\{h_i\}$ (left-hand side) coincides with the same free energy averaged over the distribution function (right-hand side) in the limit of a sufficiently large system.[3] The term self-averaging reflects this property that the value of the free energy coincides with the average of itself.

The self-averaging property allows us to calculate the free energy averaged over randomness instead of the free energy for a given set of random fields $\{h_i\}$, the latter being directly correlated to experimental situations. Since the former average is easier to evaluate than the latter, we hereafter discuss the former quantity. The average over randomness is termed the *configurational average*.

The self-averaging property is shared by many extensive quantities including the internal energy, specific heat, magnetization, and magnetic susceptibility if they are divided by the system size to give their values per spin. The reason is that these quantities are obtained by differentiation of the free energy with respect to appropriate variables.

The above discussions equally apply to randomness in the interactions, not just randomness in the external fields.

---

[3] A more accurate statement is that both sides of eqn (8.5) coincide with probability one. The distribution function of f_{tot} approaches a delta function that has a peak at the average value.

8.1.2   Mean-field theory

Let us first apply the mean-field theory of Chapter 2 to the problem of a phase transition in the presence of random fields. We would like to know how a uniformly ordered ferromagnetic phase at low temperatures is affected by random fields. The order parameter is the usual magnetization per spin m. From the self-averaging property, the magnetization is expressed as

$$m = \frac{1}{N} \sum_{i=1}^{N} \langle S_i \rangle = \left[\langle S_i \rangle \right]. \tag{8.6}$$

Here, the square brackets $[\cdots]$ denote the configurational average.[4] An application of the mean-field theory described in Section 2.1 leads to the following mean-field Hamiltonian,

$$H = N_{\mathrm{B}} J m^2 - J m z \sum_i S_i - \sum_i h_i S_i. \tag{8.7}$$

The free energy $F = -T[\log Z]$ is then straightforward to evaluate,

$$F = N_{\mathrm{B}} J m^2 - T N \left[\log 2 \cosh \beta (J m z + h_i) \right]. \tag{8.8}$$

The right-hand side is averaged over randomness using the self-averaging property. Minimization of the free energy with respect to m leads to the self-consistent equation, i.e. the equation of state,

$$m = \left[\tanh \beta (J m z + h_i) \right]. \tag{8.9}$$

We expand the right-hand side to third order in m to check the condition for ferromagnetic order to exist $m \neq 0$ following the discussions in Section 2.2,

$$m = \beta J z \left[1 - \tanh^2(\beta h_i) \right] m$$
$$- \frac{1}{3} (\beta J z)^3 \left[1 - 4 \tanh^2(\beta h_i) + 3 \tanh^4(\beta h_i) \right] m^3. \tag{8.10}$$

We have dropped odd powers of h_i because the symmetric distribution of h_i means a vanishing configurational average for odd powers.

Let us proceed with the binary distribution (8.3) as an explicit example. The configurational average of even powers like $\tanh^2(\beta h_i)$ is obtained simply by replacing h_i with h_0 in the case of the binary distribution. Then, the equation of state reads

$$m = \beta J z \left(1 - \tanh^2(\beta h_0) \right) m - \frac{1}{3} (\beta J z)^3 \left(1 - 4 \tanh^2(\beta h_0) + 3 \tanh^4(\beta h_0) \right) m^3. \tag{8.11}$$

The coefficient of the first-order term, m, on the right-hand side is small in the high-temperature region $\beta J \ll 1$ and is large at low temperatures $\beta J \gg 1$. Therefore, the

---

[4] Equation (8.6) shows that the configurational average (right-hand side) is equivalent to the spatial average (middle expression).

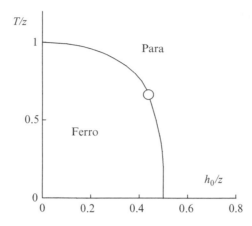

Fig. 8.2 The mean-field phase diagram of the ferromagnetic Ising model in the presence of random fields with a binary distribution. The transition is second order above the tricritical point drawn as a blank circle and is first order below. The constant J is chosen to be unity.

system undergoes a second-order transition at the temperature where the coefficient is unity

$$\beta_c Jz\big(1 - \tanh^2(\beta_c h_0)\big) = 1, \qquad (8.12)$$

and spontaneous magnetization exists below this temperature. This transition temperature (8.12) agrees with the conventional mean-field value $T_c = Jz$ when $h_0 = 0$.

Equation (8.12) shows that the transition temperature decreases as h_0 increases. Random fields gradually destroy the ferromagnetic phase. Simultaneously, the coefficient of the third-order term, m^3, in eqn (8.11) decreases from the value $(\beta Jz)^3/3$ at $h_0 = 0$ and eventually vanishes at $\tanh^2(\beta h_0) = 1/3$. The coefficient of the third-order term of the equation of state is the coefficient of the fourth-order term of the Landau free energy. Vanishing of this coefficient means a tricritical point as elucidated in Section 2.4. A further increase of h_0 beyond the tricritical point causes a first-order transition. The condition to determine the tricritical temperature T_{tc} is that the first- and third-order coefficients vanish in eqn (8.11),

$$\beta_{tc} Jz\big(1 - \tanh^2(\beta_{tc} h_0)\big) = 1, \quad \tanh^2(\beta_{tc} h_0) = \frac{1}{3}. \qquad (8.13)$$

The phase diagram on the $h_0 T$ plane has thus been determined as depicted in Fig. 8.2. As the strength of randomness h_0 increases, the temperature for the second-order transition decreases and changes over to a first-order transition beyond the tricritical point drawn as a blank circle.

For a Gaussian distribution of randomness (8.2), there is no tricritical point and the second-order transition continues to zero temperature. The mean-field theory thus predicts that qualitatively different types of transitions happen depending on the type of distribution function of random fields.

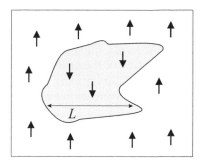

Fig. 8.3 A cluster of inverted spins of linear size L appears under the influence of random fields.

EXERCISE 8.1 Show that the random-field Ising model with a binary distribution has a first-order transition when $h_0/Jz = 1/2$ at $T = 0$ according to the mean-field theory, as shown in Fig. 8.2.

EXERCISE 8.2 Show that there exists a second-order transition at $h_{0,\mathrm{c}} = \sqrt{2/\pi}\,Jz$ at zero temperature for the Gaussian distribution function of random fields (8.2). It will be useful to take the zero-temperature limit in the equation of state and then expand the result in powers of m.

8.1.3 Lower critical dimension

It is interesting to estimate the lower and upper critical dimensions to check when the mean-field theory is applicable to random-field problems. The lower critical dimension is known to be $d_{\mathrm{lc}} = 2$ for the Ising model and $d_{\mathrm{lc}} = 4$ for continuous spins. The physical argument to derive these results is known as the *Imry–Ma argument*, which is analogous to the Peierls argument of Section 7.1.

Suppose that a cluster of down spins of approximate linear size L appears under the influence of random fields in the sea of up spins as illustrated in Fig. 8.3. Such an inversion of spin orientation causes the increase of interaction energy at the surface of the cluster, roughly of the order of JL^{d-1} in the Ising model. In the case of continuous spins the increase would be about JL^{d-2}.[5] As for the energy due to random fields, an all-up spin cluster has $-\sum_i h_i S_i = -\sum_i h_i$, the average of which vanishes and the variance is

$$\left[\left(\sum_i h_i\right)^2\right] = \sum_i [h_i^2] = h^2 L^d, \tag{8.14}$$

---

[5] The increase in energy in the continuous spin case is smaller than the one in the Ising model because the slow change of spin orientation spans the length scale of L from the center of the cluster to the boundary (surface). The change of orientation of neighboring spins $\Delta\theta$ is of the order of L^{-1} since the orientation changes by π in the length L. Then, the increase in the interaction energy from the perfectly ferromagnetic state $\Delta\theta = 0$ is $J\cos\Delta\theta - J\cos 0 \approx \mathcal{O}(L^{-2})$ for a neighboring pair of spins. This energy increase exists approximately uniformly over the area of L^d, giving the total energy increase $\mathcal{O}(L^{d-2})$.

where we have used that $[h_i h_j] = [h_i][h_j] = 0$ for $i \neq j$. The same average and variance are shared by an all-down cluster. For the Gaussian distribution (8.2), h^2 is σ^2, whereas the binary distribution (8.3) has $h^2 = h_0^2$. Equation (8.14) indicates that fluctuations of field energy with a standard deviation of the order of $\pm h L^{d/2}$ can well happen. Thus, a cluster with inverted spins as in Fig. 8.3 may exist under the influence of fluctuations in random fields, with an energy gain (decrease) of $h L^{d/2}$. The total energy change is therefore

$$JL^{d-1} - hL^{d/2} \tag{8.15}$$

for the Ising model. The second term dominates if $d < 2$, and then large-scale inversions of spin orientation due to random fields happen at many locations and the ferromagnetic state breaks down. When $d > 2$, on the other hand, the increase of the interaction energy is the dominant term, and the ferromagnetic state is stable against spin inversion by random fields. We therefore conclude that the lower critical dimension is $d_{lc} = 2$. A similar analysis leads to $d_{lc} = 4$ for continuous-spin systems.

The Imry–Ma argument is an intuitive, qualitative conjecture based on the stability analysis of a perfectly ferromagnetic state. Actually, it has been proved rigorously that the three-dimensional random-field Ising model has finite ferromagnetic order at low temperatures as long as the strength of randomness is not too large.

8.1.4   Upper critical dimension

We need an additional tool to evaluate the upper critical dimension. It is necessary to take the configurational average of the free energy $-T[\log Z]$, which is apparently a difficult task because the dependence of $\log Z$ on random fields $\{h_i\}$ is quite complicated. It is useful to remember here that each h_i appears exponentiated as $\exp(\beta \sum_i h_i S_i)$ in the partition function Z. The same is true for Z^n, where n is a natural number. We then use the identity

$$[\log Z] = \lim_{n \to 0} \frac{[Z^n] - 1}{n}, \tag{8.16}$$

which allows us to first calculate the configurational average of Z^n and then take the limit $n \to 0$. This technique is called the *replica method* since Z^n means to prepare n identical copies (replicas) of the partition function.

The operation of the limit $n \to 0$ for a quantity evaluated in terms of a natural number n may raise questions on the validity of such an 'analytical continuation'. It, nevertheless, turns out that most of the results obtained in this way for random systems have been proved or conjectured to be true. Most physicists in this field believe the validity of the replica method.

Let us recall here the derivation of the upper critical dimension $d_{uc} = 4$ for non-random systems discussed in Section 2.10. The Ginzburg criterion for consistency of the mean-field theory is written as $G(r = \xi) \ll m^2$. For a non-random ferromagnet, the Fourier transformation $\tilde{G}(q)$ of $G(r)$ is $1/(kt + bq^2)$, and hence the correlation function for $r = \xi$ is, when $t \approx \xi^{-2}$,

$$G(\xi) = \int e^{iq\xi} \frac{1}{kt + bq^2} \, dq = \xi^2 \int \frac{e^{iq\xi}}{k + b(q\xi)^2} \, dq \propto \xi^{2-d}. \tag{8.17}$$

We have multiplied the integral variable q by $1/\xi$ in the final step. This relation leads to the condition $\xi^{2-d} \ll m^2$ or equivalently $(d-2)\nu > 2\beta$ for the validity of the mean-field theory. With the mean-field exponents, this condition reduces to $d > 4$.

It is clear in the above analysis that the q dependence of $\tilde{G}(q)$ at $t = 0$, i.e. q^{-2}, determines the upper critical dimension four. If this dependence turns out to be q^{-4}, the same discussion leads to $G(\xi) \propto \xi^{4-d}$, and $d > 6$ is the condition for the validity of the mean-field theory. The random-field Ising model has this property.

We therefore study the behavior of the system in the Fourier space by taking the configurational average using the replica method. In particular, it is necessary to check the wave-number dependence of the Gaussian effective Hamiltonian (free energy), which is simply $\tilde{G}(q)^{-1}$. Using the same notation as in Section 2.9, the n-time replicated partition function is

$$Z^n = \int \left(\prod_{r} \prod_{\alpha=1}^{n} d\phi^{\alpha}(r) \right) \exp \left\{ -\sum_{\alpha=1}^{n} \left(kt \int dr \left(\phi^{\alpha}(r) \right)^2 \right. \right.$$
$$\left. \left. + b \int dr \left(\nabla \phi^{\alpha}(r) \right)^2 + \int dr \, \phi^{\alpha}(r) h(r) \right) \right\}. \tag{8.18}$$

Here, α is the replica index. We average Z^n over the distribution of random fields following the prescription of the replica method. Using the Gaussian distribution (8.2) for each $h(r)$, the random-field part of the above equation is squared,

$$[Z^n] = \int \left(\prod_{r} \prod_{\alpha=1}^{n} d\phi^{\alpha}(r) \right) \exp \left\{ -\sum_{\alpha=1}^{n} \left(kt \int dr \left(\phi^{\alpha}(r) \right)^2 \right. \right.$$
$$\left. \left. + b \int dr \left(\nabla \phi^{\alpha}(r) \right)^2 \right) + \frac{\sigma^2}{2} \int dr \sum_{\alpha,\beta=1}^{n} \phi^{\alpha}(r) \phi^{\beta}(r) \right\}. \tag{8.19}$$

The representation in the wave number by Fourier transformation is

$$[Z^n] = \int \prod_{q} \prod_{\alpha=1}^{n} d\tilde{\phi}^{\alpha}(q) \exp \left\{ -\sum_{\alpha=1}^{n} \left(kt \int dq \, \tilde{\phi}^{\alpha}(q) \tilde{\phi}^{\alpha}(-q) \right. \right.$$
$$\left. + \int dq \, bq^2 \tilde{\phi}^{\alpha}(q) \tilde{\phi}^{\alpha}(-q) \right)$$
$$\left. + \frac{\sigma^2}{2} \int dq \sum_{\alpha,\beta=1}^{n} \tilde{\phi}^{\alpha}(q) \tilde{\phi}^{\beta}(-q) \right\}. \tag{8.20}$$

The exponent on the right-hand side is the effective free energy $-F$ that corresponds to eqn (2.80).

In order to confirm that $\tilde{G}(\boldsymbol{q})$ diverges as q^{-4} at $t = 0$, we have to study the coefficient of $\tilde{\phi}(\boldsymbol{q})\tilde{\phi}(-\boldsymbol{q})$, which is now an $n \times n$ matrix with index $\{\alpha\}$ for each $\boldsymbol{q}$. Let us write the matrix as $\tilde{G}(\boldsymbol{q})^{-1}$ using the same notation as in Chapter 2. Then, at the critical point $t = 0$, we have

$$\tilde{G}(\boldsymbol{q})^{-1} = bq^2 - \frac{\sigma^2}{2}E, \tag{8.21}$$

where the first term on the right-hand side is the $n \times n$ unit matrix multiplied by bq^2 and the second term is the matrix E with all elements unity multiplied by $-\sigma^2/2$. This equation (8.21) is a generalization of the denominator of the integrand of eqn (2.85) with $t = 0$. Thus, the diagonal element of the inverse of the above matrix, $\tilde{G}^{\alpha\alpha}(\boldsymbol{q})$, gives the Fourier transformation of the correlation function $G^{\alpha\alpha}(\boldsymbol{r}) = \langle \phi^\alpha(\boldsymbol{r})\phi^\alpha(0) \rangle$, for which we check if it diverges as q^{-4} in the limit $n \to 0$. If we notice the relation $E^2 = nE$, it is straightforward to see that the inverse matrix is written as

$$\tilde{G}(\boldsymbol{q}) = \left(bq^2 - \frac{\sigma^2}{2}E \right)^{-1}$$

$$= (bq^2)^{-1} \left(1 + \sum_{j=1}^{\infty} \left(\frac{\sigma^2}{2bq^2} \right)^j n^{j-1}E \right)$$

$$= \frac{1}{bq^2} + \frac{\sigma^2 E}{bq^2(2bq^2 - n\sigma^2)}. \tag{8.22}$$

We therefore conclude that the leading divergence as $q \to 0$ is proportional to q^{-4} in the limit $n \to 0$.

8.1.5 Systems in finite spatial dimensions

It is still actively studied what type of critical behavior the random-field Ising model has between the upper and lower critical dimensions, in particular in three dimensions.[6] Problems of interest include (i) whether the structure of the mean-field phase diagram, Fig. 8.2, remains qualitatively the same in three dimensions, in particular the existence of a tricritical point, (ii) critical exponents of the second-order transition for weak random fields, and (iii) difference between the Gaussian and binary distributions. We avoid discussing these difficult, unsolved problems here and check only the relevance of random fields as a perturbation to a non-random system.

The relevance of random fields in the sense of renormalization group between the upper and lower critical dimensions can be verified by calculating the renormalization exponent y of the final term of eqn (8.19) as a perturbation to the other terms representing a non-random system ($\sigma = 0$). It is useful for this purpose to evaluate the scaling dimension of the correlation function of the operator $\phi^\alpha(\boldsymbol{r})\phi^\beta(\boldsymbol{r})$ at the

---

[6] Continuous-spin systems have $d_{lc} = 4$ and no interesting physics exists in three dimensions.

critical point $t = 0$. Since different replicas are decoupled in a non-random system, as seen in eqn (8.19) with $\sigma^2 = 0$, this correlation function behaves as

$$\langle \phi^\alpha(0)\phi^\beta(0)\phi^\alpha(\boldsymbol{r})\phi^\beta(\boldsymbol{r})\rangle_0 \propto \langle \phi^\alpha(0)\phi^\alpha(\boldsymbol{r})\rangle_0 \langle \phi^\beta(0)\phi^\beta(\boldsymbol{r})\rangle_0$$

$$= \langle \phi(0)\phi(\boldsymbol{r})\rangle_0^2 \propto r^{-4x_h}. \tag{8.23}$$

Here, $\langle \cdots \rangle_0$ is the expectation value for the non-random system and x_h is the scaling dimension of $\phi(\boldsymbol{r})$ for the non-random system, $\langle \phi(0)\phi(\boldsymbol{r})\rangle \propto r^{-2x_h}$. Thus, the scaling dimension of the random-field term $\phi^\alpha(\boldsymbol{r})\phi^\beta(\boldsymbol{r})$ of eqn (8.19) is $2x_h$ and the corresponding renormalization exponent is

$$y = d - 2x_h = d - 2(d - y_h) = 2y_h - d. \tag{8.24}$$

This y is positive for a ferromagnetic system, as can be verified from the relation $\gamma = (2y_h - d)/y_t > 0$, and the random field term is a relevant perturbation to the non-random system. It thus follows that the critical behavior between paramagnetic and ferromagnetic phases is qualitatively different in the presence of random fields from the non-random system if the ferromagnetic phase survives in the presence of random fields.

8.2 Spin glass

Let us next investigate the effects of randomness in the interactions. Interactions between spins sometimes may change sign or disappear at some places due to the randomness in the constitution of the magnetic materials. The present section will discuss the *spin glass system* in which the signs of the exchange interactions change from bond to bond.

For a spin pair with a ferromagnetic interaction, the two spins tend to align parallel, whereas they are likely to be antiparallel when the interaction is antiferromagnetic. Then, a randomly ordered phase (spin glass phase) may exist, in which there is no spatially uniform ordering as in the ferromagnetic phase but spins are apparently random in space. The spin glass phase is, nevertheless, an ordered state because the orientation of a given spin does not change significantly with time. We may therefore view a spin glass phase as a state ordered in time but random in space. Notice that the paramagnetic phase has neither temporal nor spatial order.

The goal of the spin glass theory is to clarify the conditions under which such a strange state may exist as a stable thermodynamic phase. The standard theoretical model for this purpose is the *Edwards–Anderson model*,

$$H = -\sum_{\langle ij \rangle} J_{ij} S_i S_j - h \sum_i S_i, \tag{8.25}$$

in which the interactions are quenched random variables. Experimentally, the set of interactions $\{J_{ij}\}$ is fixed (quenched) for a given sample. Correspondingly, we assume in theoretical analyses that these interactions are generated from a distribution function and fixed. It is known that a spin glass phase indeed exists within the mean-field approximation and the same is likely to be true in the three-dimensional Edwards–Anderson model with Ising spins.

We mainly give an account of the mean-field theory in the present section with some comments on finite-dimensional cases in the last part. The discussions on quenched randomness and self-averaging properties of Section 8.1.1 also apply to spin glasses without modifications.

8.2.1 Sherrington–Kirkpatrick model

The *Sherrington–Kirkpatrick model* (SK model) is the infinite-range version of the Edwards–Anderson model, and a mean-field theory of spin glasses has been developed for the SK model. Since the infinite-range model is known to give the same results as the mean-field theory for the ferromagnetic system, we expect that the SK model may be regarded as the mean-field model of spin glasses, since for the latter a direct mean-field approximation is not easy to formulate.

The SK model has the following Hamiltonian,

$$H = -\sum_{i<j} J_{ij} S_i S_j - h \sum_i S_i, \qquad (8.26)$$

where the spins are of the Ising type and the sum in the first term on the right-hand side runs over all different pairs of spins.[7] The number of terms in the first contribution on the right-hand side is $N(N-1)/2$, the number of combinations to choose two out of N. The interaction J_{ij} is a quenched random variable following the Gaussian distribution function

$$P(J_{ij}) = \frac{1}{J} \sqrt{\frac{N}{2\pi}} \exp\left\{ -\frac{N}{2J^2} \left(J_{ij} - \frac{J_0}{N} \right)^2 \right\}, \qquad (8.27)$$

which is common for all pairs (ij). This probability distribution has the average and variance,

$$[J_{ij}] = \frac{J_0}{N}, \quad [(\Delta J_{ij})^2] = \frac{J^2}{N}, \qquad (8.28)$$

both of which are proportional to $1/N$. The first relation corresponds to the factor $1/N$ in front of the interaction in the non-random infinite-range model (2.34). The second relation for the variance is necessary for extensive physical quantities, such as the free energy and its derivatives, to become proportional to N, as will be shown below.

It takes lengthy calculations to evaluate the free energy of this model. The details are given in Appendix A.14. The result is

$$-\beta f = \frac{\beta^2 J^2}{4}(1-q)^2 - \frac{1}{2}\beta J_0 m^2 + \frac{1}{\sqrt{2\pi}} \int e^{-z^2/2} \log\left(2\cosh \beta \tilde{H}(z)\right) dz, \qquad (8.29)$$

---

[7] Notice that $\sum_{i<j}$ is equivalent to $(1/2)\sum_{i\neq j}$.

where $\tilde{H}(z) = J\sqrt{q}z + J_0 m + h$ and the range of integration is from $-\infty$ to ∞. The variable m is the magnetization that characterizes the ferromagnetic phase,

$$m = [\langle S_i \rangle], \tag{8.30}$$

and q is the *spin glass order parameter* that describes a randomly frozen state,

$$q = [\langle S_i \rangle^2]. \tag{8.31}$$

If the spin state is frozen randomly, the thermal average $\langle S_i \rangle$ remains finite (which is the definition of a frozen state) but its value and sign change randomly from site to site due to the spatial randomness of the frozen state. It thus follows that the configurational average (spatial average) $[\cdots]$ in eqn (8.30) gives a vanishing value $m = 0$ due to cancellation of plus and minus signs. The average of the square in eqn (8.31) remains finite because the sign is always positive $\langle S_i \rangle^2 > 0$. Hence, the set of parameter values $q > 0$ and $m = 0$ characterize the spin glass phase with a randomly frozen state. Both of these order parameters are finite in the ferromagnetic phase.

We next derive the equations of state from the extremal condition of the free energy (8.29) following the strategy of the Landau theory. Extremization of the free energy with respect to the variable m gives

$$m = \frac{1}{\sqrt{2\pi}} \int e^{-z^2/2} \tanh \beta \tilde{H}(z) \, dz. \tag{8.32}$$

This is the equation of state for the ferromagnetic order parameter m. Comparison of eqn (8.32) with the equation for the magnetization of a single spin in a field, $m = \tanh \beta h$, suggests that the effective field $\tilde{H}(z)$ distributes according to the Gaussian distribution because of randomness in the interactions.

The extremal condition of the free energy with respect to q gives

$$\frac{\beta^2 J^2}{2}(q-1) + \frac{1}{\sqrt{2\pi}} \int e^{-z^2/2} \big(\tanh \beta \tilde{H}(z)\big) \frac{\beta J}{2\sqrt{q}} z \, dz = 0, \tag{8.33}$$

which is rewritten by integration by parts as

$$q = 1 - \frac{1}{\sqrt{2\pi}} \int e^{-z^2/2} \operatorname{sech}^2 \beta \tilde{H}(z) \, dz = \frac{1}{\sqrt{2\pi}} \int e^{-z^2/2} \tanh^2 \beta \tilde{H}(z) \, dz. \tag{8.34}$$

8.2.2 Phase diagram of the SK model

The solution of the equations of state (8.32) and (8.34) is determined by the values of the temperature T and the center of the distribution of randomness J_0. We assume that there is no external field, $h = 0$.

When the distribution is symmetric, $J_0 = 0$, we have $\tilde{H}(z) = J\sqrt{q}z$ and $\tanh \beta \tilde{H}(z)$ is an odd function of the argument. Thus, the magnetization is clearly zero $m = 0$ from eqn (8.32). There is no ferromagnetic phase. The free energy is

$$-\beta f = \frac{1}{4}\beta^2 J^2 (1-q)^2 + \frac{1}{\sqrt{2\pi}} \int e^{-z^2/2} \log \big(2\cosh(\beta J\sqrt{q}z)\big) \, dz. \tag{8.35}$$

To study the behavior of the system when the spin glass order parameter q is close to zero, we expand the right-hand side in powers of q,

$$\beta f = -\frac{1}{4}\beta^2 J^2 - \log 2 - \frac{\beta^2 J^2}{4}(1 - \beta^2 J^2)q^2 + \mathcal{O}(q^3). \tag{8.36}$$

The Landau theory suggests that the critical point is obtained by the condition that the coefficient of the quadratic term q^2 vanishes. We then have $\beta J = 1$ as the transition point between the paramagnetic and spin glass phases.

If the distribution of J_{ij} is not symmetric but the average is positive ($J_0 > 0$), there is a possibility for a ferromagnetic solution ($m \neq 0$) to exist. We expand the right-hand side of eqn (8.34) in powers of q and m and keep the leading order to find

$$q = \beta^2 J^2 q + \beta^2 J_0^2 m^2. \tag{8.37}$$

If $J_0 = 0$, the critical point is where the coefficient $\beta^2 J^2$ reaches unity, and the result agrees with the prediction given by the expansion of the free energy. When $J_0 > 0$ and $m > 0$, eqn (8.37) suggests $q = \mathcal{O}(m^2)$. We expand the right-hand side of eqn (8.32) with this fact in mind and keep only the leading-order terms to have

$$m = \beta J_0 m + \mathcal{O}(q). \tag{8.38}$$

This equation indicates that the ferromagnetic critical point is at $\beta_c J_0 = 1$ or $T_c = J_0$.

We have obtained the boundaries between the paramagnetic and spin glass phases ($\beta J = 1$) and between the paramagnetic and ferromagnetic phases ($\beta_c J_0 = 1$). The boundary between the spin glass and ferromagnetic phases can be evaluated by numerically solving eqns (8.32) and (8.34). Figure 8.4 is the phase diagram thus obtained. There exists a spin glass phase for J_0 smaller than J ($0 \leq J_0/J < 1$). Numerical solutions of eqns (8.32) and (8.34) show that the spin glass phase lies below the ferromagnetic phase in the region $J_0/J > 1$, as shown by the dashed line in Fig. 8.4. This strange behavior is called the *re-entrant transition*.

This result has been derived under the assumption of a symmetry between different replicas and is called the *replica symmetric solution*, as detailed in Appendix A.14. The re-entrant transition on the dashed line in Fig. 8.4 is actually an artefact of the replica symmetric solution and disappears if we correctly take into account the symmetry breakdown in the abstract replica space. A very intricate setup is needed to reveal this aspect and therefore we only mention here that the dashed line is actually replaced by two full lines to distinguish the ferromagnetic, mixed and spin glass phases, as drawn in Fig. 8.4. The *mixed phase*, which does not exist in the replica symmetric solution, has features both of the ferromagnetic and spin glass phases in the sense that the ferromagnetic order parameter is finite although very complicated spin states are realized there.

Failure of the replica symmetric solution at low temperatures can be verified by calculating the entropy from the free energy (8.29) for $J_0 = 0$. It turns out that the entropy is negative at low temperatures. The entropy is the logarithm of the number of possible states and should be positive or zero in systems with discrete degrees of freedom like the Ising model with or without randomness.

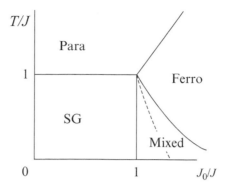

Fig. 8.4 Phase diagram of the SK model. The dashed line is the erroneous boundary between the spin glass and ferromagnetic phases suggested by the replica-symmetric solution. The correct phase boundaries are drawn in full lines.

8.2.3   Systems in finite spatial dimensions

Let us estimate the upper critical dimension to confirm how far the mean-field predictions can be trusted in more realistic systems in finite spatial dimensions. We assume $J_0 = 0$, and consequently $m = 0$, to exclude the effects of ferromagnetic order to focus our attention on the spin glass phase.

Expansion of the free energy (8.29) in powers of q has a third-order term, as suggested in eqn (8.36). This is in contrast to the infinite-range model of the ferromagnetic system or the Landau theory, in which the next term to m^2 was m^4. The symmetry of the free energy with respect to the inversion of magnetization, $f(m) = f(-m)$, meant that $f(m)$ is an even function and does not have odd powers in its expansion. The spin glass order parameter, on the other hand, is the average of the square of $\langle S_i \rangle$ and is always positive, and consequently the operation of sign inversion of the order parameter is not a proper symmetry of the system. Thus, there exist odd powers of q.

The above symmetry consideration holds irrespective of the replica method, and the problem of replica symmetry breakdown does not change the conclusion. It is therefore necessary to evaluate the renormalization-group exponent of the third-order term of the order parameter to study the stability of the Gaussian fixed point, which gives the mean-field exponent, following the prescription described in Section 4.2.1. It turns out that the same invariance argument as in Section 4.2.1 applied to the third-order term leads to the exponent $y_v = 3 - d/2$ for the coefficient v of the third-order term. This result means that the third-order term is irrelevant for $d > 6$ and the mean-field description given by the Gaussian model is stable against the non-Gaussian perturbation. This suggests that the upper critical dimension is six.

> **EXERCISE 8.3** Confirm that the value of the exponent of the third-order term is $y_v = 3 - d/2$ from invariance of the Hamiltonian under a change of scale.

The lower critical dimension is much harder to estimate than the upper critical dimension. There is no known simple theory like the Imry–Ma argument because the stability criterion of a spatially uniform ordered state does not apply to the spin glass problem. It is, nevertheless, believed mainly from numerical calculations that the lower critical dimension is somewhere between two and three.

The relevance of randomness as a perturbation to a non-random system can be studied following the method of Section 8.1.5. To see the relevance of random interactions, it is useful to separate the Hamiltonian into a non-random term H_0 and a random-interaction term H_1,

$$H_0 = -J_0 \sum_{\langle ij \rangle} S_i S_j, \tag{8.39}$$

$$H_1 = -\frac{1}{2} \sum_{r} \sum_{\delta} J_{r, r+\delta} S_r S_{r+\delta} \equiv -\sum_{r} J(r) E(r), \tag{8.40}$$

where we assume $J(r)$ to be a quenched Gaussian variable with vanishing mean and variance J^2, and δ is the vector to a neighboring site. If we take the configurational average of the n-time replicated system, the perturbation to the non-random system due to randomness is written as, like the last term of eqn (8.19),

$$\frac{J^2}{2} \int \mathrm{d}r \sum_{\alpha, \beta} E^\alpha(r) E^\beta(r). \tag{8.41}$$

The scaling dimension of this term is evaluated as $2x_t (= 2(d - y_t))$ because at $t = 0$ we have, using eqn (3.82),

$$\langle E^\alpha(0) E^\beta(0) E^\alpha(r) E^\beta(r) \rangle_0 \propto \langle E^\alpha(0) E^\alpha(r) \rangle_0^2 \propto r^{-4x_t} \quad (\alpha \neq \beta), \tag{8.42}$$

where $\langle \cdots \rangle_0$ is the average with respect to the weight of the non-random system. It has been used that different replicas decouple in the non-random system. Thus, the renormalization-group exponent that determines the relevance of the random-interaction term is

$$y = d - 2x_t = 2y_t - d = \frac{2 - d\nu}{\nu}. \tag{8.43}$$

This relation indicates that the random interaction is not a relevant perturbation for $2 < d\nu$ and the critical exponents remain unchanged. This result is called the *Harris criterion*. If we apply the hyperscaling $2 - d\nu = \alpha$ to this relation, the Harris criterion is restated by saying that the random interaction is not relevant if the critical exponent of the specific heat is negative, $\alpha < 0$.

We may interpret the Harris criterion intuitively as follows. The Harris criterion measures the effects of the response of the system to a random perturbation to the coefficient $J(r)$ of the local energy term $E(r)$. Since $J(r)$ always appears in the Boltzmann factor as a product with the inverse temperature $\beta = 1/T$, the perturbation in $J(r)$ may be regarded as a perturbation in the local temperature. The rate of change of the energy $\langle E(r) \rangle$ with respect to temperature is the specific heat, which diverges at the critical point if $\alpha > 0$. Such a divergence suggests that the system is unstable against local temperature changes when $\alpha > 0$. Therefore, the perturbation is expected to affect the critical behavior.

The Harris criterion applies not just to spin-glass-type perturbations with positive and negative signs but to general cases of quenched random interactions added to a non-random system. For example, we may apply this criterion to the randomly diluted ferromagnet discussed in the next section. The idea is to write the effective Hamiltonian using the replica method, in which the leading term due to randomness is quadratic in the local energy and the above argument applies. Higher-order terms have smaller exponents of the renormalization group.

8.3 Diluted ferromagnet and percolation

In the previous section we studied the effects of random interactions with both positive and negative signs. There exist other types of randomness in which some of the interactions or some spins vanish. For instance, some of the magnetic atoms may be replaced by non-magnetic atoms because of the effect of artificial mixing of magnetic and non-magnetic materials. These substances are called *diluted ferromagnets*. We explain the case of *site dilution* in the first part of this section, most of which applies to *bond dilution* as well. In the former case, spins in some of the sites disappear, whereas in the latter there are no interactions on some of the bonds. The bond dilution case will be described in the last part of this section. Phase transitions in diluted ferromagnets at zero temperature will be shown to be related to the geometrical phase transition called percolation.

8.3.1 Diluted ferromagnet

A diluted ferromagnet has a ferromagnetic phase as its only ordered state at low temperatures, in contrast to spin glasses.[8] Let us be more concrete and assume that a given site is occupied by a spin (magnetic moment) with probability p and unoccupied with probability $1 - p$, independently of other sites. As p decreases from unity, the ferromagnetic phase become unstable and disappears completely below a critical value p_c, as depicted in Fig. 8.5. The Harris criterion applies to the present case. The randomness, dilution, is relevant in the renormalization group sense if the specific heat exponent is positive, $\alpha > 0$, for the non-random (non-diluted) system ($p = 1$). Then, the critical behavior changes from the non-random case.

The renormalization-group flow of parameters is shown schematically in Fig. 8.5. The critical point of the non-random system at $(p = 1, T = T_c)$ is unstable along the temperature axis. If $\alpha > 0$, the same point is also unstable along the horizontal direction that decreases p from unity. In this case, there exists a *random fixed point* that controls the critical behavior of the diluted ferromagnet at some intermediate values of p and T, as indicated by a blank square in Fig. 8.5. To state it more accurately, more and more new types of parameters emerge as the renormalization steps proceed, and consequently the renormalization flow cannot be drawn on the two-dimensional phase diagram. The random fixed point is located in a multidimensional space away from the plane of Fig. 8.5, which is projected onto this figure as a blank square. There

---

[8] There exists the possibility of a special phase called the *Griffiths phase*, which is hard to detect experimentally due to its very weak singularity.

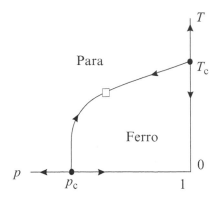

Fig. 8.5 Phase diagram of a diluted ferromagnet. The probability that a site is occupied by a spin is denoted as p. Arrows represent the flow of parameters by the renormalization group. The blank square is the projection of the random fixed point onto the two-dimensional p–T plane, which controls the critical phenomena along the phase boundary.

is no special critical point on the plane of Fig. 8.5 itself. This is a feature different from the tricritical point of Fig. 8.2.

The structure of the diagram of Fig. 8.5 suggests that there can be another unstable fixed point at zero temperature. The point marked p_c on the p-axis is this fixed point. The transition at this point along the p-axis is called the *percolation transition*, which is a geometric transition. Ising spins all align parallel in the ground state and do not play a role in the determination of the system properties. The percolation transition, nevertheless, happens because the size of the clusters drastically changes at p_c. A *cluster* is a set of occupied sites connected by bonds. Assume that a bond exists between neighboring occupied sites. In Fig. 8.6, there are four small clusters and a single large cluster percolating from the top end of the system to the bottom.

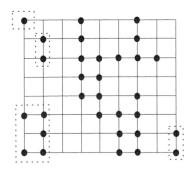

Fig. 8.6 An example of percolation. Occupied sites are denoted by black circles. Four small clusters are encircled by dotted lines. Another large cluster percolates from the top to the bottom.

Fig. 8.7 A cluster of size four in one dimension. Occupied sites are drawn in black and unoccupied sites in white.

When the probability p of site occupancy is larger than the threshold p_c, large clusters of macroscopic size connect boundaries of the system. Strictly speaking, a cluster percolates if and only if it is of infinite size. For p smaller than p_c, only finite-size clusters can exist. A percolation transition occurs at the point where a drastic change between these different states takes place.

Ising spins on clusters all point to a single direction, up state ($S_i = 1$), at zero temperature under infinitesimally small but positive field. Thus, the sum of sizes of clusters is given by $\sum_i' S_i$, where the sum runs over all occupied sites. This means that the magnetization of a diluted ferromagnet at zero temperature is closely related with the sizes of the clusters. In the thermodynamic limit, where the system size tends to infinity, finite-size clusters do not contribute to the magnetization per site. Therefore, the magnetization per site agrees with the probability P that a site belongs to an infinite-size cluster. The probability P decreases as p decreases and vanishes for $p \leq p_c$.

As will be shown below, many other quantities also show singular behavior at p_c. This phenomenon can be analyzed in a theoretical framework similar to the one developed in previous chapters if we replace $T - T_c$ by $p - p_c$.

Percolation is a geometrical concept related to configurations of sites and bonds, independent of magnetic properties of spins on occupied sites. The theory of percolation has been applied to many fields including the spread of forest fire and the search for crude oil in strata.

8.3.2 Scaling in a percolation transition

The number of finite-size clusters, with the size variable denoted as s, plays an important role in the theory of percolation transitions. As a simple example, let us consider the probability that a cluster of size four exists in one dimension, as in Fig. 8.7. The probability that four consecutive sites are occupied is p^4, and the probability that sites neighboring to the left and right of the four-site cluster (shown in white in Fig. 8.7) are both unoccupied is $(1 - p)^2$. The number of ways to place this cluster on a one-dimensional chain of size L is equal to the number of ways to put the left-most site on the chain, which is approximately L. If we ignore the boundary effects by considering an asymptotically infinitely long chain, the above number is indeed L. Then, the total number of size-four clusters is $Lp^4(1 - p)^2$, which we write as $n_4(p) = p^4(1 - p)^2$, where $n_4(p)$ is the number of size-four clusters divided by the number of sites L. Clearly, for general s, we have the cluster number density $n_s(p) = p^s(1 - p)^2 = (1 - p)^2 \, e^{-s/s_0}$, where the characteristic cluster size, $s_0 = -1/\log p$, diverges for $p \to p_c = 1$ as $s_0 \sim (p_c - p)^{-1/\sigma}$ with $\sigma = 1$.[9]

---

[9] The critical probability p_c for a general Bethe lattice with coordination number z is given by $p_c = 1/(z - 1)$, which becomes unity when $z = 2$, i.e. for the one-dimensional case.

The probability that a given site belongs to a cluster of size s is $s n_s(p)$ since the site under consideration may be any one of the s occupied sites in a cluster.

It is difficult to write the expression of $n_s(p)$ for arbitrary s in higher dimensions. Various types of clusters may exist, and it is impossible to list all of them for large s. We therefore do not try to derive the explicit form of $n_s(p)$ and instead estimate its asymptotic form for large clusters $s \gg 1$ as these large clusters should play dominant roles in the critical phenomena around $p \approx p_c$.

Our goal is to reveal the asymptotic behavior of $n_s(p)$ for p close to p_c and very large s. An analogy with the usual thermal critical phenomena suggests to assume the following scaling law,

$$n_s(p) = s^{-\tau} f\big((p - p_c)s^\sigma\big). \tag{8.44}$$

This implies that the dependence of $n_s(p)$ on s and p is described essentially as a single-variable function of the combination $(p - p_c)s^\sigma$.

We investigate the behavior of important physical quantities near p_c using eqn (8.44). The first target is the probability P that an occupied site belongs to an infinite cluster, i.e. the percolating cluster. As was mentioned in the previous section, P is essentially the zero-temperature magnetization of a diluted ferromagnet, which is positive (non-vanishing, to be more accurate) when $p > p_c$ and zero for $p \leq p_c$. Let us write β for the power by which P approaches zero toward p_c,[10]

$$P \approx (p - p_c)^\beta \quad (p > p_c). \tag{8.45}$$

We recall here that the probability for a site to be occupied is the sum of the probabilities for the site to belong to an infinite cluster and that to a finite cluster. Accordingly, we have the following relation,

$$P + \sum_{s=1}^{\infty} n_s(p)s = p. \tag{8.46}$$

This equation may be rewritten as, using $\sum_s n_s(p_c)s = p_c$ (since $P = 0$ at $p = p_c$),

$$P = \sum_s \big(n_s(p_c) - n_s(p)\big)s + (p - p_c). \tag{8.47}$$

If we assume that critical phenomena are dominated by the behavior of $n_s(p)$ for very large s, we may ignore the discreteness of s, $\Delta s = 1$, compared to the magnitude of s itself. Then, the sum in the above equation can be replaced by an integral. Using the scaling law (8.44), eqn (8.47) now reads

$$P \approx \int ds \, s^{-\tau+1} \big[f(0) - f\big((p - p_c)s^\sigma\big)\big] + (p - p_c). \tag{8.48}$$

We can ignore the second term on the right-hand side, $p - p_c$, for the purpose of investigating the singularities of P. The change of the integral variable from s to $z = (p - p_c)s^\sigma$ yields, for $p > p_c$,

---

[10] One dimension is special. In that case, $P = 0$ for $p < 1$ and $P = 1$ at $p = p_c = 1$ with $\beta = 0$.

$$P \propto (p - p_c)^{(\tau-2)/\sigma} \int dz\, z^{-1+(2-\tau)/\sigma} \big(f(0) - f(z)\big). \tag{8.49}$$

Since the left-hand side is proportional to $(p - p_c)^\beta$, it is concluded that

$$\beta = \frac{\tau - 2}{\sigma}. \tag{8.50}$$

Let us next consider the average size of finite-size clusters. The probability for a site to belong to a cluster of size s is $s n_s(p)$, and hence the expectation value of the size of a finite-size cluster, S,[11] is

$$S = \sum_s n_s(p) s^2 \approx \int ds\, s^2 n_s(p). \tag{8.51}$$

Assuming $p < p_c$ and changing the integral variable to $z = (p_c - p)s^\sigma$ using eqn (8.44), we rewrite the above equation as

$$S \propto (p_c - p)^{(\tau-3)/\sigma} \int dz\, f(-z) z^{-1+(3-\tau)/\sigma}. \tag{8.52}$$

As p increases toward p_c from below, larger and larger clusters emerge and the average size of clusters S should diverge. If we define the exponent of this divergence as γ, the above equation implies

$$\gamma = \frac{3 - \tau}{\sigma}. \tag{8.53}$$

The same relation as above holds also for $p > p_c$ if we consider finite-size clusters only, excluding infinite clusters, and see the rate of divergence of the size of finite clusters.

We write γ for the exponent of the divergence of S because S is proportional to the susceptibility of a diluted ferromagnet in the low-temperature limit. When the temperature is close to zero, almost all Ising spins in a cluster are oriented in the same direction and so a cluster can be regarded as a single isolated spin that takes values $\pm s$. Thus, the magnetization of a cluster of size s is $s \tanh(\beta s h)$ if the external field h is of the same order of magnitude as the temperature (so that βh is of order one). The total magnetization of the system is the sum of these contributions from clusters. For $p < p_c$, all clusters are of finite size, and the magnetization is

$$m = \sum_s n_s(p) s \tanh(\beta s h). \tag{8.54}$$

The susceptibility is therefore

$$\chi = \left. \frac{\partial m}{\partial h} \right|_{h \to 0} = \beta \sum_s s^2 n_s(p) = \beta S. \tag{8.55}$$

This expression shows that χ is proportional to S, apart from the p-independent factor β.[12]

---

[11] In one dimension S is determined exactly as $S = p(1+p)/(1-p) = p(p_c + p)/(p_c - p)$ and diverges as $(p_c - p)^{-\gamma}$ with a universal exponent $\gamma = 1$.

[12] This β is $1/T$, not a critical exponent.

The total number of clusters $M_0 = \sum_s n_s(p)$ also shows a singularity at p_c. If we write the singular part of M_0 as $(p_c - p)^{2-\alpha}$, it is possible to derive the following relation in a similar way to that for β and γ,

$$2 - \alpha = \frac{\tau - 1}{\sigma}. \tag{8.56}$$

Equations (8.50), (8.53) and (8.56) show that the exponents τ and σ defined in (8.44) determine the other exponents, a similar situation to the conventional critical phenomena. It is sometimes useful to remember that the exponents α, β and γ satisfy the following scaling relation,

$$\alpha + 2\beta + \gamma = 2. \tag{8.57}$$

EXERCISE 8.4 Derive eqn (8.56).

8.3.3 Fractal dimension and hyperscaling

The scaling law (8.44) may look similar to the finite-size scaling (3.98) of conventional critical phenomena. The variable $t = (T - T_c)/T_c$ corresponds to $p - p_c$ and the length scale L would correspond to the cluster size s. This identification, however, is not necessarily correct. The size s is the number of sites in a cluster, which is different from the linear size of a cluster. As the typical linear size of a cluster is the correlation length ξ, it is necessary to relate the correlation length with the size s.

The starting point of the discussion is the definitions of the correlation function $G(r)$ and the correlation length ξ. The correlation function is the probability that a site B at a distance r from a given occupied site A belongs to the same cluster as A. A very simplified example is that all sites are occupied inside a circle of radius a from site A and all sites are unoccupied outside of the same circle. Then, the correlation function from A is $G(r) = 1$ for $r < a$ and $G(r) = 0$ for $r > a$. This example should be sufficient to convince the reader that the expectation value of the size of a cluster S is the sum of $G(r)$ over all r,

$$S = \sum_r G(r). \tag{8.58}$$

Let us assume $p < p_c$, which allows us to consider only finite-size clusters. The above equation is of the same form as the relation between the susceptibility and the correlation function in spin systems. Since all clusters are of finite size, $G(r)$ decays exponentially beyond the length scale ξ, the correlation length,

$$G(r) \approx r^{-c} e^{-r/\xi}. \tag{8.59}$$

As p approaches p_c, larger and larger clusters appear, and eventually the correlation length diverges at p_c. The critical exponent ν characterizes the rate of this divergence,

$$\xi \propto (p_c - p)^{-\nu}. \tag{8.60}$$

It is, by the way, natural to assume that the borderline size

$$s_0 \approx (p_c - p)^{-1/\sigma} \tag{8.61}$$

between the regions of large and small $z(= (p - p_c)s^\sigma)$ in eqn (8.44) is related to the length ξ at which the correlation function $G(\boldsymbol{r})$ starts to decay significantly. Indeed, the change of s from values giving $|z| = |(p - p_c)s^\sigma|$ smaller than unity to values giving large $|z|$ has the same effect on $n_s(p)$ as the change of p (with fixed s) from values near p_c to values far from p_c. The system then moves from a region close to the critical point to a region far from criticality, the borderline being at $|z| \approx 1$. If we observe a cluster with a much smaller length scale than the correlation length ξ, this correlation length looks very large and the system seems as if it were in the critical region ($|z| \ll 1$). If, on the other hand, the standard of length to observe the cluster is much larger than the correlation length ξ, the finiteness of ξ is clearly recognized and the system is regarded as being outside the critical region ($|z| \gg 1$). Consequently, the s_0 corresponding to $|z| = 1$ would be identified with the cluster size (the number of sites in the cluster) where such a qualitative change of the system behavior takes place, namely the size of a cluster of linear length ξ.

At p_c, a finite-size cluster has a very complicated structure, very different from a simple sphere, characterized by a *fractal dimension D*. Discussions in the previous paragraph indicate that s_0 has a linear size ξ with a fractal dimension D defined by

$$s_0 \propto \xi^D. \tag{8.62}$$

A simple structure like a sphere has D equal to the spatial dimension d. In the case of more complicated cases, *fractal* structures, the increase of occupied sites s as ξ grows is slower than in simple sphere-like structures and thus $D < d$ holds.[13] In general, if we observe a cluster from a length scale L smaller than the correlation length ξ, the system may look critical (because the correlation length looks very long) and the cluster looks fractal-like $s \propto L^D$. As $L \to \xi$, s approaches s_0.

Equations (8.60), (8.61) and (8.62) lead to the following relation,

$$s_0 \propto \xi^D \propto (p_c - p)^{-D\nu} \propto (p_c - p)^{-1/\sigma}. \tag{8.63}$$

It then follows that

$$D\nu\sigma = 1. \tag{8.64}$$

We have thus related the critical exponent ν with the fractal dimension D. Is it possible to further establish relations between the fractal dimension D and the spatial dimension d and hyperscaling that relates critical exponents with the spatial dimension? The answer is yes with the relation

$$D = d - \frac{\beta}{\nu}. \tag{8.65}$$

---

[13] In one dimension $D = d = 1$.

By eliminating D using $D\nu\sigma = 1$, we find a hyperscaling relation. Equation (8.65) shows how much the dimension decreases from the spatial dimension d to the fractal dimension D, expressed in terms of two critical exponents.

In order to understand eqn (8.65), suppose that p is fixed to a value slightly larger than p_c and we count the number of occupied sites $M(L)$ in a finite system of linear length $L(\gg \xi)$. We divide the system into subsystems with linear length ξ. The number of such subsystems is $(L/\xi)^d$. Since the number of occupied sites in a subsystem is ξ^D according to eqn (8.62), the total number of occupied sites in the whole system is $(L/\xi)^d\xi^D = \xi^{D-d}L^d$. This number is identified with PL^d from the definition of P, and then we obtain eqn (8.65) from $P \propto (p - p_c)^\beta$.

If we eliminate D from $D\nu\sigma = 1$ and $D = d - \beta/\nu$ and use the relation $\beta + \gamma = 1/\sigma$ (which comes from eqns (8.50) and (8.53)) as well as the scaling relation (8.57), we arrive at the hyperscaling

$$2\beta + \gamma = d\nu = 2 - \alpha, \tag{8.66}$$

which is the same expression as in conventional critical phenomena. Hyperscaling holds below the upper critical dimension as before. The upper critical dimension is six for the problem of percolation, as will be explained in the next section.

8.3.4 Bond process and Potts model

It has been shown that the scaling analysis is useful to understand some aspects of critical phenomena in percolation. A direct relation between spin systems at finite temperature and percolation will further help us clarify the situation. This program is realized through the Potts model with the number of states tending to unity.

We have so far treated the *site process* of percolation in which each site is occupied or empty independently of the other sites. It is actually the *bond process*, in which each bond is occupied randomly with probability p independently of the other bonds, that has a direct correspondence with the Potts model. We will therefore consider the bond process in the present section. It should be noticed that the scaling theory developed in the previous sections applies without essential changes to the bond process since the concepts of a cluster and the cluster size (the number of occupied bonds in a cluster) can be defined essentially in the same way as in the site process.

Let us analyze the q-state Potts model on a lattice

$$H = -J\sum_{\langle ij \rangle} \delta_{S_i,S_j}, \tag{8.67}$$

introduced in Section 1.5, where S_i is a Potts spin assuming the values $1, 2, \cdots, q$ at site i. We will now show that critical phenomena of this model as a function of temperature are equivalent to critical phenomena of percolation if we take the limit $q \to 1$.

The partition function for the Hamiltonian (8.67)

$$Z = \sum_{\{S_i\}} \exp\left(K\sum_{\langle ij \rangle} \delta_{S_i,S_j}\right) = \sum_{\{S_i\}} \prod_{\langle ij \rangle} e^{K\delta_{S_i,S_j}} \tag{8.68}$$

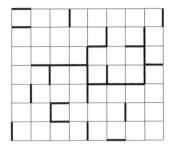

Fig. 8.8 An example of a term in the partition function of a bond process. The bold lines represent occupied bonds.

can be rewritten, using the relation $e^{K\delta_{S_i,S_j}} = 1 + (e^K - 1)\delta_{S_i,S_j}$, as

$$Z = \sum_{\{S_i\}} \prod_{\langle ij\rangle} \left(1 + u\delta_{S_i,S_j}\right) \quad (u = e^K - 1). \tag{8.69}$$

The expansion of the product on the right-hand side of this equation gives the factor 1 or $u\delta_{S_i,S_j}$ for each bond. This fact suggests that we can express the polynomial of u, obtained from the expansion of the right-hand side, graphically as follows. If the term $u\delta_{S_i,S_j}$ is chosen for a bond (ij), we take it as an occupied state of the bond (ij), and 1 is understood as an unoccupied bond. Then, each term of the expansion is represented by a graph showing the set of occupied bonds, as illustrated in Fig. 8.8.

An occupied bond has the constraint δ_{S_i,S_j}, that is, spins at sites i and j take the same value. Thus, all sites in a cluster of occupied bonds have the same spin state. Spins belonging to different clusters are uncorrelated. Consequently, the expansion of Z as a polynomial of u is written as follows,

$$Z = \sum_{\text{config}} q^{N_c} u^{N_b}. \tag{8.70}$$

Here, the sum runs over all possible configurations of occupied bonds, i.e. all possible ways to draw graphs like Fig. 8.8. N_c is the number of clusters in a given configuration of occupied bonds, and N_b is the number of occupied bonds in the given configuration.

In the bond process of percolation, the probability P_{config} for a given configuration of bond occupation is $p^{N_b}(1-p)^{N_B-N_b}$, where N_B is the total number of bonds, occupied and unoccupied. We use this probability to rewrite eqn (8.70) multiplied by $(1-p)^{N_B}$, writing $p/(1-p)$ for u, as

$$\tilde{Z}(q,p) \equiv (1-p)^{N_B} Z = (1-p)^{N_B} \sum_{\text{config}} q^{N_c} \left(\frac{p}{1-p}\right)^{N_b} = \sum_{\text{config}} q^{N_c} P_{\text{config}}. \tag{8.71}$$

In the limit $q \to 1$, $\tilde{Z}(q,p)$ represents the sum of probabilities of all configurations of occupied bonds. This suggests a close relationship between the Potts model in the limit $q \to 1$ and percolation.

We need to be a little more careful because the simple substitution of $q = 1$ in $\tilde{Z}(q, p)$ yields the sum of all possible cases, which is unity trivially. The leading correction to the limit $q \to 1$, $q = 1 + \epsilon$ for small ϵ, gives the non-trivial relation

$$\tilde{Z}(1 + \epsilon, p) = 1 + \epsilon \sum_{\text{config}} N_{\text{c}} P_{\text{config}} = 1 + \epsilon \langle N_{\text{c}} \rangle \tag{8.72}$$

for the expectation value of the number of clusters. Using the notation of the previous sections, we have, from $dq = d\epsilon$,

$$\frac{\partial}{\partial q} \log \tilde{Z}(q, p) \bigg|_{q \to 1} = \langle N_{\text{c}} \rangle = \sum_s n_s(p) = M_0. \tag{8.73}$$

Thus, M_0 is given by the $q \to 1$ limit of the free energy of the q-state Potts model, and the $q \to 1$ limit of the critical exponent α of the Potts model is the critical exponent α of percolation. Other critical exponents satisfy similar relations.

Critical points of the two systems satisfy the relation

$$\lim_{q \to 1} e^{K_{\text{c}}} - 1 = \frac{p_{\text{c}}}{1 - p_{\text{c}}}, \tag{8.74}$$

where K_{c} is the critical point of the Potts model and we have used $u = p/(1-p)$. For example, the critical point of the q-state Potts model on the square lattice is $e^{K_{\text{c}}} = \sqrt{q} + 1$ as will be shown in Exercise 10.2. Then, the above relation gives the critical probability of the bond process on the square lattice as $p_{\text{c}} = 1/2$.

Let us now consider the upper critical dimension and the mean-field theory for percolation. It is useful to check the relevance of non-quadratic terms of the Landau free energy, generalized to have a spatial dependence of the order parameter, around the Gaussian fixed point. The Potts model does not have a symmetry under inversion of spin variables except for the special case of the Ising model with $q = 2$. Hence, the Landau free energy has both even and odd terms. Consequently, a correction to the quadratic term begins with a cubic term, for which the borderline dimension of relevance is six, as was discussed for the spin glass problem in Section 8.2.3. We conclude that the upper critical dimension is six for percolation.

The mean-field critical exponent for $d > 6$ can be estimated easily from the properties of the Gaussian fixed point. As was mentioned in Section 4.2.1, critical exponents for the Gaussian fixed point satisfy

$$\alpha = 2 - \frac{d}{2}, \quad \beta = \frac{d-2}{4}, \quad \gamma = 1, \quad \nu = \frac{1}{2}, \quad \eta = 0. \tag{8.75}$$

By inserting the value of the upper critical dimension $d = 6$, which is the limiting dimension for the hyperscaling to be applicable, we find the mean-field exponents as follows,

$$\alpha = -1, \quad \beta = 1, \quad \gamma = 1, \quad \nu = \frac{1}{2}, \quad \eta = 0. \tag{8.76}$$

These values coincide with the direct solution of the percolation problem on the Bethe lattice, a lattice with special structure on which the Bethe approximation gives the exact solution.

The final remark is on the order of the transition. The Potts model with $q > 2$ has a first-order transition in the mean-field theory since the Landau free energy has a cubic term. The order changes at $q = 2$ because the coefficient of the cubic term changes the sign there, and consequently the problem of percolation corresponding to $q \to 1$ undergoes a second-order transition.

EXERCISE 8.5 Let us confirm that the correlation function of the Potts model reduces to the correlation function of percolation in the limit $q \to 1$. We first recall the definition of the correlation function of the Ising model, $G(\boldsymbol{r}) = \langle S_0 S_{\boldsymbol{r}} \rangle$, which decays exponentially to zero as $r \to \infty$ in the paramagnetic phase. Physically, this behavior reflects the fact that each spin takes both values $S_i = 1$ and $S_i = -1$ with equal probability, leading to the vanishing average. This observation suggests that the correlation function of the Potts model is to be defined in terms of the product of $\delta_{S_i,1} - q^{-1}$ (for which the simple average vanishes),

$$G(\boldsymbol{r}) = \langle (\delta_{S_0,1} - q^{-1})(\delta_{S_{\boldsymbol{r}},1} - q^{-1}) \rangle. \tag{8.77}$$

We now reinterpret this quantity in terms of percolation by using the correspondence explained in this section. (1) Show that the average of $(\delta_{S_0,1} - q^{-1})(\delta_{S_{\boldsymbol{r}},1} - q^{-1})$ vanishes if site 0 and $\boldsymbol{r}$ belong to different clusters. (2) Show that the average of $(\delta_{S_0,1} - q^{-1})(\delta_{S_{\boldsymbol{r}},1} - q^{-1})$ is $(q-1)/q^2$ if site 0 and $\boldsymbol{r}$ belong to the same cluster. (3) Set $q = 1 + \epsilon$ and show that the coefficient of ϵ of the expansion of the correlation function of the Potts model $G(\boldsymbol{r})$ coincides with the correlation function of percolation.

9
Exact solutions and related topics

Only a limited number of models of phase transitions and critical phenomena can be solved exactly. These examples, nevertheless, play important roles in many aspects including the verification of the accuracy of approximation theories such as the mean-field theory and renormalization group. Also, mathematical methods to solve such examples are interesting in their own right and constitute an important subfield of mathematical physics. In particular, the exact solution of the two-dimensional Ising model occupies an outstanding status as one of the founding studies of the modern theory of phase transitions and critical phenomena. We elucidate in the present chapter simple but typical examples of exact solutions such as the one-dimensional classical spin system, the spherical model, the one-dimensional quantum XY model and the two-dimensional Ising model. An account of the Yang–Lee theory on the zeros of the partition function will also be given as a set of basic rigorous results on phase transitions.

9.1 One-dimensional Ising model

We have already studied the one-dimensional Ising model in the context of a real-space renormalization group. It is, nevertheless, illuminating to explain the well-established method of its solution here partly because the solution of the two-dimensional Ising model is closely related to some aspects of the one-dimensional solution. Also, it is one of those exactly solvable models where the independence of thermodynamic quantities on the boundary conditions, in the thermodynamic limit $N \to \infty$, can be rigorously established. We will study the free and periodic boundary conditions cases.

9.1.1 Free boundary condition

The Hamiltonian of the one-dimensional Ising model at zero field, $h = 0$, under free boundary conditions reads

$$H = -\sum_{i=1}^{N-1} J_i S_i S_{i+1}. \tag{9.1}$$

The interactions are considered to depend on site index i for later convenience. A superscript (F) for the partition function Z will denote the free boundary condition,

$$Z_N^{(\mathrm{F})} = \sum_{S_1 = \pm 1} \cdots \sum_{S_N = \pm 1} \exp\left(\beta \sum_{i=1}^{N-1} J_i S_i S_{i+1} \right), \tag{9.2}$$

where $\beta = 1/T$ is the inverse temperature. We first carry out the sum over S_N, the spin on the edge, to evaluate $Z_N^{(\mathrm{F})}$. If we separate the part that includes S_N, we have

$$Z_N^{(\mathrm{F})} = \sum_{S_1, S_2, \cdots, S_{N-1}} e^{\beta J_1 S_1 S_2 + \cdots + \beta J_{N-2} S_{N-2} S_{N-1}} \cdot \sum_{S_N = \pm 1} e^{\beta J_{N-1} S_{N-1} S_N}. \quad (9.3)$$

The sum over S_N can be performed easily to yield a factor

$$\sum_{S_N = \pm 1} e^{\beta J_{N-1} S_{N-1} S_N} = 2 \cosh(\beta J_{N-1} S_{N-1}) = 2 \cosh(\beta J_{N-1}). \quad (9.4)$$

We have used here that cosh is an even function and thus $\cosh(\beta J_{N-1} S_{N-1})$ is independent of $S_{N-1} (= \pm 1)$. We therefore have a recursion relation

$$Z_N^{(\mathrm{F})} = 2 \cosh(\beta J_{N-1}) \cdot Z_{N-1}^{(\mathrm{F})} \quad (9.5)$$

between $Z_N^{(\mathrm{F})}$ and $Z_{N-1}^{(\mathrm{F})}$. Repeated applications of this recursion relation leads to

$$Z_N^{(\mathrm{F})} = 2 \cosh(\beta J_{N-1}) \cdot 2 \cosh(\beta J_{N-2}) \cdots 2 \cosh(\beta J_1) \cdot 2 = 2^N \prod_{i=1}^{N-1} \cosh(K_i), \quad (9.6)$$

where the final factor 2 comes from the sum over S_1 and $K_i = \beta J_i$. This is the solution for the partition function.

Equation (9.6) immediately gives physical quantities such as the free energy F, energy E, and specific heat C from the logarithm and its derivatives. The entropy is $S = (-F + E)/T$. For a uniform system $J_i = J$, for which $Z_N^{(\mathrm{F})} = 2(2 \cosh K)^{N-1}$,

$$F = -T(\log 2 + (N-1) \log(2 \cosh K)) \approx -TN \log(2 \cosh K) \quad (9.7)$$

$$E = -J(N-1) \tanh K \approx -JN \tanh K \quad (9.8)$$

$$C = \frac{K^2(N-1)}{\cosh^2 K} \approx \frac{K^2 N}{\cosh^2 K}, \quad (9.9)$$

where we have taken the large-N limit to replace $N-1$ by N and chosen to ignore $\log 2$ in comparison with $N \log(2 \cosh K)$. The dependence of the energy and specific heat on the temperature is shown in Fig. 9.1. The specific heat in the low-temperature limit $(K \gg 1)$ is given as, using $\cosh^2 K \approx e^{2K}/4$,

$$C \approx 4NK^2 e^{-2K}. \quad (9.10)$$

This function vanishes exponentially as K increases (T decreases).

Next, the correlation function is defined as

$$\langle S_i S_{i+r} \rangle = \frac{\sum_{S_1, \cdots, S_N} S_i S_{i+r} \, e^{\beta(J_1 S_1 S_2 + \cdots + J_{N-1} S_{N-1} S_N)}}{Z_N^{(\mathrm{F})}}. \quad (9.11)$$

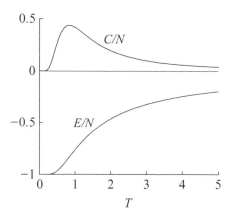

Fig. 9.1 Energy and specific heat per spin for the one-dimensional Ising model. The unit of energy is set to $J = 1$. The specific heat has a peak around $T = 1$.

The numerator can be expressed as a derivative of the denominator $Z_N^{(\mathrm{F})}$: Using $S_j^2 = 1$ and eqn (9.6), we rewrite the above equation as follows,

$$
\langle S_i S_{i+r} \rangle \cdot Z_N^{(\mathrm{F})}
$$
$$
= \sum_{S_1, \cdots, S_N} (S_i S_{i+1}) \cdot (S_{i+1} S_{i+2}) \cdots (S_{i+r-1} S_{i+r}) \, e^{\beta(J_1 S_1 S_2 + \cdots + J_{N-1} S_{N-1} S_N)}
$$
$$
= \frac{\partial}{\partial(\beta J_i)} \frac{\partial}{\partial(\beta J_{i+1})} \cdots \frac{\partial}{\partial(\beta J_{i+r-1})} Z_N^{(\mathrm{F})}
$$
$$
= (2 \cosh \beta J_1 \cdot 2 \cosh \beta J_2 \cdots 2 \cosh \beta J_{i-1})
$$
$$
\cdot (2 \sinh \beta J_i \cdot 2 \sinh \beta J_{i+1} \cdots 2 \sinh \beta J_{i+r-1})
$$
$$
\cdot (2 \cosh \beta J_{i+r} \cdots 2 \cosh \beta J_{N-1}) \cdot 2 = Z_N^{(\mathrm{F})} \prod_{j=i}^{i+r-1} \tanh K_j. \tag{9.12}
$$

Thus, the correlation function for the case of uniform interactions ($J_i = J$) is

$$
\langle S_i S_{i+r} \rangle = (\tanh K)^r = \exp\big(-r(-\log \tanh K)\big), \tag{9.13}
$$

which is independent of i and N. The correlation length is then

$$
\xi = -\frac{1}{\log \tanh K}. \tag{9.14}
$$

In the low-temperature limit, the correlation length diverges exponentially since $\xi \approx e^{2K}/2$ from $\tanh K \approx 1 - 2e^{-2K}$.

The evaluation of the susceptibility starts from its expression in terms of the sum of correlation functions as described in Appendix A.2,

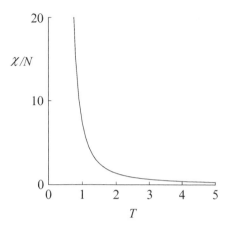

Fig. 9.2 Susceptibility of the one-dimensional Ising model with the energy unit fixed to $J = 1$. It diverges exponentially as $T \to 0$.

$$\chi = \beta \sum_{i,j=1}^{N} \langle S_i S_j \rangle. \tag{9.15}$$

We first notice that the correlation function $\langle S_i S_j \rangle$ depends only on the distance between the two sites, $|i - j|$, as seen in eqn (9.13), not on i and j separately. If we simplify the discussion by ignoring boundary effects (justified in the limit of large N), we may replace the above equation with the following,

$$\frac{\chi}{N} = \beta \left(1 + 2 \sum_{k=1}^{N} \langle S_i S_{i+k} \rangle \right). \tag{9.16}$$

Here, the first term on the right-hand side comes from the terms with $i = j$ in eqn (9.15). The factor 2 on the second term comes from the two cases of $j (\equiv i + k)$ larger than i ($k > 0$) and smaller than i ($k < 0$) in eqn (9.15). The double sum over i and j in eqn (9.15) is replaced by a single sum over k in eqn (9.16) with the factor N (the number of possible is) extracted to the outside, and the sum over j in eqn (9.15) is rewritten in terms of the sum over k in eqn (9.16). We now insert eqn (9.13) into eqn (9.16) and carry out the sum to find

$$\lim_{N \to \infty} \frac{\chi}{N} = \beta \cdot \frac{1 + \tanh K}{1 - \tanh K}. \tag{9.17}$$

This susceptibility shows a strong exponential divergence proportional to βe^{2K} at low temperatures, as depicted in Fig. 9.2. This feature reflects the fact that the lower critical dimension is one with $T_c = 0$.

9.1.2 Periodic boundary condition

A periodic boundary condition does not allow us to trace out spin variables from the boundary site one by one as in the free-boundary case. It is, nevertheless, possible to evaluate the partition function by a powerful technique called the *transfer matrix method*. The transfer matrix method is a common tool in many contexts and is used also in the solution of the two-dimensional Ising model, and so we explain this technique in some detail.

Let us first write the definition of the partition function for the uniform system in the presence of an external field h,

$$Z_N^{(\mathrm{P})} = \sum_{S_1,\cdots,S_N} \exp\big(KS_1S_2 + hS_1 + KS_2S_3 + hS_2 + \cdots + KS_NS_1 + hS_N\big), \quad (9.18)$$

where the superscript (P) denotes a periodic boundary. It is useful to introduce the following notation,

$$T(S_1, S_2) = \exp\left(KS_1S_2 + \frac{h(S_1 + S_2)}{2}\right). \quad (9.19)$$

Then, eqn (9.18) is written as follows,

$$Z_N^{(\mathrm{P})} = \sum_{S_1,\cdots,S_N} T(S_1, S_2)T(S_2, S_3)T(S_3, S_4)\cdots T(S_{N-1}, S_N)T(S_N, S_1). \quad (9.20)$$

We note here that $T(S_i, S_{i+1})$ takes four values depending on S_i and S_{i+1}, and is thus regarded as a 2×2 matrix (the *transfer matrix*),[1]

$$T = \begin{pmatrix} T(1,1) & T(1,-1) \\ T(-1,1) & T(-1,-1) \end{pmatrix} = \begin{pmatrix} e^{K+h} & e^{-K} \\ e^{-K} & e^{K-h} \end{pmatrix}. \quad (9.21)$$

Then, the sums over S_2 to S_N in eqn (9.20) (excluding the sum over S_1) may be identified with the diagonal element $T^N(S_1, S_1)$ of the product of N matrices T^N,

$$T^N(S_1, S_1) = \sum_{S_2,\cdots,S_N} T(S_1, S_2)T(S_2, S_3)T(S_3, S_4)\cdots T(S_{N-1}, S_N)T(S_N, S_1). \quad (9.22)$$

The final sum over S_1 is equivalent to the trace of the matrix T^N,

$$Z_N^{(\mathrm{P})} = \mathrm{Tr}\, T^N. \quad (9.23)$$

Consequently, we have, by writing $\lambda_\pm$ for the two eigenvalues of T,

$$Z_N^{(\mathrm{P})} = \lambda_+^N + \lambda_-^N. \quad (9.24)$$

The eigenvalues of T are easily calculated as

$$\lambda_\pm = \frac{e^{K+h} + e^{K-h} \pm \sqrt{(e^{K+h} + e^{K-h})^2 - 4\,e^{2K} + 4\,e^{-2K}}}{2}. \quad (9.25)$$

This completes the evaluation of the partition function.

In the thermodynamic limit $N \to \infty$, we expect the present result to agree with the corresponding result for the free boundary condition, eqn (9.7), because the boundary

---

[1] Do not confuse temperature T with the transfer matrix T.

effects may be neglected. Since $\lambda_+ > \lambda_-$, we have

$$Z_N^{(\mathrm{P})} = \lambda_+^N \left\{ 1 + \left(\frac{\lambda_-}{\lambda_+} \right)^N \right\} \to \lambda_+^N, \tag{9.26}$$

from which the free energy per spin in the absence of fields ($h = 0$) is

$$\lim_{N \to \infty} \frac{F}{N} = -T \log \lambda_+ = -T \log(2 \cosh K). \tag{9.27}$$

This agrees with eqn (9.7). The derivatives of the free energy, i.e. energy and specific heat, also do not depend on the boundary condition.

We next write the definition of the correlation function to evaluate it under the periodic boundary condition, assuming $h = 0$ for simplicity. To be concrete, the example of the correlation between S_2 and S_4 will be explained,

$$\langle S_2 S_4 \rangle \cdot Z_N^{(\mathrm{P})} = \sum_{S_1, \cdots, S_N} T(S_1, S_2)$$

$$\cdot S_2 \cdot T(S_2, S_3) T(S_3, S_4) \cdot S_4 \cdot T(S_4, S_5) \cdots T(S_N, S_1). \tag{9.28}$$

As already mentioned, the trace over spin variables can be regarded as the trace of the product of transfer matrices. We therefore calculate the expectation value of the product from $T(S_1, S_2)$ to $T(S_N, S_1)$ in the summand of the above equation, the expectation value being taken with respect to the eigenvectors of T. The normalized eigenvectors for $h = 0$ are

$$|\pm\rangle = \frac{1}{\sqrt{2}} \begin{pmatrix} 1 \\ \pm 1 \end{pmatrix}, \tag{9.29}$$

where the signs correspond to the two eigenvalues $\lambda_\pm$. We first take the expectation value of the summand of eqn (9.28) over $|+\rangle$. The application of $|+\rangle$ from the right leads to simple multiplication by λ_+ as each of $T(S_N, S_1)$ to $T(S_4, S_5)$ is applied to $|+\rangle$. Similarly, $\langle +|$ applied to $T(S_1, S_2)$ from the left yields $\lambda_+ \langle +|$. Consequently,

$$z_+ \equiv$$

$$\langle +|T(S_1, S_2) \cdot S_2 \cdot T(S_2, S_3) T(S_3, S_4) \cdot S_4 \cdot T(S_4, S_5) \cdots T(S_N, S_1)|+\rangle$$

$$= \lambda_+ \langle +|S_2 T(S_2, S_3) T(S_3, S_4) S_4|+\rangle \lambda_+^{N-3}. \tag{9.30}$$

Next, we notice that $T(S_3, S_4) S_4$ is obtained from $T(S_3, S_4)$ by changing the sign of the second column, and therefore

$$T(S_3, S_4) S_4 |+\rangle = \begin{pmatrix} e^K & -e^{-K} \\ e^{-K} & -e^K \end{pmatrix} \cdot \frac{1}{\sqrt{2}} \begin{pmatrix} 1 \\ 1 \end{pmatrix} = \frac{\lambda_-}{\sqrt{2}} \begin{pmatrix} 1 \\ -1 \end{pmatrix} = \lambda_- |-\rangle. \tag{9.31}$$

Similarly, one finds

$$\langle +|S_2 T(S_2, S_3) = \frac{1}{\sqrt{2}} (1 \quad 1) \begin{pmatrix} e^K & e^{-K} \\ -e^{-K} & -e^K \end{pmatrix} = \frac{\lambda_-}{\sqrt{2}} (1 \quad -1) = \lambda_- \langle -|. \tag{9.32}$$

Then, it follows that

$$z_+ = \lambda_+^{N-2} \lambda_-^2 . \tag{9.33}$$

The expectation value z_- obtained by using $|-\rangle$ can be evaluated in a similar fashion. Simple replacements of λ_+ by λ_- and $|+\rangle$ by $|-\rangle$ are sufficient. The result is

$$z_- = \lambda_-^{N-2} \lambda_+^2 . \tag{9.34}$$

We therefore arrive at the following relation, using eqn (9.24),

$$\langle S_2 S_4 \rangle = \frac{\lambda_+^{N-2} \lambda_-^2 + \lambda_-^{N-2} \lambda_+^2}{\lambda_+^N + \lambda_-^N} . \tag{9.35}$$

From $\lambda_+ > \lambda_-$, the correlation function in the thermodynamic limit is

$$\lim_{N \to \infty} \langle S_2 S_4 \rangle = \left(\frac{\lambda_-}{\lambda_+} \right)^2 = \tanh^2 K . \tag{9.36}$$

It is straightforward to apply the same method to an arbitrary correlation function. The result is, for $r \le N/2$,

$$\langle S_i S_{i+r} \rangle = \left(\frac{\lambda_-}{\lambda_+} \right)^r = (\tanh K)^r . \tag{9.37}$$

This expression coincides with eqn (9.13) for a free boundary. We therefore conclude that the correlation function does not depend on the boundary condition in the thermodynamic limit.

It is interesting to notice here that eqn (9.37) can be rewritten as

$$\langle S_i S_{i+r} \rangle = \exp \left(-r \log \frac{\lambda_+}{\lambda_-} \right) . \tag{9.38}$$

The correlation length is written in terms of the logarithm of the ratio of two eigenvalues of the transfer matrix,

$$\xi = \frac{1}{\log \dfrac{\lambda_+}{\lambda_-}} . \tag{9.39}$$

It is established in higher dimensions as well that the correlation length can be expressed as a function of the ratio of the two largest eigenvalues of the transfer matrix.

EXERCISE 9.1 Evaluate the partition function of the three-state Potts model in one dimension,

$$\beta H = -K \sum_i \delta(S_i, S_{i+1}) \quad (S_i = 0, 1, 2) \tag{9.40}$$

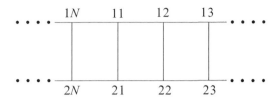

Fig. 9.3 A two-leg ladder.

with a free boundary condition. Solve the same problem also for a periodic boundary condition. Confirm that the free energy per spin does not depend on the boundary condition in the thermodynamic limit $N \to \infty$.

EXERCISE 9.2 Evaluate the partition function of the two-leg ladder Ising model,

$$\beta H = -K_1 \sum_{i=1}^{N} S_{1i} S_{1i+1} - K_1 \sum_{i=1}^{N} S_{2i} S_{2i+1} - K_2 \sum_{i=1}^{N} S_{1i} S_{2i} \tag{9.41}$$

with periodic boundary conditions, i.e. $S_{1N+1} = S_{11}$, and $S_{2N+1} = S_{21}$, see Fig. 9.3. Write also the correlation function $\langle S_{1i} S_{1i+r} \rangle$ in terms of the transfer matrix. Hint: The transfer matrix is a 4×4 matrix.

9.2   One-dimensional n-vector model

The *n-vector model* is a system of coupled spins $\{\boldsymbol{S}_i\}$, where $\boldsymbol{S}_i$ is an n-component classical vector at site i, $\boldsymbol{S}_i = (S_{i1}, S_{i2}, \cdots, S_{in})$, normalized to unity, $|\boldsymbol{S}_i| = 1$. The components of $\boldsymbol{S}_i$ can take continuous values as long as the normalization condition $|\boldsymbol{S}_i| = 1$ is satisfied. It is also called the $O(n)$ *model*. The Hamiltonian is, in the absence of external fields,

$$H = -J \sum_{\langle ij \rangle} \boldsymbol{S}_i \cdot \boldsymbol{S}_j. \tag{9.42}$$

This model includes the Ising model when $n = 1$ and the XY and Heisenberg models for $n = 2$ and $n = 3$, respectively. The n-vector model with $n = 4$ does not directly represent real physical systems. It is, nevertheless, often useful to discuss general-n cases for theoretical analyses. In particular, the limit $n \to \infty$ is known as the *spherical model*, for which the exact solution can be derived in any dimension, as will be shown in Section 9.3.

The mean-field theory for the n-vector model yields critical phenomena (critical exponents) independent of n. This result is not applicable to dimensions lower than the upper critical dimension, where the critical exponents depend on n. In the present section we solve the one-dimensional n-vector model under a free boundary condition and discuss its properties.

Let us start with the definition of the partition function of the n-vector model in an N sites chain with free boundary,

$$Z_N^{(\mathrm{F})} = \int \prod_{i=1}^{N} \mathrm{d}\boldsymbol{S}_i \, \exp\left(K(\boldsymbol{S}_1 \cdot \boldsymbol{S}_2 + \boldsymbol{S}_2 \cdot \boldsymbol{S}_3 + \cdots + \boldsymbol{S}_{N-1} \cdot \boldsymbol{S}_N)\right). \tag{9.43}$$

The integration is carried out constrained to the normalization condition $|\boldsymbol{S}_i| = 1$ for each i.

Now, we first perform the above integration for $\boldsymbol{S}_N$ and then for $\boldsymbol{S}_{N-1}$, following the prescription for the Ising model. The part involving $\boldsymbol{S}_N$ is extracted for this purpose,

$$Z_N^{(\mathrm{F})} = \int \prod_{i=1}^{N-1} \mathrm{d}\boldsymbol{S}_i \, \exp\left(K(\boldsymbol{S}_1 \cdot \boldsymbol{S}_2 + \boldsymbol{S}_2 \cdot \boldsymbol{S}_3 + \cdots + \boldsymbol{S}_{N-2} \cdot \boldsymbol{S}_{N-1})\right)$$
$$\cdot \int \mathrm{d}\boldsymbol{S}_N \exp(K\boldsymbol{S}_{N-1} \cdot \boldsymbol{S}_N). \tag{9.44}$$

The range of integration for $\boldsymbol{S}_N$ is over the unit sphere of the n-dimensional space. Since the unit sphere is isotropic, we may choose any direction as the first axis of the Cartesian coordinate. We therefore choose the direction of the vector $\boldsymbol{S}_{N-1}$ as the first axis because this facilitates the calculation significantly. The product $\boldsymbol{S}_{N-1} \cdot \boldsymbol{S}_N$ then has only the first component S_{N1}, and the $\boldsymbol{S}_N$-integral of eqn (9.44), to be denoted as $G(K)$, is written as

$$G(K) = \int \mathrm{d}\boldsymbol{S}_N \, \mathrm{e}^{KS_{N1}}$$
$$= \int_{-\infty}^{\infty} \mathrm{d}S_{N1} \cdots \mathrm{d}S_{Nn} \, \delta\left((S_{N1})^2 + (S_{N2})^2 + \cdots (S_{Nn})^2 - 1\right) \mathrm{e}^{KS_{N1}}. \tag{9.45}$$

This integral can be carried out as a Gaussian integral if we use the Fourier representation of the delta function as detailed in Appendix A.15. The result is

$$G(K) = c \left(\frac{K}{2}\right)^{1-n/2} I_{n/2-1}(K), \tag{9.46}$$

where c is a trivial constant and $I_{n/2-1}(K)$ is the modified Bessel function of the first kind.

The partition functions (9.43) and (9.44) thus satisfy the recursion relation $Z_N^{(\mathrm{F})} = G(K) \cdot Z_{N-1}^{(\mathrm{F})}$. Repeated applications of this recursion relation lead to the following solution,

$$Z_N^{(\mathrm{F})} = G(K)^{N-1} \cdot \mathrm{const.} \tag{9.47}$$

We can now calculate the free energy and its derivatives. The energy per site is

$$E_0 \equiv \lim_{N \to \infty} \frac{E}{N} = -J \frac{\mathrm{d}}{\mathrm{d}K} \log G(K) = -J \frac{I_{n/2}(K)}{I_{n/2-1}(K)}, \tag{9.48}$$

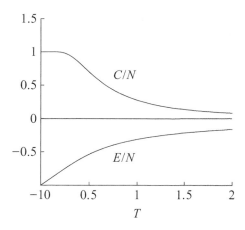

Fig. 9.4 The energy and specific heat of the one-dimensional (classical) Heisenberg model. J is set to unity.

where the final equality was derived using the following identities satisfied by the modified Bessel function,

$$I'_m(K) = \frac{1}{2}\left(I_{m+1}(K) + I_{m-1}(K)\right), \quad I_m(K) = \frac{K}{2m}\left(I_{m-1}(K) - I_{m+1}(K)\right). \quad (9.49)$$

As an example, let us choose $n = 3$ corresponding to the Heisenberg model. The modified Bessel functions applicable to this case can be written, in terms of elementary functions, as

$$I_{3/2}(K) = \sqrt{\frac{2}{\pi K}}\left(\cosh K - K^{-1}\sinh K\right), \quad I_{1/2}(K) = \sqrt{\frac{2}{\pi K}}\sinh K. \quad (9.50)$$

The energy per site E_0 and its temperature derivative, the specific heat per site C_0, are

$$E_0 = J\left(\frac{1}{K} - \frac{1}{\tanh K}\right), \quad C_0 = K^2\left(\frac{1}{K^2} - \frac{1}{\sinh^2 K}\right). \quad (9.51)$$

These functions are drawn in Fig. 9.4. The low-temperature limit of the specific heat is

$$C_0 \approx (1 - 4K^2 e^{-2K}), \quad (9.52)$$

which approaches a finite value at $T \to 0$, a behavior significantly differently from the Ising model, see Fig. 9.1. This qualitative difference originates in the symmetry; the Ising model has a global discrete $\mathbb{Z}_2$ symmetry, whereas the model with $n \geq 2$ has a continuous symmetry, i.e. an invariance of the Hamiltonian under a simultaneous rotation of all spins by the same angle.

To understand this behavior, it is helpful to differentiate the free energy with respect to temperature to obtain the entropy. The entropy of the classical Heisenberg model turns out to behave as $S \approx \log T$ in the low-temperature limit and diverges toward $-\infty$. This is the same phenomenon as observed in the entropy of a classical ideal gas. In a classical ideal gas the low-temperature entropy behaves unphysically due to the continuous translational invariance of the Hamiltonian. The same mechanism works in the classical Heisenberg model. In real materials with continuous symmetries, quantum effects come into play at low temperatures, preventing physical quantities from diverging. Such an example will be shown in Section 9.4.

9.3   Spherical model

The n-vector model can be solved exactly in any spatial dimension d if we take the limit $n \to \infty$. The result shows a finite-temperature phase transition for $d > 2$, and the low-temperature phase has finite spontaneous magnetization. The critical exponents are functions of d in the range $2 < d < 4$, whereas the same exponents take the mean-field values above four dimensions $d > 4$. At $d = 4$ critical exponents coincide with the mean-field results but with logarithmic corrections. The n-vector model in the limit $n \to \infty$ has been studied in detail due to these notable properties expected for more conventional systems with finite n, which include the existence of upper and lower critical dimensions and the d dependence of critical exponents between the upper and lower critical dimensions.

The free energy of the n-vector model in the limit $n \to \infty$ is known to be identical to a model called the *spherical model*. Spins $\{S_i\}$ of the spherical model take arbitrary real values under the constraint $\sum_{i=1}^{N} S_i^2 = N$, where N is the number of lattice sites. The Hamiltonian has the conventional expression

$$H = -J \sum_{\langle ij \rangle} S_i S_j - h \sum_i S_i. \tag{9.53}$$

We can choose an arbitrary spatial dimension and lattice structure. The Ising model also satisfies the constraint $\sum_{i=1}^{N} S_i^2 = N$, but a spin of the spherical model S_i can take values other than ± 1. The constraint $\sum_{i=1}^{N} S_i^2 = N$ represents an N-dimensional sphere, from which the name comes. In the present section we derive the solution of the n-vector model in the $n \to \infty$ limit, which is often also called the spherical model.

9.3.1   Partition function and free energy

The Hamiltonian of the n-vector model in the presence of an external field reads

$$\beta H = -K \sum_{\langle ij \rangle} \boldsymbol{S}_i \cdot \boldsymbol{S}_j - \boldsymbol{h} \cdot \sum_i \boldsymbol{S}_i$$

$$= -K \sum_{\langle ij \rangle} \sum_{a=1}^{n} S_{ia} S_{ja} - h \sum_i \sum_{a=1}^{n} S_{ia} \quad \left(\sum_{a=1}^{n} S_{ia}^2 = n, \; \forall i \right), \tag{9.54}$$

where $\boldsymbol{S}_i$ is a vector with n components. In order to have a non-trivial result in the limit $n \to \infty$, we normalize $\boldsymbol{S}_i$ as $|\boldsymbol{S}_i|^2 = n$, not as $|\boldsymbol{S}_i| = 1$ employed in the previous section. The external field h is assumed to be applied along all axes with the same amplitude. For simplicity, we consider the d-dimensional hypercubic lattice and the interactions only exist between nearest neighbors. Periodic boundaries will be assumed for translational invariance.

The partition function of this model is written as

$$
Z = \int_{-\infty}^{\infty} \prod_{i=1}^{N} \mathrm{d}\boldsymbol{S}_i \, \exp\left(K \sum_{\langle ij \rangle} \sum_{a=1}^{n} S_{ia} S_{ja} + h \sum_{i} \sum_{a=1}^{n} S_{ia} \right)
$$
$$
\cdot \prod_{i=1}^{N} \delta\left(n - \sum_{a=1}^{n} S_{ia}^2 \right). \tag{9.55}
$$

From the Fourier representation of the delta function, we find

$$
Z = \int_{-\infty}^{\infty} \prod_{i=1}^{N} \mathrm{d}\boldsymbol{S}_i \cdot \frac{1}{(2\pi \mathrm{i})^N} \int_{-\mathrm{i}\infty}^{\mathrm{i}\infty} \prod_{i=1}^{N} \mathrm{d}z_i
$$
$$
\cdot \prod_{a=1}^{n} \exp\left(K \sum_{\langle ij \rangle} S_{ia} S_{ja} + h \sum_{i} S_{ia} + \sum_{i} z_i (1 - S_{ia}^2) \right). \tag{9.56}
$$

It is convenient to perform the integration over $\boldsymbol{S}_i$ first. The integral over $\boldsymbol{S}_i$ can be performed independently over each component, $a = 1, 2, \cdots, n$, and the result does not depend on the index a. Thus, we may simply raise the integral over a single component to the nth power,

$$
Z = \frac{1}{(2\pi \mathrm{i})^N} \int_{-\mathrm{i}\infty}^{\mathrm{i}\infty} \prod_{i=1}^{N} \mathrm{d}z_i \left\{ \int_{-\infty}^{\infty} \prod_{i} \mathrm{d}S_i \right.
$$
$$
\left. \cdot \exp\left(K \sum_{\langle ij \rangle} S_i S_j + h \sum_{i} S_i + \sum_{i} z_i (1 - S_i^2) \right) \right\}^n. \tag{9.57}
$$

The multiple integral over $\{S_i\}$ is a Gaussian integral and can be evaluated explicitly. The result is a function of $\{z_i\}$, which we have to further integrate over $\{z_i\}$. A significant simplification takes place here in the limit $n \to \infty$ since the integral over $\{S_i\}$ is raised to the nth power, which allows us to apply the saddle-point method to the z-integral. The result of the integral over $\{z_i\}$ is simply the maximum of the integrand. This is why the spherical model can be solved exactly. This program will be carried out explicitly.

Since all sites are equivalent, we may assume that the saddle point of z_i is independent of i. We therefore set $z_i = z$ and focus our attention on the evaluation of the integral

$$
I \equiv \int_{-\infty}^{\infty} \prod_{i} \mathrm{d}S_i \, \exp\left(K \sum_{\langle ij \rangle} S_i S_j - z \sum_{i} S_i^2 + h \sum_{i} S_i \right). \tag{9.58}
$$

This is the multiple Gaussian integral detailed in Appendix A.16, where the following formula is derived,

$$\int_{-\infty}^{\infty} e^{-\frac{1}{2}\,{}^t x \cdot Cx + i\,{}^t x \cdot q}\, \mathrm{d}x = \frac{(2\pi)^{N/2}}{(\det C)^{1/2}} e^{-\frac{1}{2}\sum_{n,l} q_n q_l (C^{-1})_{nl}}. \tag{9.59}$$

Here, x and q are N-dimensional vectors and C is an $N \times N$ matrix. We choose C in this equation as follows,

$$C_{ii} = 2z \ \ (\text{diagonal}), \quad C_{i,i+\delta} = -K \ \ (\text{nearest neighbor}), \quad \text{other } C_{ij} = 0, \tag{9.60}$$

and set $iq = {}^t(h, h, \cdots, h)$ to express eqn (9.58). Then, we can apply eqn (9.59) to the present problem if we appropriately evaluate various quantities in this equation. We start with the determinant of C in the denominator by writing the eigenvalues of C, which can be obtained following the method of eqn (A.263),

$$C(\mathbf{k}) = 2z - 2K\lambda(\mathbf{k}), \quad \lambda(\mathbf{k}) = \sum_{j=1}^{d} \cos k_j. \tag{9.61}$$

This leads to

$$\det C = (2K)^N \prod_{\mathbf{k}} (\tilde{z} - \lambda(\mathbf{k})), \tag{9.62}$$

where $\tilde{z} = z/K$. The quadratic form of q in the exponent of eqn (9.59) can be rewritten as follows, using the translational invariance of the lattice Green function $G(= C^{-1})$, $iq = {}^t(h, h, \cdots, h)$, and eqn (9.61),

$$-\frac{1}{2}\sum_{n,l} q_n q_l G_{nl} = \frac{h^2}{2}\sum_{n,l} G_{nl} = \frac{h^2 N}{2}\sum_{l} G_{nl}$$

$$= \frac{h^2}{2} NG(\mathbf{k}=0) = \frac{h^2 N}{2C(\mathbf{k}=0)} = \frac{h^2 N}{4(z - Kd)}. \tag{9.63}$$

We are ready to use eqn (9.59) to write the result of the integral (9.58) explicitly,

$$\frac{1}{N}\log I = \frac{1}{2}\log 2\pi - \frac{1}{2}\log 2K - \frac{1}{2N}\sum_{\mathbf{k}}\log(\tilde{z} - \lambda(\mathbf{k})) + \frac{h^2}{4K(\tilde{z} - d)}. \tag{9.64}$$

The free energy per spin f is then written as, using the saddle point z_0 (or $\tilde{z}_0 - d = z_0/K - d \equiv u$),

$$\beta f(u) = -\lim_{N,n \to \infty} \frac{1}{Nn}\log Z = -\frac{1}{2}\log\frac{\pi}{K} - \frac{h^2}{4Ku} - (Ku + Kd)$$

$$+ \frac{1}{2(2\pi)^d}\int_0^{2\pi}\log(u + d - \lambda(\mathbf{k}))\,\mathrm{d}\mathbf{k} \tag{9.65}$$

in the limit $N, n \to \infty$. We further need the explicit properties of the saddle point z_0 (or $u = z_0/K - d$), which will be analyzed in the next section.

9.3.2 Solution of the saddle-point equation and critical exponents

The minimization condition of f as a function of u is

$$H(u) \equiv \frac{1}{(2\pi)^d} \int_0^{2\pi} \frac{1}{u + d - \lambda(\boldsymbol{k})} \, d\boldsymbol{k} = -\frac{h^2}{2Ku^2} + 2K. \tag{9.66}$$

Let us investigate the condition for this equation $H(u) = 2K$ to have a solution in the case of $h = 0$. We first note that $H(u)$ is a monotone decreasing function of u, as can be verified by differentiation. The function $H(u)$ has three different types of behavior around $u = 0$ depending on the dimensionality d as follows.

(i) $d \leq 2$.   The behavior of $H(u)$ as $u \to 0$, whether it diverges or not, is determined by the properties of the integrand around the origin $k = |\boldsymbol{k}| \approx 0$. We may therefore adopt the approximation, using $\lambda(\boldsymbol{k}) \approx d - k^2/2$,

$$H(u) \approx \int_0 \frac{k^{d-1}}{u + k^2/2} \, dk, \tag{9.67}$$

where the upper limit is omitted as it has no relevance for the divergence at the lower limit. This integral diverges as $u \to 0$ if $d \leq 2$. The other limit $u \to \infty$ gives $H(u) \to 0$. Consequently, $H(u)$ takes all positive values, which means that the equation $H(u) = 2K$ always has a solution for arbitrary $K(> 0)$. It is also clear that $H(u)$ has no singularities at any positive u. Hence, u as the solution of $H(u) = 2K$ is an analytic function of K and $f(u)$ is not singular as a function of K. Thus, the system does not undergo a phase transition for $d \leq 2$.

(ii) $2 < d < 4$.   As seen in eqn (9.67), $H(u)$ tends to a finite value as $u \to 0$ for $2 < d < 4$. The first-order derivative of $H(u)$, however, diverges as $u \to 0$. To see this, we set $k = \sqrt{2u}\,x$ and rewrite eqn (9.67) as

$$H(u) \approx u^{d/2-1} \int_0 \frac{x^{d-1}}{1 + x^2} \, dx \propto u^{d/2-1} + \text{const.} \tag{9.68}$$

This equation shows that the first-order derivative diverges as $u^{d/2-2}$ for $2 < d < 4$.

The function $H(u)$ thus behaves as $H(u) \approx H(0) - cu^{d/2-1}$ for $u \approx 0$. It also monotonically decreases from $H(0)$ at $u = 0$ to $H \to 0$ as $u \to \infty$. Accordingly, the saddle-point equation $H(u) = 2K$ has a solution in the high-temperature region satisfying $K < H(0)/2 \equiv K_{\rm c}$ but not in the low-temperature region $K > K_{\rm c}$. The free energy therefore changes drastically at $K_{\rm c}$, a phase transition.[2]

---

[2] No saddle point exists for $K > K_{\rm c}$ in the sense that eqn (9.66) does not have a solution and $f(u)$ is not stationary anywhere. We may, nevertheless, consider that the stationary point stays at $u = 0$ since $f(u)$ is minimum at $u = 0$, that is, the argument of the parentheses $\{\cdot\}$ of eqn (9.57) is maximum at $u = 0$.

An expansion of the saddle-point equation (9.66)($h = 0$) near the transition point yields, using $H(0) = 2K_c$,

$$2K_c - cu^{d/2-1} = 2K, \tag{9.69}$$

from which we have $\Delta K \equiv K_c - K \propto u^{d/2-1}$. Since the integral in the free energy (9.65) is given as the u-integral of $H(u)$, the singular part is proportional to $u^{d/2}$ from $H(u) \approx H(0) - cu^{d/2-1}$. Thus, the singular part of the free energy is

$$f \propto u^{d/2} \propto (\Delta K)^{d/(d-2)}. \tag{9.70}$$

A comparison of this equation with the definition of the critical exponent α, $f \approx (\Delta K)^{2-\alpha}$, yields

$$\alpha = -\frac{d}{d-2} + 2 = \frac{d-4}{d-2}. \tag{9.71}$$

Consequently, the specific heat does not diverge for $2 < d < 4$ since $\alpha < 0$. The exponent α diverges as $d \to 2$ and tends to vanish as $d \to 4$. It is interesting to remember that $\alpha = 0$ is the mean-field value.

Since u stays constant for $K > K_c$, the temperature dependence of the free energy (9.65) exists only in the terms explicitly dependent on K. The second-order derivative of this equation with $h = 0$ then shows that the specific heat is a constant. We therefore conclude that the specific heat has the temperature dependence as depicted in Fig. 9.5. A notable feature is that the specific heat stays finite as $T \to 0$, similarly to the one-dimensional n-vector model, a problem specific to continuous classical systems.

The susceptibility is the second-order derivative of f with respect to h and is therefore divergent, as follows according to eqn (9.65),

$$\chi \propto u^{-1} \propto (\Delta K)^{-2/(d-2)}. \tag{9.72}$$

The exponent is then $\gamma = 2/(d-2)$.[3] The limiting behaviors are $\gamma \to \infty$ as $d \to 2$ and $\gamma \to 1$ as $d \to 4$, the latter reproducing the mean-field result.

These two critical exponents are sufficient to fix the other values from the scaling relation,

$$\boxed{\alpha = \frac{d-4}{d-2}, \ \beta = \frac{1}{2}, \ \gamma = \frac{2}{d-2}, \ \delta = \frac{d+2}{d-2}, \ \nu = \frac{1}{d-2}, \ \eta = 0} \tag{9.73}$$

Expansions of these expressions to second order in ϵ with $4 - d = \epsilon$ agree with the ϵ-expansion results of Section 4.2.2 in the limit $n \to \infty$. All these exponents approach the mean-field values as $d \to 4$. Exponents other than β and η diverge as $d \to 2$, indicating that the rate of divergence is larger than power laws at the lower critical dimension. This

---

[3] In eqn (9.65) u is also a function of h through eqn (9.66) but this dependence can be ignored in the evaluation of the derivative in the limit $h \to 0$. This fact can be confirmed if we take the second-order derivative of eqn (9.56) with respect to h, set $h \to 0$ and reproduce the following argument.

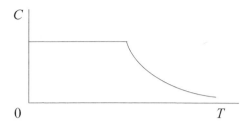

Fig. 9.5 A qualitative description of the temperature dependence of the specific heat of the spherical model for $d > 2$.

is in agreement with the strong exponential divergence observed in the two-dimensional XY model and the one-dimensional Ising model.

(iii) $d \geq 4$. Since $H(u) = H(0) - cu$ as $u \approx 0$ above four dimensions, we can apply the discussions for $2 < d < 4$ to the present case simply by setting $d \to 4$. It of course means that the mean-field theory gives the exact values for the critical exponents. At $d = 4$, logarithmic corrections appear, reflecting the logarithmic divergence of physical quantities.

EXERCISE 9.3 Show that the spherical model does not have a phase transition in any dimension if $h \neq 0$. A finite field erases a phase transition, as in Fig. 1.3. It will be useful to check if the saddle-point equation (9.66) has a solution for arbitrary K and if the solution has a singularity.

EXERCISE 9.4 Evaluate the spontaneous magnetization of the spherical model near the transition point for $2 < d < 4$. Differentiate eqn (9.65) with respect to h and eliminate u using eqn (9.66).

9.4   One-dimensional quantum XY model

The present book discusses topics of phase transitions and critical phenomena in classical statistical mechanics. A main reason is that phase transitions and critical phenomena are macroscopic phenomena involving very many degrees of freedom and quantum effects, which usually appear prominently in microscopic length scales, do not play significant roles. Quantum fluctuations, nevertheless, become dominant at low temperatures, where thermal fluctuations become irrelevant, and phase transitions caused by quantum effects sometimes appear.

We do not discuss quantum phase transitions in general. Nevertheless, one of the simplest examples of quantum spin systems, the one-dimensional quantum XY model, will be studied in the present section. This model is interesting not just because the exact solution can be derived by relatively straightforward calculations but also since the solution has common aspects with the two-dimensional Ising model to be discussed in the next section.

The Hamiltonian of the one-dimensional quantum XY model is

$$H = -J \sum_{j=1}^{N} (S_j^x S_{j+1}^x + S_j^y S_{j+1}^y) - h \sum_{j=1}^{N} S_j^z, \tag{9.74}$$

where $\boldsymbol{S}_j$ is a spin-1/2 quantum operator, whose algebra was already presented in Section 4.3.1, and a field h is applied along the z-axis.[4] The boundary condition is periodic. The set of spin-1/2 operators along a one-dimensional chain, $\{\boldsymbol{S}_j\}$, can be written in terms of Fermionic operators $\{a_j, a_j^\dagger\}$ by means of the *Jordan–Wigner transformation* explained in Appendix A.17. The raising and lowering operators of spins

$$S_j^\pm = S_j^x \pm \mathrm{i} S_j^y \tag{9.75}$$

and the z-component S_j^z have the following expressions in terms of a_j, and $a_j^\dagger$,

$$S_j^+ = (1 - 2n_1)(1 - 2n_2) \cdots (1 - 2n_{j-1}) a_j^\dagger$$
$$S_j^- = (1 - 2n_1)(1 - 2n_2) \cdots (1 - 2n_{j-1}) a_j$$
$$S_j^z = a_j^\dagger a_j - \frac{1}{2}. \tag{9.76}$$

Here, $n_j = a_j^\dagger a_j$ is the number operator of Fermions with eigenvalues 0 and 1. From the trivial relation $(1 - 2n_j)^2 = 1$, we obtain the following Fermionic representations of the products of neighboring spin operators,

$$S_j^+ S_{j+1}^+ = a_j^\dagger (1 - 2n_j) a_{j+1}^\dagger = a_j^\dagger a_{j+1}^\dagger$$
$$S_j^- S_{j+1}^- = a_j (1 - 2n_j) a_{j+1} = -a_j a_{j+1} = a_{j+1} a_j$$
$$S_j^+ S_{j+1}^- = a_j^\dagger (1 - 2n_j) a_{j+1} = a_j^\dagger a_{j+1}$$
$$S_j^- S_{j+1}^+ = a_j (1 - 2n_j) a_{j+1}^\dagger = -a_j a_{j+1}^\dagger = a_{j+1}^\dagger a_j. \tag{9.77}$$

Rewriting the x- and y-components of the spin operators in the Hamiltonian (9.74) by eqn (9.75) and using the above equations, we find the representation of interactions in terms of Fermionic operators,

$$S_j^x S_{j+1}^x = \frac{1}{4}(a_j^\dagger a_{j+1}^\dagger + a_{j+1} a_j + a_j^\dagger a_{j+1} + a_{j+1}^\dagger a_j)$$

$$S_j^y S_{j+1}^y = -\frac{1}{4}(a_j^\dagger a_{j+1}^\dagger + a_{j+1} a_j - a_j^\dagger a_{j+1} - a_{j+1}^\dagger a_j). \tag{9.78}$$

---

[4] An application of fields along the x- or y-axis precludes the model to be exactly solved. Try it.

The Hamiltonian is therefore[5]

$$H = -\frac{J}{2} \sum_{j=1}^{N} (a_j^\dagger a_{j+1} + a_{j+1}^\dagger a_j) - h \sum_{j=1}^{N} \left(a_j^\dagger a_j - \frac{1}{2} \right). \tag{9.79}$$

This Hamiltonian represents a set of free Fermions (i.e. Fermions without interactions) hopping from a site to a neighboring site, for which we can derive the eigenvalues by using translational invariance.

It is useful to change the basis from the real-space representation to the space of wave numbers by Fourier transformation, as is usually done in the analysis of translationally invariant systems. We then define

$$a_j = \frac{1}{\sqrt{N}} \sum_q e^{iqj} a_q, \quad a_j^\dagger = \frac{1}{\sqrt{N}} \sum_q e^{-iqj} a_q^\dagger, \tag{9.80}$$

where the operators $\{a_q, a_q^\dagger\}$ are also Fermionic. The Hamiltonian (9.79) is transformed into

$$H = -\sum_q (J \cos q + h) a_q^\dagger a_q + \frac{hN}{2}. \tag{9.81}$$

Here, each wave number q is independent, i.e decoupled, and thus the problem has been solved. Since the eigenvalues of $a_q^\dagger a_q$ are 0 and 1, the partition function is

$$Z = e^{-\beta hN/2} \prod_q \left(1 + e^{\beta J \cos q + \beta h} \right). \tag{9.82}$$

The free energy and energy per spin in the thermodynamic limit are calculated from this partition function as

$$f = \frac{h}{2} - \frac{T}{2\pi} \int_{-\pi}^{\pi} \log \left(1 + e^{\beta J \cos q + \beta h} \right) dq \tag{9.83}$$

$$E_0 = \frac{h}{2} - \frac{1}{2\pi} \int_{-\pi}^{\pi} \frac{J \cos q + h}{1 + e^{-\beta J \cos q - \beta h}} dq. \tag{9.84}$$

This is the exact solution of the one-dimensional quantum XY model.

Let us investigate the properties of this quantum system at zero and low temperatures, where the system behaves quite differently from the corresponding classical system due to quantum fluctuations. The zero-temperature limit of the energy for the case $h = 0$ is derived by setting $h = 0$ in eqn (9.84) and taking the limit $\beta \to \infty$. The $\cos q$ term in the denominator of the integrand has positive and negative values

---

[5] The periodic boundary condition of spin operators does not directly correspond to the periodic boundary of Fermionic operators. However, if we are interested only in the macroscopic quantities such as the energy and magnetization in the thermodynamic limit, the boundary condition does not affect the result. We therefore do not go into the details of the problem of boundaries. See Appendix A.17 and Section 9.5.2 for more details.

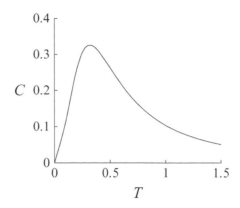

Fig. 9.6 Specific heat of the one-dimensional quantum XY model with the unit of energy $J = 1$.

depending on the range of q. Owing to the denominator in eqn (9.84), only the positive range contributes to the zero-temperature limit and we have

$$E_0 = -\frac{J}{2\pi} \int_{-\pi/2}^{\pi/2} \cos q \, dq = -\frac{J}{\pi} \tag{9.85}$$

as the exact ground-state energy per spin. Non-trivial quantum effects manifest themselves as the factor π. The specific heat per spin in the absence of external field

$$C = \frac{\beta^2 J^2}{2\pi} \int_{-\pi}^{\pi} \frac{\cos^2 q \, e^{-\beta J \cos q}}{(1 + e^{-\beta J \cos q})^2} \, dq \tag{9.86}$$

behaves in the limit $T \to 0$ as

$$C \to \frac{\pi T}{3J}, \tag{9.87}$$

and thus approaches 0 linearly in temperature.

EXERCISE 9.5 Derive eqn (9.87) by taking the low-temperature limit of the specific heat C, eqn (9.86).

The specific heat of the quantum XY model vanishes at $T = 0$, in contrast to the classical n-vector model, although the quantum system share the same invariance properties under uniform rotation of the spins as the classical case. The temperature dependence of the specific heat (9.86) is shown in Fig. 9.6.

The ground-state energy expression for non-vanishing h depends on whether $h(> 0)$ is larger or smaller than J. The result is

$$E_0 = \begin{cases} -\dfrac{h}{2} & (h > J) \\ \dfrac{h}{2} - \dfrac{\sqrt{J^2 - h^2}}{\pi} - \dfrac{h}{\pi} \arccos\left(-\dfrac{h}{J}\right) & (h < J) \end{cases}. \tag{9.88}$$

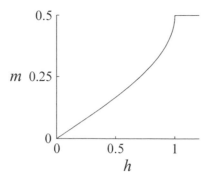

Fig. 9.7 Ground-state magnetization along the z-axis for the one-dimensional quantum XY model. The unit of energy is chosen as $J = 1$.

EXERCISE 9.6 Derive eqn (9.88) by taking the low-temperature limit of eqn (9.84).

The magnetic susceptibility along the z-axis is obtained by differentiation of the free energy with respect to $-h$. The final answer for $T = 0$, derived in a manner similar to the ground-state energy, is

$$m = -\frac{1}{2} + \frac{1}{\pi} \arccos\left(-\frac{h}{J}\right), \tag{9.89}$$

for $h < J$. This function is plotted in Fig. 9.7. When $h > J$, all spins align in the z-direction and the interactions in the XY plane become ineffective. For $h < J$ the interplay between the field and the interactions determines the spin state, which may be understood classically that the spin orientation gradually becomes parallel to the XY plane as the field along the z-axis is reduced. Actually, quantum fluctuations make the system state more complex.

We may regard the sudden change of the state at $h = J$ as a quantum phase transition. Any finite temperature, however small, destroys this singularity.

EXERCISE 9.7 Evaluate the partition function of the following Fermionic Hamiltonian chain,

$$\beta H = -4K_1 \sum_{j=1}^{N-1} n_j n_{j+1} - 4K_2 \sum_{j=1}^{N-2} n_j n_{j+2} + 4(K_1 + K_2) \sum_{j=1}^{N} n_j, \tag{9.90}$$

with a free boundary condition. Here, $n_j = a_j^\dagger a_j$ is the number operator with the eigenvalues 0 and 1. Since this Hamiltonian is actually written only in terms of the classical numbers $\{n_j\}$, it is convenient to map this Hamiltonian to a next-nearest-neighbor Ising chain (up to an irrelevant constant and boundary terms)

$$\beta H = -K_1 \sum_{j=1}^{N-1} S_j S_{j+1} - K_2 \sum_{j=1}^{N-2} S_j S_{j+2} \quad (S_j = \pm 1). \tag{9.91}$$

Compute the partition function of this equivalent Ising chain. Determine also the correlation function $\langle S_j S_{j+1} \rangle$ in the thermodynamic limit $N \to \infty$.

9.5 Two-dimensional Ising model

The next example is the two-dimensional Ising model. The exact solution of the two-dimensional Ising model is considered an outstanding milestone in the theory of phase transitions and critical phenomena. There have been a number of methods of solution proposed. In this section we explain the one that uses Majorana fields because it involves a relatively smaller amount of computation than other methods.

9.5.1 Construction of the transfer matrix

It is convenient to reanalyze the transfer matrix method used for the one-dimensional Ising model as a preparation for the two-dimensional case. Equation (9.21) shows that the transfer matrix from site i to site $i + 1$ is, in the absence of external fields,

$$T(S_i, S_{i+1}) = \begin{pmatrix} e^K & e^{-K} \\ e^{-K} & e^K \end{pmatrix}. \tag{9.92}$$

The effect of this matrix is to add an interaction between spins S_i and S_{i+1} and extend the length of the one-dimensional system, see Fig. 9.8. This transfer matrix T is represented by a Pauli matrix (see Section 4.3.1), using the fact that the diagonal element is e^K and the off-diagonal e^{-K} in eqn (9.92), as

$$\begin{aligned} T &= e^K + e^{-K}\sigma^x = e^K(1 + e^{-2K}\sigma^x) \\ &= e^K(1 + \tanh K^* \sigma^x) = e^K(\cosh K^*)^{-1} e^{K^* \sigma^x} \\ &\equiv g(K) e^{K^* \sigma^x}. \end{aligned} \tag{9.93}$$

Here, K^* is a function of K defined by $e^{-2K} = \tanh K^*$ and discussed in Chapter 10 in relation to duality. Also, we defined $g(K) = e^K / \cosh K^* = (2 \sinh 2K)^{1/2}$.

We now consider the two-dimensional case in which the process of extending the system size, by adding columns one by one as in Fig. 9.9, is expressed by the operations V_1 (for the addition of interactions between two columns) and V_2 (for the addition of interactions within a column). We first notice that the addition of interactions between columns represented by horizontal dashed lines in Fig. 9.9 can be performed at each site independently of other sites. This operation of adding the interaction to the site neighboring to the right of a given site is exactly the same as in the transfer matrix for

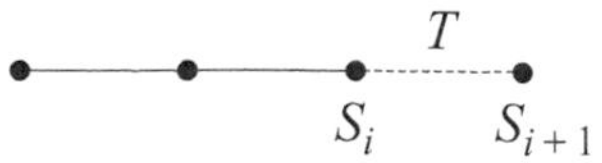

Fig. 9.8 The transfer matrix of the one-dimensional Ising model adds a spin to the existing system.

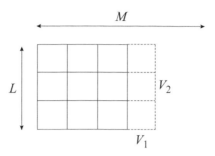

Fig. 9.9 The process to add columns by the operation of the transfer matrix. There are M columns and L rows, both of which have periodic boundaries.

the one-dimensional model in Fig. 9.8. For the jth site $(j = 1, 2, \cdots, L)$, the operation is expressed as $g(K)\mathrm{e}^{K^* \sigma_j^x}$, according to eqn (9.93). The total operation of this type then amounts to

$$V_1 = g(K)^L \exp\left(K^* \sum_{j=1}^{L} \sigma_j^x \right). \tag{9.94}$$

The next operation of adding interactions within a column is written as follows using the Pauli operator σ^z,

$$V_2 = \exp\left(K \sum_{j=1}^{L} \sigma_j^z \sigma_{j+1}^z \right). \tag{9.95}$$

Alternate operations of V_1 and V_2, M times, and then taking the trace reproduce the partition function with periodic boundary conditions,

$$Z = \mathrm{Tr}\left(V_2 V_1 \right)^M = \mathrm{Tr}\left(V_2^{1/2} V_1 V_2^{1/2} \right)^M \equiv \mathrm{Tr} V^M. \tag{9.96}$$

Here, we have introduced the symmetric operator $V = V_2^{1/2} V_1 V_2^{1/2}$ because of its useful properties such as the diagonalizability and its real-valued eigenvalues. Given the structure of V_1 and V_2 this transformation realizes a classical–quantum mapping from a classical $d = 2$ Ising model to an effective $d = 1$ quantum problem.

The problem has now been reduced to the evaluation of the largest eigenvalue of the transfer matrix V. This matrix has a very large dimension $2^L \times 2^L$ but can be diagonalized as described in the following sections.

9.5.2 Representation in terms of Majorana fields

The transfer matrix V can be diagonalized by the Jordan–Wigner transformation, which rewrites spin operators in terms of Fermionic operators, as we showed in the one-dimensional quantum XY model. We adopt here a related but slightly simpler approach that uses *Majorana fields*. Let us first define a set of operators $\psi_1(j)$, $\psi_2(j)$ as

$$\psi_1(j) = \frac{1}{\sqrt{2}} \sigma_1^x \sigma_2^x \cdots \sigma_{j-1}^x \sigma_j^y \tag{9.97}$$

$$\psi_2(j) = \frac{1}{\sqrt{2}} \sigma_1^x \sigma_2^x \cdots \sigma_{j-1}^x \sigma_j^z. \tag{9.98}$$

These $\psi_1(j), \psi_2(j)$ are Hermitian operators because they are defined by the x-, y-, z-components of the Pauli operators. It is straightforward to verify that they satisfy the following anticommutation relations, using the commutation relations of the Pauli operator,

$$\left[\psi_a(j), \psi_b(l)\right]_+ = \psi_a(j)\psi_b(l) + \psi_b(l)\psi_a(j) = \delta_{a,b}\delta_{j,l}. \tag{9.99}$$

The set of operators satisfying these properties are Majorana fields. The matrices V_1 and V_2 in eqns (9.94) and (9.95) are expressed as follows in terms of Majorana fields

$$V_1 = g(K)^L \exp\left(-2iK^* \sum_{j=1}^{L} \psi_1(j)\psi_2(j)\right) \tag{9.100}$$

$$V_2 = \exp\left(2iK \sum_{j=1}^{L} \psi_1(j)\psi_2(j+1)\right), \tag{9.101}$$

as can be confirmed by rewriting ψ_1 and ψ_2 in these equations using eqns (9.97) and (9.98).

Here, a comment on the boundary condition is in order.[6] The product of Majorana fields at boundary sites reads

$$\psi_1(L)\psi_2(1) = \frac{i}{2} \sigma_1^x \cdots \sigma_L^x \sigma_L^z \sigma_1^z, \tag{9.102}$$

according to eqns (9.97) and (9.98). Correspondingly, if we want the boundary term $\sigma_L^z \sigma_1^z$ appearing in the Pauli-operator representation of V_2, eqn (9.95), to be expressed as in eqn (9.101), we have to impose an antiperiodic boundary condition $\psi_2(L+1) = -\psi_2(1)$ in the subspace U_+, where $\sigma_{\text{prod}} \equiv \sigma_1^x \cdots \sigma_L^x$ results to be $+1$. In the subspace U_- with $\sigma_{\text{prod}} = -1$, in contrast, a periodic boundary $\psi_2(L+1) = \psi_2(1)$ is appropriate. Consequently, the wave number q after Fourier transformation has the following different sets of values according to the sign of $e^{iqL}(=\pm 1)$,

$$U_+ : q = \pm\frac{\pi}{L}, \pm\frac{3\pi}{L}, \cdots, \pm\frac{L-1}{L}\pi \tag{9.103}$$

$$U_- : q = 0, \pm\frac{2\pi}{L}, , \cdots, \pm\frac{L-2}{L}\pi, \pi, \tag{9.104}$$

where L has been assumed to be even. This subtle difference of the two subspaces is essential for the evaluation of correlation functions as the difference between the

---

[6] The reader may skip this paragraph on first reading.

largest and second largest eigenvalues of the transfer matrix is related to the difference of these two subspaces. This is the same situation as in the one-dimensional model described in Section 9.1.2, where the ratio of the largest and second largest eigenvalues was directly related to the correlation length. However, if we are interested only in the free energy per spin in the thermodynamic limit, this subtlety does not come into play, which is indeed the case in the following discussions.

9.5.3 Fourier representation in terms of Fermionic operators

The system is translationally invariant due to periodic boundary conditions. We thus use Fourier transformation to represent the Majorana fields in terms of the Fermionic operators $C_1(q), C_1^\dagger(q), C_2(q), C_2^\dagger(q)$ for positive q,

$$\psi_i(j) = \frac{1}{\sqrt{L}} \sum_{q \geq 0} \left(e^{iqj} C_i(q) + e^{-iqj} C_i^\dagger(q) \right) \quad (i = 1, 2). \tag{9.105}$$

Here, the sum runs over all non-negative q in eqn (9.103) or eqn (9.104). It is not difficult to confirm that $\psi_i(j)$ is a Hermitian operator satisfying the anticommutation relation of the Majorana field (9.99). Now, V_1 and V_2 in eqns (9.100) and (9.101) can be rewritten in terms of the Fermionic operators as[7]

$$V_1 = g(K)^L \exp \left[-2iK^* \sum_{q \geq 0} \left(C_1(q) C_2^\dagger(q) + C_1^\dagger(q) C_2(q) \right) \right] \tag{9.106}$$

$$V_2 = \exp \left[2iK \sum_{q \geq 0} \left(e^{-iq} C_1(q) C_2^\dagger(q) + e^{iq} C_1^\dagger(q) C_2(q) \right) \right]. \tag{9.107}$$

The quadratic forms in the exponents commute with each other for different q, and thus we can decompose the transfer matrix $V = V_2^{1/2} V_1 V_2^{1/2}$ into a product of matrices for different q as

$$V = g(K)^L \prod_{q \geq 0} V(q), \quad V(q) = V_2(q)^{1/2} V_1(q) V_2(q)^{1/2}. \tag{9.108}$$

Here, $V_1(q)$ and $V_2(q)$ are the exponential matrices obtained by removing the summation symbols in the exponents of eqns (9.106) and (9.107), respectively. The problem has been reduced to the diagonalization of $V(q)$.

9.5.4 Eigenvalues and the free energy

It is convenient to adopt the states $|n_1 n_2\rangle$, which are eigenstates of the number operators $C_1^\dagger(q) C_1(q)$ and $C_2^\dagger(q) C_2(q)$, as the basis to diagonalize $V(q)$. $n_1 (= 0, 1)$

---

[7] The number of Fermions is conserved in this representation because C and $C^\dagger$ appear in pairs. This is in contrast to the conventional method using the Jordan–Wigner transformation, which makes the problem slightly more complicated than in the present formulation.

is the eigenvalue of the first Fermion number operator $C_1^\dagger(q)C_1(q)$, and $n_2(=0,1)$ is for $C_2^\dagger(q)C_2(q)$.

Consider the two-dimensional space spanned by $|00\rangle$ and $|11\rangle(= C_1^\dagger(q)C_2^\dagger(q)|00\rangle)$. The operators in the exponents of eqns (9.106) and (9.107) for $V_1(q)$ and $V_2(q)$ have only a vanishing eigenvalue in this subspace,

$$C_1(q)C_2^\dagger(q)|11\rangle = C_1^\dagger(q)C_2(q)|00\rangle = 0. \tag{9.109}$$

Hence, $|00\rangle$ and $|11\rangle$ are both eigenstates of $V_1(q)$ and $V_2(q)$ with eigenvalue 1, and consequently of $V(q)$. The matrix $V(q)$ has a doubly degenerate eigenvalue 1 in this two-dimensional subspace.

Next, we study the subspace spanned by $|+\rangle \equiv |01\rangle(= C_2^\dagger|00\rangle)$ and $|-\rangle \equiv |10\rangle(= C_1^\dagger|00\rangle)$. The operators satisfy

$$-C_1(q)C_2^\dagger(q)|+\rangle = 0, \quad -C_1(q)C_2^\dagger(q)|-\rangle = |+\rangle, \tag{9.110}$$

which allows us to regard $-C_1(q)C_2^\dagger(q)$ as a raising operator from state $|-\rangle$ to state $|+\rangle$. Consequently, we can rewrite $-C_1(q)C_2^\dagger(q)$ using a new Pauli operator τ,

$$-C_1(q)C_2^\dagger(q) = \tau^+ = \frac{\tau^x + i\tau^y}{2}. \tag{9.111}$$

Therefore, the operators $\tilde{V}_1(q)$ and $\tilde{V}_2(q)$, obtained by restricting $V_1(q)$ and $V_2(q)$ in eqns (9.106) and (9.107) to the present two-dimensional subspace, are expressed as

$$\tilde{V}_1(q) = \exp\left(2iK^*(\tau^+ - \tau^-)\right) = \exp(-2K^*\tau^y) \tag{9.112}$$

$$\tilde{V}_2(q) = \exp\left(-2iK(\tau^+ e^{-iq} - \tau^- e^{iq})\right) = \exp\left(2K(\tau^y \cos q - \tau^x \sin q)\right). \tag{9.113}$$

To facilitate the diagonalization of $\tilde{V}_2(q)^{1/2}$, it is useful to rotate the spin space by an angle q around the z-axis,

$$\tilde{V}_1(q) = \exp\left(-2K^*(\tau^y \cos q + \tau^x \sin q)\right) \tag{9.114}$$

$$\tilde{V}_2(q) = \exp(2K\tau^y). \tag{9.115}$$

We further apply a rotation by $\pi/2$ around the x-axis, $(\tau^x, \tau^y, \tau^z) \to (\tau^x, -\tau^z, \tau^y)$,

$$\tilde{V}_1(q) = \exp\left(2K^*(\tau^z \cos q - \tau^x \sin q)\right) \tag{9.116}$$

$$\tilde{V}_2(q) = \exp(-2K\tau^z). \tag{9.117}$$

Then, the exponential in $\tilde{V}_1(q)$ is expanded using $(\tau^z \cos q - \tau^x \sin q)^2 = 1$ to derive the following matrix representation,

$$\tilde{V}_1(q) = C^* + (\tau^z \cos q - \tau^x \sin q)S^*$$
$$= \begin{pmatrix} C^* + \cos q\, S^* & -\sin q\, S^* \\ -\sin q\, S^* & C^* - \cos q\, S^* \end{pmatrix} \tag{9.118}$$

$$\tilde{V}_2(q)^{1/2} = \begin{pmatrix} e^{-K} & 0 \\ 0 & e^K \end{pmatrix}, \tag{9.119}$$

where $C^* = \cosh 2K^*$, $S^* = \sinh 2K^*$. The resulting two-dimensional matrix is therefore written explicitly as

$$\tilde{V}_2(q)^{1/2}\tilde{V}_1(q)\tilde{V}_2(q)^{1/2} = \begin{pmatrix} \mathrm{e}^{-2K}\left(C^* + \cos q\, S^*\right) & -\sin q\, S^* \\ -\sin q\, S^* & \mathrm{e}^{2K}\left(C^* - \cos q\, S^*\right) \end{pmatrix}. \qquad (9.120)$$

The characteristic equation of this two-dimensional matrix reveals that the product of the two eigenvalues is 1 and their sum is $2\cosh 2K \cosh 2K^* - 2\cos q$, according to the relation between the solutions and coefficients of an algebraic equation of second order. The two eigenvalues can be expressed as $\mathrm{e}^{\pm\epsilon(q,K)}$ since their product is unity. Their sum $\mathrm{e}^{\epsilon(q,K)} + \mathrm{e}^{-\epsilon(q,K)}$ satisfies

$$\cosh \epsilon(q, K) = \cosh 2K \cosh 2K^* - \cos q = \cosh 2K \coth 2K - \cos q. \qquad (9.121)$$

The four eigenvalues of $V(q)$ have been evaluated as $1, 1, \mathrm{e}^{\pm\epsilon(q,K)}$. The partition function is therefore

$$Z = g(K)^{LM} \prod_{q\geq 0} \mathrm{Tr}\left(V(q)\right)^M$$

$$= g(K)^{LM} \prod_{q\geq 0} \left(2 + \mathrm{e}^{M\epsilon(q,K)} + \mathrm{e}^{-M\epsilon(q,K)}\right). \qquad (9.122)$$

Since $\epsilon(q, K) > 0$, only the eigenvalue $\mathrm{e}^{M\epsilon(q,K)}$ survives in the large-M limit in the above parentheses, and the free energy per spin is

$$-\beta f = \lim_{L,M\to\infty} \frac{1}{LM} \log Z = \frac{1}{2}\log(2\sinh 2K) + \frac{1}{2\pi}\int_0^\pi \epsilon(q, K)\,\mathrm{d}q, \qquad (9.123)$$

where $\epsilon(q, K)$ is the positive solution of eqn (9.121). This is the exact solution of the two-dimensional Ising model.

Equation (9.123) can be rewritten in a more transparent form using the identity

$$\int_0^\pi \log(2\cosh \epsilon - 2\cos x)\,\mathrm{d}x = \pi\epsilon \qquad (9.124)$$

as

$$-\beta f = \frac{1}{2}\log(2\sinh 2K)$$
$$+ \frac{1}{2\pi^2}\int_0^\pi \mathrm{d}q \int_0^\pi \mathrm{d}x\, \log\left(2\cosh \epsilon(q, K) - 2\cos x\right)$$
$$= \frac{1}{2}\log(2\sinh 2K)$$
$$+ \frac{1}{2\pi^2}\int_0^\pi \mathrm{d}q \int_0^\pi \mathrm{d}x\, \log\left(2\cosh 2K \coth 2K - 2\cos q - 2\cos x\right)$$
$$= \log(2\cosh 2K)$$
$$+ \frac{1}{2\pi^2}\int_0^\pi \mathrm{d}\omega_1 \int_0^\pi \mathrm{d}\omega_2\, \log\left(1 - k_1^2 \cos \omega_1 \cos \omega_2\right), \qquad (9.125)$$

where

$$k_1^2 = \frac{2 \sinh 2K}{\cosh^2 2K}. \tag{9.126}$$

This last expression appears more often than eqn (9.123) in the literature.

9.5.5 Logarithmic divergence of the specific heat

The celebrated logarithmic divergence of the specific heat is derived in this section from eqn (9.123). Let us write the right-hand side of eqn (9.121) as $u(q, K)$ to have

$$\epsilon(q, K) = \log\left(u(q, K) + \sqrt{u(q, K)^2 - 1}\right). \tag{9.127}$$

$u(q, K)$ assumes its minimum as a function of K at K_c, which is the solution of $K = K^*$. As shown below, this is the transition temperature with the value $K_c = 0.4407$ or $T_c = 2.269$ according to $\mathrm{e}^{-2K_c} = \tanh K_c$, which comes from $K = K^*$. The expansion of u under the condition $K \approx K_c, q \approx 0$ is

$$u(q, K) \approx 1 + \frac{q^2}{2} + 8(\Delta K)^2, \tag{9.128}$$

where $\Delta K = K - K_c$. The singular part of $\epsilon(q, K)$ is therefore $\sqrt{q^2 + 16(\Delta K)^2}$. The insertion of this relation into eqn (9.123) and performing the integration yields the following expression for the singular part of the free energy,

$$\int_0^\pi \sqrt{q^2 + 16(\Delta K)^2} \, \mathrm{d}q$$

$$= \frac{1}{2}\left[q\sqrt{q^2 + 16(\Delta K)^2} + 16(\Delta K)^2 \log\left|q + \sqrt{q^2 + 16(\Delta K)^2}\right|\right]_0^\pi$$

$$= -8(\Delta K)^2 \log|\Delta K| + (\text{regular part}). \tag{9.129}$$

We see clearly that the specific heat, the second-order derivative of the free energy with respect to temperature, has a logarithmic singularity at $\Delta K = 0$. It is concluded that the two-dimensional Ising model has a phase transition with a logarithmically divergent specific heat and a critical exponent $\alpha = 0$. Equation (9.129) shows that the critical amplitudes have the same value above and below the transition temperature. Figure 9.10 depicts the temperature dependence of the specific heat.

EXERCISE 9.8 Solve the one-dimensional transverse-field Ising model defined by

$$H = -J\sum_{j=1}^L \sigma_j^z \sigma_{j+1}^z - h\sum_{j=1}^L \sigma_j^x, \tag{9.130}$$

where $\sigma_j^{x,y,z}$ are Pauli operators. The method to diagonalize the transfer matrix of the two-dimensional Ising model applies to this case almost without modifications. Derive the expressions of the free energy and ground-state energy. From the latter, show that the system undergoes a zero-temperature phase transition at $h = J$.

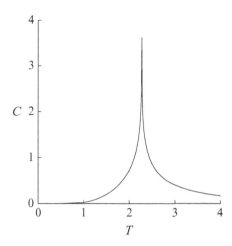

Fig. 9.10 Temperature dependence of the specific heat of the two-dimensional Ising model. The energy unit is $J = 1$.

9.6 Zeros of the partition function

This section discusses the *Yang–Lee theory* of phase transitions. This theory relates singularities of the free energy as a function of the field h with zeros of the partition function in the complex field plane, often termed the *Yang–Lee* (or *Lee–Yang*) *zeros*. The partition function Z is usually positive as it is a sum of exponentials but can vanish if we consider complex values of the field h. The vanishing of the partition function is directly reflected in the singularity of the free energy $F = -T \log Z$ as can be realized from the fact that the logarithmic function is singular only at the origin. The Yang–Lee theory rigorously justifies this intuition and provides a unique point of view on phase transitions for the Ising model and related problems such as gas–liquid transitions.

The following theorems are the central results.

THEOREM 9.1 *Suppose that the partition function of the Ising model with uniform (i.e. bond independent) interactions as a function of the complex field h is free from zeros, $Z(h) \neq 0$, in a region R that contains a segment of the real axis. Assume also that the thermodynamic limit is taken appropriately such that the number of sites on the surface is kept sufficiently smaller than the total number of sites N. Then, $N^{-1} \log Z$ converges to a limit f uniformly in R as $N \to \infty$. As a consequence, the free energy per spin f is not singular in R in the thermodynamic limit.*

THEOREM 9.2[Circle theorem] *Consider the Ising model with two-body ferromagnetic interactions. Zeros of its partition function all lie on the imaginary axis in the complex field h plane, or equivalently, on the unit circle in the plane of $z = \mathrm{e}^{-2\beta h}$.*

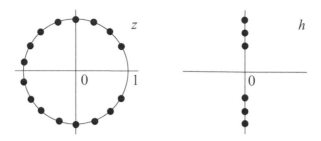

Fig. 9.11 Distribution of zeros of a finite-size ferromagnetic Ising model in the complex-z plane (left panel) and complex-h plane (right panel).

A proof of Theorem 9.1 is given in Appendix A.18. A proof of Theorem 9.2 is more complicated and we refer the interested reader to the original paper.[8] Theorem 9.1 justifies the naive conjecture that the singularity of the free energy is caused by the zeros of the partition function. Theorem 9.2 holds only for models with two-body, ferromagnetic interactions, but the interaction strengths need not be translationally invariant in Theorem 9.2, as long as they are ferromagnetic.

To see the significance of Theorem 9.2, it is convenient to write the partition function Z as a function of $z = \mathrm{e}^{-2\beta h}$ to see that Z is essentially a polynomial of z,

$$
\tilde{Z} \equiv \mathrm{e}^{-\beta h N} Z = \sum_{\{S_i\}} \exp\left(K \sum_{\langle ij \rangle} S_i S_j + \beta h \sum_i (S_i - 1) \right)
$$
$$
= \tilde{Z}_0 + z\tilde{Z}_1 + z^2 \tilde{Z}_2 + \cdots + z^N \tilde{Z}_N, \tag{9.131}
$$

where $\tilde{Z}_k$ stands for

$$
\tilde{Z}_k = {\sum_{\{S_i\}}}' \exp\left(K \sum_{\langle ij \rangle} S_i S_j \right). \tag{9.132}
$$

Here, the summation is over the spin configurations with k down spins. Since $\tilde{Z}_k > 0$, Z cannot vanish for real, positive z and hence the zeros or roots of Z all lie away from the positive real axis in the complex-z plane as long as the system size N is finite, as illustrated in Fig. 9.11. The number of roots increases with N. The zero closest to $z = 1$ ($h = 0$) approaches $z = 1$ ($h = 0$) as the system size grows if the temperature is lower than the critical value.

> **EXERCISE 9.9** Find the locations of the zeros of the partition functions for the single-spin and two-spin Ising models,
>
> $$
> H = -hS_1, \quad H = -JS_1 S_2 - h(S_1 + S_2) \tag{9.133}
> $$

[8] C. N. Yang and T. D. Lee, *Phys. Rev.* **87** (1952) 404 and 410.

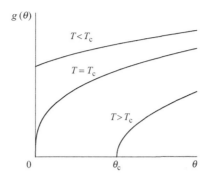

Fig. 9.12 Density of Lee–Yang zeros of the ferromagnetic Ising model along the imaginary axis on the complex field plane. θ_c is the edge of the density above the critical temperature.

in the complex-z plane.

As there are N zeros on the unit circle in the complex-z plane, according to Theorem 9.2 and eqn (9.131), the partition function is expressed as

$$Z = e^{\beta h N} \prod_{k=1}^{N} (z - e^{i\theta_k}). \tag{9.134}$$

Correspondingly, the free energy per spin is written as

$$\frac{F}{N} = -h - \frac{T}{N} \sum_{k=1}^{N} \log(z - e^{i\theta_k}) \longrightarrow -h - T \int_{-\pi}^{\pi} \log(z - e^{i\theta}) g(\theta) \, d\theta, \tag{9.135}$$

where the thermodynamic limit has been taken in the last step, and the density of zeros in this limit has been denoted as $g(\theta)$. This function is normalized as

$$\int_{-\pi}^{\pi} g(\theta) \, d\theta = 1. \tag{9.136}$$

Using the symmetry $g(\theta) = g(-\theta)$, which comes from the symmetry $Z(h) = Z(-h)$, we can rewrite the above equation as

$$f = -h - \frac{T}{2} \int_{-\pi}^{\pi} \log(1 - 2z \cos\theta + z^2) g(\theta) \, d\theta. \tag{9.137}$$

The magnetization is then

$$m(h) = 1 - 2z \int_{-\pi}^{\pi} \frac{z - \cos\theta}{1 - 2z \cos\theta + z^2} g(\theta) \, d\theta$$

$$= \sinh 2\beta h \int_{-\pi}^{\pi} \frac{g(\theta)}{\cosh 2\beta h - \cos\theta} \, d\theta. \tag{9.138}$$

The spontaneous magnetization is therefore

$$m(0+) = \int_{-\pi}^{\pi} 2\pi\delta(\theta)g(\theta)\,\mathrm{d}\theta = 2\pi g(0), \qquad (9.139)$$

where we used the relation

$$\lim_{x\to 0+} \frac{\sinh x}{\cosh x - \cos\theta} = 2\pi\delta(\theta). \qquad (9.140)$$

This last relation can be verified by the expansions of $\sinh x$ and $\cosh x$ for small x and $\cos\theta$ for small θ. Equation (9.139) indicates that the density of zeros at the origin in the complex field plane is directly proportional to the spontaneous magnetization. Thus, $g(0)$ is finite for $T < T_c$, whereas it vanishes for $T > T_c$. In the latter high-temperature region, the density $g(\theta)$ vanishes for $\theta < \theta_c$, where the threshold value θ_c is called the *edge* of the density. The edge approaches the origin $\theta = 0$ as the temperature decreases toward the critical point, $\theta_c \to 0+$ as $T \to T_c + 0$. The overall qualitative behavior of the density is depicted in Fig. 9.12.

In general, the Yang–Lee theory provides a framework to characterize a phase transition but does not establish analytically the presence or lack of a singularity in the partition function in the thermodynamic limit.

EXERCISE 9.10 Show that the density of zeros $g(\theta)$ behaves as $\theta^{1/\delta}$ near the origin $\theta = 0$ at the critical point $T = T_c$. For this purpose, first expand $\sinh(\cdot)$, $\cosh(\cdot)$ and $\cos(\cdot)$ in eqn (9.138) for their small arguments assuming that the singular behavior of m for small h is dominated by the values of the integrand at small θ as this is where the integrand becomes largest. Then, insert the functional form $g(\theta) = \theta^a$ to the integrand and verify that $m \propto h^{1/\delta}$ results only when $a = 1/\delta$.

10
Duality

Exact solutions of model systems are the most reliable source of information in the theory of phase transitions and critical phenomena, as typically exemplified in the two-dimensional Ising model. However, there are not many model systems that can be exactly solved and, consequently, many approximate schemes have been developed. It is sometimes possible, though, to extract exact information without directly solving those model systems. In two dimensions in particular, arguments using duality transformations make it possible to derive the exact location of the phase-transition point and the exact value of the energy at the transition point. These remarkable results can be obtained by much simpler arguments than the direct solutions. Duality not only determines the exact location of the transition point of the two-dimensional Ising model and related models but also is useful to rewrite the XY model into a different form, which reveals new physical aspects of the system.

10.1 Classical duality

Duality in classical statistical mechanics usually means a transformation that relates the partition functions of two distinct model systems under the replacement of the value of temperature by another value. When the two model systems involved are the same, the duality transformation is known as *self-duality*. Self-dualities are mappings between the high- and low-temperature phases of the model that allow us, for instance, to determine the location of the phase transition point when there is a unique singularity in the free energy.

It is convenient to write the partition function, a function of $K = J/T$, as $Z(K)$. If we write the conclusion first, the duality for the two-dimensional Ising model on the square lattice with periodic boundaries without external field means that the partition function satisfies the following relation

$$\frac{Z(K)}{2^N (\cosh K)^{2N}} = \frac{Z(K^*)}{2\,\mathrm{e}^{2NK^*}},$$

(10.1)

or, equivalently,[1]

$$\frac{Z(K)}{(\sinh 2K)^{N/2}} = \frac{Z(K^*)}{2(\sinh 2K^*)^{N/2}}.$$

(10.2)

---

[1] The factor 2 in the denominator of the right-hand side introduces a sort of asymmetry in the self-dual relation, and is related to the particular periodic boundary conditions used in both spatial directions. Boundary effects play no role in the thermodynamic limit, $N \to \infty$.

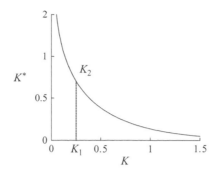

Fig. 10.1 The dual coupling K^* as a function of the original coupling K. The value of the partition function at some K, K_1 for example, is equal to the value of the partition function at K_2 apart from a trivial factor.

Here, N is the number of lattice sites, $2N$ is the number of bonds (equal to the number of nearest-neighbor interactions), and the *dual coupling* K^* is a monotonically decreasing function of K defined by

$$e^{-2K^*} = \tanh K. \tag{10.3}$$

See Fig. 10.1. This is indeed an example of a self-dual mapping since it relates the same Ising model, i.e. the partition function of the same Ising model at two different temperatures.

The duality transformation of the coupling (interaction strength), eqn (10.3), transforms the high-temperature region (small K) to the low-temperature region (large K^*) and vice versa. Thus, the duality relation (10.1) implies that the values of the partition function are essentially equal at high and low temperatures, apart from the trivial factors in the denominators. For example, two systems, one with coupling K_1 and the other with $K_2(= K^*(K_1))$ of Fig. 10.1, have essentially the same partition function. This fact leads to important consequences for the singularity of the free energy.

Let us take the logarithm of both sides of eqn (10.1) and divide the result by the number of spins,

$$\frac{1}{N} \log Z(K) = \frac{1}{N} \log Z(K^*(K)) + (\text{regular part}). \tag{10.4}$$

In the thermodynamic limit $N \to \infty$, the function $\log Z/N$ has a singularity at the phase transition point as this quantity is essentially equal to the free energy. Equation (10.4) indicates that, if the left-hand side is singular at K_c, so is the right-hand side at $K^*(K_c)$. Then, if it happens that $K_c \neq K^*(K_c)$, the function $\lim_{N\to\infty} \log Z(K)/N$ is singular at K_c and $K^*(K_c)$, two different singular points. Therefore, we conclude that $K_c = K^*(K_c)$ as long as the system is singular at a unique transition point. The transition point of the two-dimensional Ising model is therefore given by

$$e^{-2K_c} = \tanh K_c, \tag{10.5}$$

which is eqn (10.3) with $K = K^* = K_c$. This relation is solved for e^{-2K_c} as $e^{-2K_c} = \sqrt{2} - 1$, in agreement with the direct exact solution in Section 9.5.

The duality relation allows us to evaluate the energy at the transition point. The logarithm of both sides of eqn (10.1) divided by N is

$$\frac{1}{N} \log Z(K) - \log 2 - 2 \log \cosh K = \frac{1}{N} \log Z(K^*) - 2K^*, \tag{10.6}$$

where the additional term $-(1/N) \log 2$ has been ignored as it vanishes in the thermodynamic limit. A differentiation of this equation with respect to K yields

$$\frac{1}{N} \frac{Z'(K)}{Z(K)} - 2 \tanh K = \left(\frac{1}{N} \frac{Z'(K^*)}{Z(K^*)} - 2 \right) \frac{\mathrm{d}K^*}{\mathrm{d}K}. \tag{10.7}$$

Since the logarithmic derivative of the partition function $Z'(K)/Z(K)$ is equal to minus the energy $-E(K)/J (\equiv -NE_0(K))$, we have

$$-E_0(K) - 2 \tanh K = \left(-E_0(K^*) - 2 \right) \frac{\mathrm{d}K^*}{\mathrm{d}K}. \tag{10.8}$$

If the fixed point of the duality, the self-dual point, coincides with the transition point, $K = K^* = K_c$, and if the energy is a continuous function of K at this point, then $E_0(K)$ and $E_0(K^*)$ in the above relation should share the same value $E_0(K_c) \equiv E_{0c}$. At the fixed point we have $\tanh K_c = \sqrt{2} - 1$ and $\mathrm{d}K^*/\mathrm{d}K = -1$ from eqn (10.3). These relations together with eqn (10.8) lead to $E_{0c} = -\sqrt{2}$, in agreement with the exact solution, as can be verified from the exact free energy given in Section 9.5.

It is also possible to prove from duality that the specific heat is either divergent or continuous at the transition point, which excludes the possibility that the specific heat has a jump, as in the mean-field theory of Chapter 2. A further differentiation of eqn (10.7) with respect to K leads to

$$\frac{1}{N} \left\{ \frac{Z''(K)}{Z(K)} - \left(\frac{Z'(K)}{Z(K)} \right)^2 \right\} - \frac{2}{\cosh^2 K}$$

$$= \frac{1}{N} \left\{ \frac{Z''(K^*)}{Z(K^*)} - \left(\frac{Z'(K^*)}{Z(K^*)} \right)^2 \right\} \left(\frac{\mathrm{d}K^*}{\mathrm{d}K} \right)^2 + \left(\frac{1}{N} \frac{Z'(K^*)}{Z(K)} - 2 \right) \frac{\mathrm{d}^2 K^*}{\mathrm{d}K^2}. \tag{10.9}$$

The quantity in curly brackets $\{\cdots\}$ appearing on both sides is the K- or K^*-derivative of the energy per spin (the minus sign of it, strictly speaking) and is equal to the specific heat per spin C_0 (the T-derivative of the energy) multiplied by T^2. We collect those quantities related to the specific heat in the left-hand side,

$$T^2 C_0(K) - (T^*)^2 C_0(K^*) \left(\frac{\mathrm{d}K^*}{\mathrm{d}K} \right)^2 = \frac{2}{\cosh^2 K} - (2 + E_0(K^*)) \frac{\mathrm{d}^2 K^*}{\mathrm{d}K^2}. \tag{10.10}$$

The quantities on the right-hand side have the following values at the transition point, $\cosh^2 K_c = (1 + \sqrt{2})/2$, $E_0(K_c) = -\sqrt{2}$ and $\mathrm{d}^2 K^*/\mathrm{d}K^2 = 2\sqrt{2}$. Thus, the right-hand side vanishes. The quantity $(\mathrm{d}K^*/\mathrm{d}K)^2$ on the left-hand side is unity at the transition point. Therefore, as the temperature approaches the transition point, there exist two possibilities, either $C_0(K)$ approaches $C_0(K^*)$ (continuous specific heat) or both diverge. It is forbidden that $C_0(K)$ and $C_0(K^*)$ approach different values from both sides of the transition point, as in the mean-field theory or the Bethe approximation.

It is impossible to determine the value of the critical exponent from the duality relation. We are, nevertheless, able to show that the critical exponents and critical amplitudes at both sides of the transition point coincide. According to eqn (10.4) the singular part of the free energy per spin f_s is an invariant quantity under the duality transformation,

$$f_s(T) = f_s(T^*(T)) + (\text{regular part}), \qquad (10.11)$$

where T^* is J/K^*. Let us assume that this singular part behaves near the transition point $(T \approx T_c + ct \ (t = (T - T_c)/T_c))$ as

$$f_s(T) \approx A_\pm |t|^{2-\alpha_\pm}. \qquad (10.12)$$

Here, $A_\pm$ and $\alpha_\pm$ are the critical amplitudes and critical exponents of the specific heat above and below the transition point, respectively. The dual temperature behaves around the transition point as

$$T^*(T) \approx T^*(T_c + ct) \approx T_c + c(T^*)'_c t, \qquad (10.13)$$

where $(T^*)'_c$ is the derivative of $T^*(T)$ at T_c and it is -1 according to eqn (10.3). From eqns (10.13), (10.11) and (10.12) we find

$$A_+ |t|^{2-\alpha_+} \approx A_- |t|^{2-\alpha_-} \qquad (10.14)$$

assuming $t > 0$. This equation shows $A_+ = A_-$ and $\alpha_+ = \alpha_-$.

Therefore, as has been shown, duality is an important mathematical tool to obtain information on the system properties. In the rest of this chapter we prove the duality relation (10.1) and its generalization to models other than the Ising model. The next section will introduce a graphical discussion on the derivation of duality using high- and low-temperature series expansions. More general arguments will be developed in later sections. The last section introduces duality relations in quantum systems.

10.2 High- and low-temperature series expansions

The technique of series expansions can be applied to arbitrary discrete spin models, and together with fast computers, it is a powerful means to study critical phenomena. The main idea behind those expansions is to start from an exactly known limit and expand in terms of graphs around that limit. The low-temperature series expansion starts from the ground state and includes the low-energy excitations, while the high-temperature series starts from the totally disordered state. The high-temperature series expansion can also be easily applied to continuous-spin systems. In that case, one basically expands the partition function in powers of β

$$Z = \text{Tr } e^{-\beta H} = \text{Tr} \left(1 - \beta H + \frac{(\beta H)^2}{2} + \cdots \right), \qquad (10.15)$$

or, similarly, the expectation value of any arbitrary physical quantity.

In this section we are interested in applying the series-expansion technique to the Ising model but in the context of deriving the self-duality of the two-dimensional model, eqn (10.1).[2] The latter duality relation can be derived by the correspondence between high-temperature and low-temperature expansions of the partition function. A more general framework will be developed in the next section, based on the technique of Fourier transforms, in which the result of the present section is included. It should, nevertheless, be useful to learn the basic knowledge of series expansions. Also, the present graphical derivation is easier to understand intuitively than the formal algebraic discussions.

The first topic is the high-temperature expansion of the partition function. Since $S_i S_j = \pm 1$ for Ising spins, the identity

$$e^{KS_i S_j} = \cosh K + S_i S_j \sinh K \tag{10.16}$$

holds, from which the following expression of the partition function results,

$$
\begin{aligned}
Z(K) &= \sum_{\{S_i\}} e^{K \sum_{\langle ij \rangle} S_i S_j} \\
&= \sum_{\{S_i\}} \prod_{\langle ij \rangle} (\cosh K + S_i S_j \sinh K) \\
&= (\cosh K)^{2N} \sum_{\{S_i\}} \prod_{\langle ij \rangle} (1 + S_i S_j \tanh K).
\end{aligned}
\tag{10.17}
$$

We have used $\prod_{\langle ij \rangle} \cosh K = (\cosh K)^{2N}$ in the final equality since we assume periodic boundary conditions on the square lattice, where the total number of bonds $\langle ij \rangle$ is $2N$. The expansion of the product over nearest-neighbor pairs in the final expression of eqn (10.17) yields either 1 or $S_i S_j \tanh K$ for each pair $\langle ij \rangle$. Let us draw a bold line between site i and site j in the latter case for $S_i S_j \tanh K$ and do nothing in the former case of 1. Then, each term of the expansion of the product is expressed as a combination of bold lines, as in Fig. 10.2. Each bold line carries the weight $\tanh K$, and we may order the graphs by the number of bold lines in the graphs, which results in a series expansion of the partition function in powers of $\tanh K$.

Another factor to consider is the sum over spin configurations, $\sum_{\{S_i\}}$, in eqn (10.17). If a term in the expansion has an even power of S_i, it does not vanish since $S_i^{2n} = 1$, with n an integer, and hence $\sum_{\{S_i\}} S_i^{2n}$ is positive. A term of odd power, in contrast, vanishes since $S_i^{2n+1} = S_i$ and $\sum_{\{S_i\}} S_i = 0$. For example, all terms but the 0th order in Fig. 10.2 have a factor S_i at the end of a bold line and thus vanish. Non-vanishing contributions come only from closed graphs, as illustrated in Fig. 10.3, in which each spin S_i appears an even number of times (none, twice, or four times).

The series expansion in powers of $\tanh K$ is the *high-temperature expansion* because smaller K (higher temperature) corresponds to smaller $\tanh K$. Any S_i appears an even number of times in a closed graph and contributes with unity, giving

[2] Hereafter, high- and low-temperature series expansions will be called high- and low-temperature expansions for simplicity.

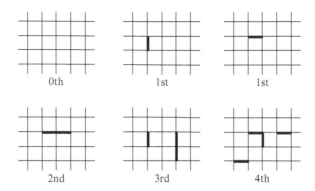

Fig. 10.2 A graphical representation of some of the terms appearing in the expansion of the partition function of the Ising model. A bold line is for the factor $S_i S_j \tanh K$. All graphs drawn here but the 0th order do not contribute to the high-temperature expansion. The order of the expansion is given by the power of $\tanh K$.

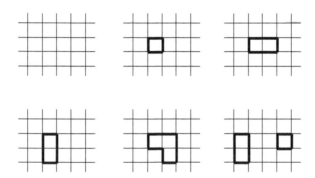

Fig. 10.3 Closed graphs that contribute to the high-temperature expansion of the partition function. A bold line expresses the factor $\tanh K$.

$\sum_{\{S_i\}} 1 = 2^N$. Hence, the high-temperature expansion is symbolically written as

$$\frac{Z(K)}{2^N (\cosh K)^{2N}} = \sum_{n=0,2,3,\cdots} (\text{the number of closed graphs with } 2n \text{ bold lines})(\tanh K)^{2n}.$$

(10.18)

The term $n = 1$ is missing in the summation because any closed graph has more than two bold lines. For example, the coefficient of $(\tanh K)^4$ is the number of ways to draw unit squares (plaquettes) on a square lattice. Since each unit square can be represented by the site on the left-lower corner, the number of unit squares is equal to the number of sites N.[3]

---

[3] Corrections are necessary if we use free boundary conditions. The number of unit squares is slightly smaller than the number of sites in such a case. We do not go into those details because we

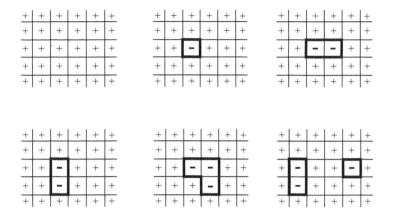

Fig. 10.4 Representation of the low-temperature expansion by closed graphs. Each bold line separates neighboring antiparallel spins and carries the factor e^{-2K}.

We next consider the *low-temperature expansion* of the partition function. To write the result first, a graphical representation of the low-temperature expansion has exactly the same set of graphs as in the high-temperature expansion. This makes it possible to find a one-to-one correspondence between the terms of the two types of series expansions. We can then equate the partition functions written in terms of $\tanh K$ (high-temperature expansion) and of e^{-2K} (low-temperature expansion), the latter being the small parameter of the low-temperature expansion, leading to the duality relation of eqn (10.1). To show this result, it is useful to choose an appropriate order in taking the summation in the definition of the partition function

$$Z(K) = \sum_{\{S_i\}} e^{K \sum_{\langle ij \rangle} S_i S_j}. \tag{10.19}$$

Let us first pick the all-up configuration ($S_i = 1$, $\forall i$), whose contribution is e^{2NK}, corresponding to the upper-left graph of Fig. 10.4. The correct factor is indeed twice this, $2e^{2NK}$, because the all-down configuration ($S_i = -1$, $\forall i$) gives the same factor. These are ground-state configurations. The next configuration to be considered in the sum of eqn (10.19) is a state with a single spin flipped from the all-up state, the upper-middle graph of Fig. 10.4. The inverted spin has the interaction energy raised from $-J(= -J \cdot 1 \cdot 1)$ to $J(= -J \cdot 1 \cdot (-1))$ for each bond, and there are four of them around a site on the square lattice. Thus, the factor $e^{-4 \cdot 2K}$ multiplies the ground-state Boltzmann factor. Since there are N ways to flip a single spin, the first two terms of the low-temperature expansion are[4]

$$Z(K) = 2\,e^{2NK}(1 + Ne^{-8K} + \cdots). \tag{10.20}$$

will be interested in the system behavior in the thermodynamic limit, $N \to \infty$, in which boundary effects are expected to play no role.

[4] The overall factor 2 is the one responsible for the asymmetry in eqn (10.2).

Each term inside the parentheses has its corresponding term in the high-temperature expansion of eqn (10.18). A graphical expression of the low-temperature expansion helps us understand this correspondence. A term of the low-temperature expansion is expressed as a graph with inverted spins surrounded by up spins. If we draw bold lines between neighboring antiparallel spins, we obtain a set of closed graphs, as in Fig. 10.4. It should be clear that a graph in Fig. 10.4 has its unique counterpart in Fig. 10.3. Then, if we replace e^{-2K} with $\tanh K$ in each term of the low-temperature expansion of $Z(K)/(2e^{2NK})$, eqn (10.20), we obtain the high-temperature expansion of $Z(K)/(2^N(\cosh K)^{2N})$, eqn (10.18). This completes the proof of the duality relation (10.1) and the law of change of the temperature parameter (10.3).

Series expansions are often formulated not for the partition function but for its logarithm, the free energy, and its derivatives like the susceptibility. Although it is often difficult to evaluate the coefficients of high-order terms of an expansion, methods to extrapolate the result of finite-order calculations to infinite order have been developed and have turned out to be powerful tools to estimate the critical points and critical exponents. Types of graphs in these expansions of the free energy and its derivatives are known to be a little different from the present case of the partition function.

EXERCISE 10.1 Determine the coefficient of the sixth-order term $(\tanh K)^6$ in the high-temperature expansion of the partition function of the Ising model on the square lattice.

We next apply these techniques to the Z_2 lattice gauge theory introduced in Section 7.7, again in the context of duality.[5] The Hamiltonian is the sum of four-spin interactions,

$$H = -J \sum_\square S_j S_k S_l S_m \qquad (10.21)$$

with each Ising spin residing on a bond, as depicted in Fig. 7.5. The lattice may not necessarily be the square lattice. The summation extends over all plaquettes (unit squares) of the lattice. The partition function can be expanded as in eqn (10.17),

$$Z(K) = \sum_{\{S_i\}} e^{K \sum_\square S_j S_k S_l S_m}$$

$$= (\cosh K)^{N_p} \sum_{\{S_i\}} \prod_\square (1 + S_j S_k S_l S_m \tanh K), \qquad (10.22)$$

where N_p is the total number of plaquettes. Similarly to the high-temperature expansion of the usual Ising model, the spin variables should appear an even number of times if the summation over the spin values $\{S_i = \pm 1\}$ is to give a non-vanishing contribution. Since spins are located on bonds, this constraint for finite contributions

---

[5] This part can be skipped on a first reading unless the reader is interested in the generalization of duality techniques to systems with many-body interactions in higher dimensions.

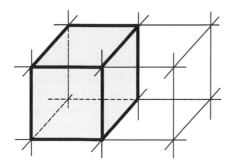 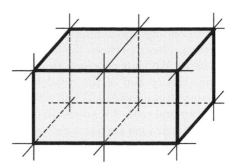

Fig. 10.5 Two leading-order graphs of the high-temperature expansion of the three-dimensional Z_2 gauge theory. A bold line represents a bond where spin variables appear twice to give unity, $S_j^2 = 1$. A shaded plaquette corresponds to a bond drawn bold in Fig. 10.3 and carries the weight $\tanh K$.

requires any bond to appear an even number of times in the expansion of the product of eqn (10.22). In two dimensions, the expansion of eqn (10.22) can be expressed by graphs of the type of Fig. 10.3, where each plaquette inside the closed bold lines represents a factor $S_j S_k S_l S_m \tanh K$. Spin variables are located on all bold lines of the diagrams and thus the summation over their values, ± 1, gives zero. Spin variables on bonds inside a closed graph (written as thin lines in Fig. 10.3) appear twice, yielding $S_j^2 = 1$, because a bond on the square lattice is shared by two neighboring plaquettes. In this way, any graph in Fig. 10.3, except for the trivial one at the upper left corner, vanishes. Therefore, the partition function of the two-dimensional Z_2 gauge theory is trivially given by

$$Z(K) = (\cosh K)^{N_p} \cdot 2^{2N}. \tag{10.23}$$

The corresponding free energy has no singularity and the system does not have a phase transition at finite temperature.

In three dimensions, non-trivial contributions exist in the high-temperature expansion. The leading term comes from the product of four-spin interactions over six plaquettes surrounding a unit cube, as depicted in the left graph of Fig. 10.5. The next term is the right graph of Fig. 10.5, in which ten plaquettes form a product over two unit cubes. The essential difference from the two-dimensional case is the possibility to generate closed objects surrounded by plaquettes. They are analogous to the closed polygons in the high-temperature expansion of the usual Ising model illustrated in Fig. 10.3. The weight of the left graph of Fig. 10.5 is $(\tanh K)^6$ since there are six plaquettes surrounding the unit cube. The right graph has a weight $(\tanh K)^{10}$.

Each of these graphs in the high-temperature expansion of the three-dimensional Z_2 gauge theory has its counterpart in the low-temperature expansion of the usual Ising model in three dimensions. The left graph of Fig. 10.5, for instance, corresponds to a down spin at the center of the cube surrounded by up spins everywhere else (dual lattice). In general, the sites of the dual Ising model are located at the centers of the cubes. The right graph, similarly, has two down spins inside the closed object in the corresponding low-temperature expansion of the Ising model. The analogy with the correspondence between the high- and low-temperature expansions of Fig. 10.4 should be clear. We thus conclude that the partition functions of the three-dimensional Z_2 gauge theory and of the usual Ising model are related by a duality relation similar to eqns (10.1) and (10.3) but with different partition functions on the two sides of the equality and with $3N$ bonds. In other words, the Ising and Z_2 gauge theory models in three dimensions are not self-dual models. Since the three-dimensional Ising model has a phase transition at finite temperature, i.e. a singularity in the free energy, it readily follows that the three-dimensional Z_2 gauge theory also undergoes a phase transition at finite temperature.

10.3 Duality by Fourier transformation

The graphical derivation of duality relations is intuitively appealing and relatively easy to understand, but it is not straightforward to generalize to models other than the Ising model. Another formulation based on Fourier transformation, albeit a little abstract, is more suitable to apply systematically to a wide class of model systems including the Ising model. We explain this method in the present section.

10.3.1 General form of the partition function

Suppose that a spin variable $\xi_i(=1,2,\cdots q)$ is assigned to site i. The Boltzmann factor $u(\xi_i,\xi_j)$ for the neighboring pair $\langle ij \rangle$ will be assumed to be a function only of the difference between ξ_i and ξ_j and periodic with period q,

$$u(\xi_i,\xi_j) = u(\xi_i - \xi_j) \ (\text{mod } q). \tag{10.24}$$

The partition function is written as

$$Z = \sum_{\{\xi_i\}} \prod_{\langle ij \rangle} u(\xi_i - \xi_j). \tag{10.25}$$

As an example, the q-state Potts model

$$H = -J\sum_{\langle ij \rangle} \delta(\xi_i,\xi_j) \tag{10.26}$$

has $u(0) = e^K$ and $u(\xi_i - \xi_j) = 1 \ (\xi_i - \xi_j \neq 0)$. The q-state clock model

$$H = -J\sum_{\langle ij \rangle} \cos(\theta_i - \theta_j) \ \left(\theta_i = \frac{2\pi\xi_i}{q}\right) \tag{10.27}$$

has the Boltzmann factor $u(\xi_i - \xi_j) = \exp\left(K\cos(2\pi(\xi_i - \xi_j)/q)\right)$. Both of these models reduce to the Ising model when $q = 2$. Equation (10.25) is therefore a generalization of the Ising model.

Duality of the partition function (10.25) is derived by the Fourier transformation of the Boltzmann factor at each bond. As a preparation, it is convenient to Fourier

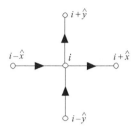

Fig. 10.6 The sign of the difference of spin variables $\xi_i - \xi_j$ is determined by the sense of arrows pointing up and right.

transform $u(\xi_i - \xi_j)$ using its periodicity as

$$u(\xi_i - \xi_j) = \frac{1}{q} \sum_{\eta_{ij}=1}^{q} \exp\left(2\pi i \frac{\xi_i - \xi_j}{q} \eta_{ij}\right) \lambda(\eta_{ij}). \tag{10.28}$$

An outstanding advantage of this expression is that the dependence on ξ is very simple on the right-hand side and we can sum the Boltzmann factor over spin variables easily. The partition function can then be written as a product of λs after summation over ξ, and this turns out to be simply the duality transformation.

10.3.2 Duality transformation

To be concrete, let us consider the case of the square lattice to illustrate the procedure and consequences of the summation over ξ. For a given i, the variable ξ_i appears in the Boltzmann factors for interactions with four neighboring sites. As shown in Fig. 10.6, we choose the sense of arrows as up and right, and assign the sign such that i is at the head of an arrow and j is at the tail in $\xi_i - \xi_j$.[6] Then, ξ_i appears as follows according to eqn (10.28),

$$\exp \frac{2\pi i}{q} \Big\{ (\xi_{i+\hat{x}} - \xi_i)\eta_{i+\hat{x},i} + (\xi_i - \xi_{i-\hat{x}})\eta_{i,i-\hat{x}}$$

$$+ (\xi_{i+\hat{y}} - \xi_i)\eta_{i+\hat{y},i} + (\xi_i - \xi_{i-\hat{y}})\eta_{i,i-\hat{y}} \Big\}, \tag{10.29}$$

where $\hat{x}$ and $\hat{y}$ are the unit vectors along the x- and y-axes, respectively.[7] This expression permits us to take the sum over $\xi_i (= 1, 2, \cdots, q)$ easily to give the constraint

$$-\eta_{i+\hat{x},i} + \eta_{i,i-\hat{x}} - \eta_{i+\hat{y},i} + \eta_{i,i-\hat{y}} = 0 \quad (\text{mod } q). \tag{10.30}$$

If all the Fourier variables η satisfy this constraint, the sum over ξ_i for a given i is just the sum of 1 over $\xi_i = 1, 2, \cdots, q$, which gives q, and the total result is the factor q^N. Consequently, the partition function is represented as the sum of the product of

---

[6] It is not essential how to choose the sense of arrows. Nevertheless, it is useful to assign the sense systematically in order to keep the discussions transparent.

[7] The lattice constant is chosen to be unity.

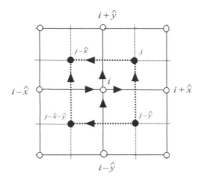

Fig. 10.7 New spin variables are defined on dual lattice sites shown in black dots and the original variable η_{ij} is written in terms of the difference of neighboring spins on the dual lattice, which is also a square lattice.

$\lambda(\eta_{ij})$ with the restriction of eqn (10.30),

$$Z = q^{N-2N} \sum_{\{\eta_{ij}\}}{}' \prod_{\langle ij \rangle} \lambda(\eta_{ij}). \tag{10.31}$$

The term $-2N$ in the exponent of q comes from the factor $1/q$ on the right-hand side of eqn (10.28), applied to all $2N$ bonds.[8] The prime ($'$) denotes the restriction that the sum over η includes only those terms satisfying eqn (10.30) for all i.

Figure 10.6 allows us to present a transparent interpretation of the set of η variables satisfying the constraint of eqn (10.30). η_{ij} is a bond variable with the sense from site i to site j. Equation (10.30) indicates that the sum of incoming η_{ij} is equal to the sum of outgoing η_{ij}. If we regard η_{ij} as an electric current flowing along the bond $\langle ij \rangle$, eqn (10.30) suggests that the current is conserved at each site and hence the field is free of divergence. As is well known in vector analysis, a divergence-free field can be represented in terms of the curl of another field. The present field is defined on lattice sites and assumes only integer values. It, nevertheless, has similar properties to the case of continuous fields.

To show this fact, we place a site of the *dual lattice* at the center of a plaquette (a unit square) and define a variable $\mu_j(= 1, 2, \cdots, q)$ there, see Fig. 10.7. The two sets of variables η and μ are chosen to be related as

$$\eta_{i+\hat{x},i} = \mu_j - \mu_{j-\hat{y}}, \quad \eta_{i+\hat{y},i} = \mu_{j-\hat{x}} - \mu_j,$$

$$\eta_{i,i-\hat{x}} = \mu_{j-\hat{x}} - \mu_{j-\hat{x}-\hat{y}}, \quad \eta_{i,i-\hat{y}} = \mu_{j-\hat{x}-\hat{y}} - \mu_{j-\hat{y}}, \tag{10.32}$$

all of which are valid modulus q. The sign of the difference of the variables μ on the right-hand sides is determined as follows. We first rotate the arrow of the original bond η by $+90°$ and subtract the μ at the tail of the arrow from the μ at the head. It is straightforward to check that the divergence-free condition (10.30) is

[8] We are assuming periodic boundary conditions.

satisfied. Equation (10.32) implies that the curl of the two-dimensional field μ gives the field η.[9]

The constraint on η, eqn (10.30), is automatically satisfied if we rewrite η in terms of μ at all sites, and the partition function

$$Z = q^{-1-N} \sum_{\{\mu_i\}} \prod_{\langle ij \rangle} \lambda(\mu_i - \mu_j) \tag{10.33}$$

results.[10] Here, $\langle ij \rangle$ stands for the neighboring pair on the dual lattice. The partition function has thus been shown to have two different expressions, eqn (10.25) and its dual eqn (10.33). This is the generalized form of duality.

It is possible to apply the above argument to two-dimensional lattices other than the square lattice. We first place a dual lattice site at the center of a unit polygon surrounded by bonds. After fixing the sense of bonds on the original lattice, we determine the sense of dual bonds by rotating the original bonds by $+90°$. The difference of dual variables is fixed by the sense of the bonds. The argument of the dual Boltzmann factor $\lambda(\eta_{ij})$ is replaced by the difference $\mu_i - \mu_j$, and the total Boltzmann factor of the dual system is represented by the product of local dual Boltzmann factors. The partition function is the sum of this total Boltzmann factor over the dual variables μ, which coincides with the original partition function up to a trivial factor.

Apparently, in many instances, the essence of duality is the Fourier transformation of the local Boltzmann factor. The lattice structure is arbitrary. Regular lattices like the square, triangular and hexagonal lattices are of course included, and irregular lattices with non-uniform coordination number are also allowed. It should, however, be remembered that the lattice structure generally changes after a duality transformation. The square lattice is one of the exceptions, in which the dual lattice is another square lattice. The dual of the triangular lattice is the hexagonal lattice. The boundary condition also changes after duality, except for the case of periodic boundaries. The concept of duality applies also to other spatial dimensions. For example, in three dimensions, a bond of the original lattice corresponds to a plaquette of the dual lattice. As a consequence, the dual system is not the usual Ising model but the Z_2 gauge theory as discussed at the end of the previous section.

10.3.3 Ising model

Let us confirm that the general duality relation

$$Z = \sum_{\{\xi_i\}} \prod_{\langle ij \rangle} u(\xi_i - \xi_j) = q^{-1-N} \sum_{\{\mu_i\}} \prod_{\langle ij \rangle} \lambda(\mu_i - \mu_j) \tag{10.34}$$

---

[9] Equation (10.32) indicates that the x-component of η, $\eta_{i+\hat{x},i}$, is the difference of μ along the y-axis, $\mu_j - \mu_{j-\hat{y}}$, and the y-component of η, $\eta_{i+\hat{y},i}$, is the minus sign of the difference of μ along the x-axis, $-(\mu_j - \mu_{j-\hat{x}})$. This is the discrete, or lattice, curl in two dimensions, the continuum limit of which is $\partial_y \mu$ for the x-component and $-\partial_x \mu$ for the y-component of the vector $\boldsymbol{\eta}$, and hence the field is divergence free, $\partial_x \eta_x + \partial_y \eta_y = 0$.

[10] The overall factor q^{-1} is there to remove the extra degree of freedom in μ to change all μs simultaneously by a constant according to eqn (10.32), which makes the sum over μ in eqn (10.33) redundant by the factor q.

reproduces the duality for the Ising model ($q = 2$) derived previously by the high- and low-temperature series expansions.

The interaction between two neighboring spins takes the values $\pm J$, and accordingly the local Boltzmann factor results $u(0) = e^K$ for the spin-parallel ($S_i = S_j$, i.e. $\xi_i - \xi_j = 0 \pmod 2$) and $u(1) = e^{-K}$ for the antiparallel cases ($S_i = -S_j$, $\xi_i - \xi_j = 1 \pmod 2$). Application of the inverse Fourier transformation to eqn (10.28)

$$\lambda(\eta_{ij}) = \sum_{\xi_{ij}=1}^{q} \exp\left(-2\pi i \frac{\xi_{ij}}{q} \eta_{ij}\right) u(\xi_{ij}) \tag{10.35}$$

for the Ising ($q = 2$) case gives

$$\lambda(0) = u(0) + u(1) = e^K + e^{-K}, \quad \lambda(1) = u(0) - u(1) = e^K - e^{-K}. \tag{10.36}$$

Thus, the dual Boltzmann factor $\lambda(0)$ for the same state of the neighboring dual variables is $e^K + e^{-K}$, and $\lambda(1) = e^K - e^{-K}$ for different states. Since the ratio of the Boltzmann factors for the two states of neighboring spins (parallel and antiparallel) is $u(1)/u(0) = e^{-2K}$ on the original lattice, it is reasonable to define the dual coupling K^* by $\lambda(1)/\lambda(0) \equiv e^{-2K^*}$. Then, the relation $e^{-2K^*} = \tanh K$ follows, which is the duality relation of eqn (10.3).

Equation (10.1) for the partition function can also be derived from the general framework. We first replace the mod 2 variable ξ_i with the conventional $S_i(= \pm 1)$ and similarly for the dual variable, $\sigma_i(= \pm 1)$ instead of μ_i. Then, eqn (10.34) reads

$$\sum_{\{S_i\}} e^{K \sum_{\langle ij \rangle} S_i S_j} = 2^{-1-N} \sum_{\{\sigma_i\}} e^{\sum_{\langle ij \rangle} (K^* \sigma_i \sigma_j + a)}. \tag{10.37}$$

The constant a comes from the ambiguity of a multiplicative factor in the dual Boltzmann factor that was defined only by the ratio of $\lambda(1)$ and $\lambda(0)$, $\lambda(1)/\lambda(0) = e^{-2K^*}$. To specify this constant we note that $\lambda(0) = e^K + e^{-K}$ is equal to e^{K^*+a},[11]

$$e^a = \frac{2\cosh K}{e^{K^*}}. \tag{10.38}$$

Then, eqn (10.37) is rewritten as

$$\sum_{\{S_i\}} e^{K \sum_{\langle ij \rangle} S_i S_j} = 2^{-1-N} \sum_{\{\sigma_i\}} e^{K^* \sum_{\langle ij \rangle} \sigma_i \sigma_j} \left(\frac{2\cosh K}{e^{K^*}}\right)^{N_B^*}, \tag{10.39}$$

where N_B^* is the number of bonds on the dual lattice.

The above discussion applies not only to the square lattice but also to arbitrary lattices. In particular, the square lattice is self-dual and has $N_B^* = 2N$, from which we conclude that eqn (10.39) reduces to eqn (10.1). The self-duality manifests itself as the same function Z on both sides of eqn (10.1). Non-self-dual lattices such as the triangular lattice will have a different function Z^* on the right-hand side from Z on the left-hand side. The duality relation holds in such cases as well but it relates different functions,

---

[11] This also means that $\lambda(1) = e^K - e^{-K}$ is equal to e^{-K^*+a}.

Fig. 10.8 The triangular and hexagonal lattices are mutually dual.

$$Z(K) = \frac{(2\cosh K)^{N_{\mathrm{B}}}}{2^{1+N} \mathrm{e}^{N_{\mathrm{B}} K^*}} Z^*(K^*). \tag{10.40}$$

It is therefore impossible, from eqn (10.40), to identify the unique transition point using the argument of Section 10.1.

A similar theory can be developed for the Potts model to derive the transition point on the square lattice.

EXERCISE 10.2 Apply the general theory of duality to the Potts model and derive the duality relations corresponding to eqns (10.1) and (10.3). Identify the transition point on the square lattice.

EXERCISE 10.3 Derive a relation between the critical exponents above and below the transition point, $\alpha_{\pm}$, for the three-state Potts model on the square lattice. Also, derive a relation of the same type for the critical amplitudes $A_{\pm}$.

EXERCISE 10.4 Consider Ising models on the triangular and hexagonal lattices. Derive a relation between the critical exponents above and below the transition temperature, $\alpha_{\pm}$, the former for the triangular lattice and the latter for the hexagonal lattice. Do the same for the critical amplitudes, $A_{\pm}$. Notice that the triangular lattice is dual to the hexagonal lattice and vice versa. See Fig. 10.8.

EXERCISE 10.5 Let us determine the transition point of the Ising model on the triangular lattice by the introduction of a *star–triangle transformation* in combination with the duality relation. If $Z(K)$ on the left-hand side of eqn (10.40) is the partition function for the triangular lattice, then $Z^*(K^*)$ on the right-hand side is the partition function for the hexagonal lattice. Since the function Z is different from Z^*, the duality relation (10.40) is insufficient to identify a unique singularity. We can, nevertheless, reduce the Ising model on the hexagonal lattice to the Ising model on a triangular lattice by taking a partial trace, as depicted in Fig. 10.9, from which the (dual) partition function for the Ising model on the hexagonal lattice is transformed to the partition function of another Ising model on the triangular lattice, $Z(\tilde{K})$. In combination with the duality relation, we then have a relation connecting the Ising models on triangular lattices, which allows us to apply the argument in Section 10.1 and identify the fixed point with the singularity.

Now comes the problem: The central black spin S_0 in Fig. 10.9 has interactions with neighboring white spins with the Boltzmann factor $B = \mathrm{e}^{K^* S_0 (S_1 + S_2 + S_3)}$.

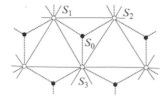

Fig. 10.9 Trace over alternating spins on the hexagonal lattice, shown in black dots, generates two-body interactions between the remaining spins (marked white), effectively realizing a new triangular lattice.

Suppose we perform a trace over spin S_0 and write the result as a set of interactions between white spins as $Ae^{\tilde{K}(S_1 S_2 + S_2 S_3 + S_3 S_1)}$. This is the star–triangle transformation. Write $\tilde{K}$ explicitly as a function of K^*. Use this result to find the transition point of the Ising model on the triangular lattice from the fixed-point condition $K = \tilde{K}$.

10.3.4 Villain model and roughening transition

The concept of duality applies also to systems with continuous degrees of freedom like the XY model. As elucidated in detail in Chapter 7, the two-dimensional XY model has two physically relevant types of excitations, the spin wave that represents a slowly changing state and the vortices that correspond to rapid changes of the spin orientation around specific points. The existence of vortices is intimately related with the periodicity of the system manifested by a Hamiltonian invariant under a rotation by 2π. We therefore often treat the *Villain model*, whose local Boltzmann factor is a periodic version of the spin-wave Boltzmann factor $e^{-K\psi_{ij}^2/2}$ (which is derived by expanding the cosine interaction of the original Boltzmann factor $e^{K\cos\psi_{ij}}$ ($\psi_{ij} = \phi_i - \phi_j$),

$$\exp\left(V(\psi_{ij})\right) \equiv \sum_{m=-\infty}^{\infty} \exp\left(-\frac{K}{2}(\psi_{ij} - 2\pi m)^2\right). \tag{10.41}$$

This is the sum of the spin-wave Boltzmann factors with period 2π and is sometimes called the *periodic Gaussian model*. The Villain model has both vortex and spin-wave excitations and consequently has a KT phase similarly to the XY model. Since the Villain model is essentially the same as the XY model in its properties concerning the phase transition, but is easier to analyze than the XY model itself, the theory of the KT transition is often developed for the Villain model instead of the XY model. The second half of Chapter 7 was essentially for the Villain model.

We first write the partition function to study the duality of the Villain model,

$$Z = \int_0^{2\pi} \prod_i \mathrm{d}\phi_i \, \exp\left(\sum_{\langle ij \rangle} V(\phi_i - \phi_j)\right). \tag{10.42}$$

The initial step of a duality transformation is to Fourier transform the local Boltzmann factor, which in the present case consists of multiplying eqn (10.41) by $e^{-il\psi_{ij}}$ and then integrate the result over ψ_{ij} from 0 to 2π. The integration range of ψ_{ij} then expands from $-\infty$ to ∞ due to the infinite sum over m, and we can carry out integration easily. In this way, we have the Fourier series expression

$$\exp\left(V(\psi_{ij})\right) = \frac{1}{\sqrt{2\pi K}} \sum_{l=-\infty}^{\infty} \exp\left(-\frac{l^2}{2K} + il\psi_{ij}\right). \tag{10.43}$$

Similarly to the case of discrete degrees of freedom explained in Section 10.3.2, we give a sense of arrow to each bond and collect the factors of the form $e^{il_{ij}(\phi_i - \phi_j)}$ involving ϕ_i for each i. Then, the integration over ϕ_i from 0 to 2π shows that l is a divergence-free field. This field is next expressed by the curl of a field μ on the dual lattice. Equation (10.32) remains valid for the square lattice, the only difference being in the absence of a constraint of modulus q. Equation (10.33) is now replaced by

$$Z = \text{const} \cdot \sum_{\{\mu_i = -\infty\}}^{\infty} \exp\left(-\frac{1}{2K} \sum_{\langle ij \rangle} (\mu_i - \mu_j)^2\right). \tag{10.44}$$

The sum over μ_i runs over all integers, from $-\infty$ to ∞, which is different from eqn (10.33) where $\mu_i = 1, 2, \cdots, q$. This is the dual representation of the Villain model. Since eqn (10.44) is written in terms of discrete variables in contrast to the original representation of the Villain model (10.42) or the XY model, the system is not self-dual even on the square lattice.

The dual representation of the Villain model (10.44) may be regarded as a *solid-on-solid (SOS) model* of *roughening transitions*, in which the surface of a solid suddenly changes its smoothness at a transition temperature. Suppose that μ_i atoms stack up at lattice site i on the surface of a solid. It is reasonable to assume that a site with a higher stack is unstable. This fact is expressed by writing the energy as $(\mu_i - \mu_j)^2$, which is higher for larger differences of neighboring heights. The partition function (10.44) corresponds to this model.

The Villain model has a KT transition and the free energy shows a singularity there. The dual representation (10.44) implies that there exists a roughening transition in the corresponding model of solid surfaces. For small K, the system is dominated by states with small $(\mu_i - \mu_j)^2$ and the surface is smooth. As K grows, larger differences of height are likely to appear due to thermal fluctuations,[12] and the surface suddenly roughens at the transition temperature.

Let us return to the XY model. The dual representation of the Villain model (10.44) is of the same form as the spin-wave approximation discussed in Section 7.2, if we disregard the discreteness of μ_i. By taking into account this discreteness, we can derive the energy of vortices (7.48) of Section 7.5 as follows.

---

[12] Here, K is identified with the temperature, not the inverse temperature, since it appears in the denominator of eqn (10.44).

The *Poisson summation formula*

$$\sum_{\mu=-\infty}^{\infty} g(\mu) = \sum_{n=-\infty}^{\infty} \int_{-\infty}^{\infty} e^{2\pi i \phi n} g(\phi) \, d\phi \qquad (10.45)$$

is useful to replace the discrete variable μ_i with a continuous counterpart ϕ_i. See Appendix A.19 for the derivation of this formula. Using this formula, we can rewrite eqn (10.44) as

$$Z = \sum_{\{n_j=-\infty\}}^{\infty} \int_{-\infty}^{\infty} \prod_j d\phi_j \, \exp\left(2\pi i \sum_j \phi_j n_j - \frac{1}{2K} \sum_{\langle ij \rangle} (\phi_i - \phi_j)^2\right) \qquad (10.46)$$

up to a trivial constant. This multiple Gaussian integral over $\{\phi_i\}$ can be performed as detailed in Appendix A.16 to give the result

$$Z = \sum_{\{n_j=-\infty\}}^{\infty} \exp\left(-2\pi^2 K \sum_{j,l} n_j n_l G(j-l)\right) \qquad (10.47)$$

up to a trivial multiplicative factor. Here, $G(j-l)$ is the *lattice Green function*,

$$G(\boldsymbol{j}-\boldsymbol{l}) = \frac{1}{2(2\pi)^2} \int_{-\pi}^{\pi} \frac{e^{i\boldsymbol{q}\cdot(\boldsymbol{j}-\boldsymbol{l})}}{2 - \cos q_x - \cos q_y} \, dq_x dq_y, \qquad (10.48)$$

where we have recovered the vector notation for two-dimensional vectors. It should be remembered that the neutrality condition $\sum_j n_j = 0$ applies to $\{n_j\}$ in eqn (10.47) as in Section 7.4. The asymptotic behavior of the lattice Green function for large $|\boldsymbol{j}-\boldsymbol{l}| \equiv r$ is analyzed in Appendix A.16 with the result

$$G(0) - G(r) \approx \frac{1}{2\pi} \log r. \qquad (10.49)$$

Then, the partition function (10.47) may be written as

$$Z = \sum_{\{n_j\}} \exp\left(\pi K \sum_{j \neq l} n_j n_l \log |j-l|\right). \qquad (10.50)$$

This is in accordance with the energy of vortices in eqn (7.48). We have reached the same expression from a more systematic approach than in Section 7.5, where an intuitive argument assuming the dominant role of vortices was employed.

EXERCISE 10.6 Let us study the roughening transition using the dual representation of the Villain model (10.44). We write T_r for K and call it the temperature since K appears in the position of temperature in eqn (10.44). In the limit of low temperature, $(\mu_i - \mu_j)^2$ takes only the smallest possible value, 0, and hence all μ_i assume a common value to be denoted as k. This corresponds to a perfectly smooth surface. As the temperature rises, some of the variables become deviated from k to $k \pm 1$. This is an Ising-like discrete excitation, and therefore long-range order exists at low temperatures, a smooth phase. In the limit of high temperature, on the other hand, we may approximate the discrete variable μ_i by continuous values since the minimum change of $(\mu_i - \mu_j)^2/(2T_r)$ is very small.

Problem: Show in the high-temperature phase that fluctuations of μ_i diverge in the long-range limit, which implies a rough surface. It will be useful to evaluate the expectation value of $(\mu_i - \mu_{i+r})^2$ and analyze its large-r limit.

10.4   Quantum duality

It has been shown in Exercise 9.8 that the transverse-field Ising model in one dimension undergoes a quantum phase transition, i.e. a qualitative change of the ground-state correlations, as the field strength h is varied. The critical properties of this transition are identical to those of the two-dimensional classical Ising model where the control parameter is the temperature T. This fact suggests an essential equivalence between these two model systems. This is indeed the case, as can be shown by a quantum-classical mapping. *Quantum dualities* are unitary mappings between quantum Hamiltonians that preserve the quasi-local character of their interaction terms. In some cases, known as *quantum self-dualities*, those mappings also conserve the form of the Hamiltonian operator. Physically, through this mapping one relates the weak-coupling phase of the model to its strong-coupling phase. Our goal in this section is first to prove a quantum duality, indeed a self-duality, relation for the one-dimensional transverse-field Ising model, from which we can locate the quantum phase transition point. Secondly, we show how classical and quantum dualities are simply related by a *quantum-classical mapping*.

10.4.1   Duality in the transverse-field Ising chain

Let us consider the transverse-field Ising model on a chain with free boundary conditions,

$$H[J, h] = -J \sum_{j=1}^{N-1} \sigma_j^z \sigma_{j+1}^z - h \sum_{j=2}^{N} \sigma_j^x. \tag{10.51}$$

Notice that the transverse field h is not applied to site 1. This special arrangement makes it possible to formulate an exact quantum duality in a compact manner.

The dual spin operators are defined as follows,

$$\tilde{\sigma}_1^x = \sigma_1^z \sigma_2^z, \ \tilde{\sigma}_2^x = \sigma_2^z \sigma_3^z, \ \cdots, \ \tilde{\sigma}_{N-1}^x = \sigma_{N-1}^z \sigma_N^z, \ \tilde{\sigma}_N^x = \sigma_N^z \tag{10.52}$$

$$\tilde{\sigma}_1^z = \sigma_1^x, \ \tilde{\sigma}_2^z = \sigma_1^x \sigma_2^x, \ \tilde{\sigma}_3^z = \sigma_1^x \sigma_2^x \sigma_3^x, \ \cdots, \ \tilde{\sigma}_N^z = \sigma_1^x \sigma_2^x \sigma_3^x \cdots \sigma_N^x. \tag{10.53}$$

These dual operators satisfy the usual anticommutation relations for Pauli operators,

$$\tilde{\sigma}_j^x \tilde{\sigma}_j^z = -\tilde{\sigma}_j^z \tilde{\sigma}_j^x. \tag{10.54}$$

The reason is that $\tilde{\sigma}_j^x$ and $\tilde{\sigma}_j^z$ have a common site index only in σ_j^z and σ_j^x. Hence, the anticommutation relation of the dual operators reflects that of the original operators. The y-component of the dual operator is constructed by the usual rule, e.g.

$$\tilde{\sigma}_1^y = -\mathrm{i}\,\tilde{\sigma}_1^z \tilde{\sigma}_1^x = -\sigma_1^y \sigma_2^z. \tag{10.55}$$

This set of dual operators $\{\tilde{\sigma}_j^x, \tilde{\sigma}_j^y, \tilde{\sigma}_j^z\}$ are themselves Pauli operators.

The exchange interaction term of the Hamiltonian (10.51) can be expressed in terms of the dual operators as

$$-J \sum_{j=1}^{N-1} \sigma_j^z \sigma_{j+1}^z = -J \sum_{j=1}^{N-1} \tilde{\sigma}_j^x, \tag{10.56}$$

while the field term becomes an exchange interaction,

$$-h \sum_{j=2}^{N} \sigma_j^x = -h \sum_{j=1}^{N-1} \tilde{\sigma}_j^z \tilde{\sigma}_{j+1}^z. \tag{10.57}$$

Therefore, the total Hamiltonian is rewritten as

$$H[\tilde{J}, \tilde{h}] = -\tilde{J} \sum_{j=1}^{N-1} \tilde{\sigma}_j^z \tilde{\sigma}_{j+1}^z - \tilde{h} \sum_{j=1}^{N-1} \tilde{\sigma}_j^x \tag{10.58}$$

with the dual parameters defined as $\tilde{J} = h$ and $\tilde{h} = J$. This hermitian operator is of the same form as the original Hamiltonian (10.51) with a slight difference in the site index; the site without field is $j = 1$ in eqn (10.51), whereas it is $j = N$ in eqn (10.58). This difference can be eliminated by the change of site numbering backward, $(1, 2, \cdots, N) \rightarrow (N, N-1, \cdots, 2, 1)$. Thus, the form of the original Hamiltonian and its dual is the same and the present duality becomes a self-duality.

The transformation of the Hamiltonian eqn (10.51) into the dual eqn (10.58) preserves the eigenvalue spectrum and is unitary. Mathematically, a quantum self-duality corresponds to[13]

$$H[J, h] = U H[h, J] U^\dagger, \tag{10.59}$$

where U is the unitary operator performing the duality. This, in turn, means that the eigenvalues E_n of the Hamiltonian and its dual are related by

$$E_n(J, h) = E_n(h, J). \tag{10.60}$$

Then, it readily follows that the quantum critical point, if a singularity exists in the ground-state energy, should be located at the fixed point of the duality, $h = J$, in agreement with the direct solution.[14] The quantum critical point is known as the *self-dual point* of the duality mapping.

Notice that the exact quantum duality mapping can be performed over a finite-N chain. Care must be exercised when one performs the transformation in the thermodynamic limit, $N \rightarrow \infty$, since appropriate boundary terms must be kept. For

[13] A general quantum duality corresponds to $H[J, h] = U H^*[h, J] U^\dagger$, where H^* is the dual Hamiltonian.

[14] If there is more than one singularity and the model is self-dual, the phase transition point does not necessarily coincide with the self-dual point. This is the case in the p-clock model when $p \geq 5$, and the self-dual point is in-between the two transitions. See Exercise 7.4 in which the existence of an intermediate phase, and thus of two phase transitions, is shown.

instance, if we consider the same Hamiltonian of eqn (10.51) but with an additional boundary term, $h\,\sigma_1^x$, i.e.

$$H = -J \sum_{j=1}^{N-1} \sigma_j^z \sigma_{j+1}^z - h \sum_{j=1}^{N} \sigma_j^x, \qquad (10.61)$$

then there is no *exact* quantum self-duality.

The transformation of eqns (10.52) and (10.53) can be systematized so that it is applicable to more general problems, which is, however, beyond the scope of this book.

10.4.2   Relation between classical and quantum duality

A natural question that emerges is whether there is any relation between the quantum and classical dualities, i.e. relations between the partition functions that were presented in previous sections. The answer is affirmative and the result is that both are two sides of the same coin. To address this issue we will illustrate the main ideas with the transverse-field Ising model. However, the reader must keep in mind that the above is a very general and deep statement.

The main idea consists in finding a representation of the trace of the exponential of the quantum Hamiltonian operator $H[J, h]$ of eqn (10.51)

$$\mathrm{Tr}\ e^{-H[J,h]} \propto \sum_{\{\sigma_i\}} e^{-\beta H_c[J,h]} \qquad (10.62)$$

in terms of a sum over classical configurations $\{\sigma_i\}$ with weights that can be associated to a classical model system with energy function $H_c[J, h]$ and fictitious temperature $T = 1/\beta$. Let us start by rewriting the quantum Hamiltonian

$$H[J, h] = H_z + H_x, \qquad (10.63)$$

where H_z represents the exchange interaction, which is diagonal in the σ^z-basis, and H_x is the transverse-field component. Notice that

$$e^{-H[J,h]} \neq e^{-H_z} e^{-H_x}, \qquad (10.64)$$

since these two operators do not commute, i.e. $[H_z, H_x] \neq 0$. However, one can use the *Suzuki–Trotter–Lie decomposition* of an exponential, whose argument is the sum of two bounded operators,

$$e^{-(H_z+H_x)} = \lim_{\mathcal{N}\to\infty} \left[e^{-\frac{1}{\mathcal{N}} H_z} e^{-\frac{1}{\mathcal{N}} H_x} \right]^{\mathcal{N}} = \lim_{\mathcal{N}\to\infty} \left[e^{-\frac{1}{\mathcal{N}} H_x} e^{-\frac{1}{\mathcal{N}} H_z} \right]^{\mathcal{N}}. \qquad (10.65)$$

It is not difficult to understand this formula by taking the leading-order contribution of $\mathcal{O}(1/\mathcal{N})$ on the right-hand side as

$$\lim_{\mathcal{N}\to\infty} \left[e^{-\frac{1}{\mathcal{N}} H_x} e^{-\frac{1}{\mathcal{N}} H_z} \right]^{\mathcal{N}} = \lim_{\mathcal{N}\to\infty} \left(1 - \frac{H_x}{\mathcal{N}} - \frac{H_z}{\mathcal{N}} \right)^{\mathcal{N}} = e^{-(H_z+H_x)}. \qquad (10.66)$$

Therefore, the trace can be rewritten as

$$
\mathrm{Tr}\, \mathrm{e}^{-H[J,h]} = \lim_{\mathcal{N}\to\infty} \sum_{\{\sigma_1\}} \langle\sigma_1| \mathrm{e}^{-\frac{1}{\mathcal{N}}H_x} \mathrm{e}^{-\frac{1}{\mathcal{N}}H_z} \cdot \mathbb{1} \cdot \mathrm{e}^{-\frac{1}{\mathcal{N}}H_x} \mathrm{e}^{-\frac{1}{\mathcal{N}}H_z} \cdot \mathbb{1}\cdots \mathrm{e}^{-\frac{1}{\mathcal{N}}H_x} \mathrm{e}^{-\frac{1}{\mathcal{N}}H_z} |\sigma_1\rangle
$$

$$
= \lim_{\mathcal{N}\to\infty} \sum_{\{\sigma_1,\cdots,\sigma_{\mathcal{N}}\}} \langle\sigma_1| \mathrm{e}^{-\frac{1}{\mathcal{N}}H_x} \mathrm{e}^{-\frac{1}{\mathcal{N}}H_z} |\sigma_{\mathcal{N}}\rangle \langle\sigma_{\mathcal{N}}| \mathrm{e}^{-\frac{1}{\mathcal{N}}H_x} \mathrm{e}^{-\frac{1}{\mathcal{N}}H_z} |\sigma_{\mathcal{N}-1}\rangle
$$

$$
\cdots \langle\sigma_2| \mathrm{e}^{-\frac{1}{\mathcal{N}}H_x} \mathrm{e}^{-\frac{1}{\mathcal{N}}H_z} |\sigma_1\rangle
$$

$$
= \lim_{\mathcal{N}\to\infty} \sum_{\{\sigma_1,\cdots,\sigma_{\mathcal{N}}\}} \langle\sigma_1| \mathrm{e}^{-\frac{1}{\mathcal{N}}H_x} |\sigma_{\mathcal{N}}\rangle \mathrm{e}^{-\frac{1}{\mathcal{N}}H_z(\sigma_{\mathcal{N}})} \langle\sigma_{\mathcal{N}}| \mathrm{e}^{-\frac{1}{\mathcal{N}}H_x} |\sigma_{\mathcal{N}-1}\rangle \mathrm{e}^{-\frac{1}{\mathcal{N}}H_z(\sigma_{\mathcal{N}-1})}
$$

$$
\cdots \mathrm{e}^{-\frac{1}{\mathcal{N}}H_z(\sigma_2)} \langle\sigma_2| \mathrm{e}^{-\frac{1}{\mathcal{N}}H_x} |\sigma_1\rangle \mathrm{e}^{-\frac{1}{\mathcal{N}}H_z(\sigma_1)}, \quad (10.67)
$$

with $l = 1, 2, \cdots, \mathcal{N}$, and where we have intercalated $\mathcal{N} - 1$ resolutions of the identity

$$
\mathbb{1} = \sum_{\sigma_l} |\sigma_l\rangle\langle\sigma_l|. \quad (10.68)
$$

This sum over σ_l represents a sum over the complete set of 2^N configurations of the one-dimensional transverse-field Ising model in the σ^z-basis. One such configuration is, for instance, $|-1-1+1+1\cdots-1+1\rangle$, which is an eigenstate of σ_1^z with eigenvalue -1, and of $\sigma_{\mathcal{N}}^z$ with eigenvalue $+1$. Thus, the index l labels an *extra* dimension. Note that because of its origin in a trace, eqn (10.67), the boundary condition in that extra dimension is periodic, while the boundary condition in the other direction is free.

Since we have chosen the eigenstates of σ^z as the basis, it implies that the exponential of the exchange interaction term H_z becomes diagonal since

$$
\langle\sigma_l| \mathrm{e}^{J\sigma_j^z\sigma_{j+1}^z} |\sigma_l\rangle = \mathrm{e}^{K^z(J)\,\sigma_{j,l}\sigma_{j+1,l}}, \quad (10.69)
$$

with $K^z(J) = J$, and classical Ising spins $\sigma_{j,l} = \pm 1$. It only remains to compute the matrix elements of the exponential of H_x in that basis. This is simple to accomplish because of the mathematical identity

$$
\langle\sigma_{l+1}| \mathrm{e}^{h\sigma_j^x} |\sigma_l\rangle = \langle\sigma_{l+1}|(\cosh h + \sinh h\; \sigma_j^x) |\sigma_l\rangle = A(h)\, \mathrm{e}^{K^x(h)\,\sigma_{j,l}\sigma_{j,l+1}}, (10.70)
$$

where

$$
K^x(h) = -\frac{1}{2}\log\tanh(h), \qquad A(h)^2 = \frac{1}{2}\sinh(2h). \quad (10.71)
$$

The validity of eqns (10.70) and (10.71) can be checked by setting $\sigma_l = \sigma_{l+1}$ in eqn (10.70) (in which case $\cosh h = A(h)\mathrm{e}^{K^x(h)}$) and $\sigma_l = -\sigma_{l+1}$ (in which case $\sinh h = A(h)\mathrm{e}^{-K^x(h)}$). Finally, we have accomplished the task to effectively map the original quantum transverse-field Ising chain in $d = 1$ into a classical statistical mechanical problem in $d = 2$ dimensions

$$
\mathrm{Tr}\, \mathrm{e}^{-H[J,h]} = \lim_{\mathcal{N}\to\infty} A(K_h^z)^{\mathcal{N}(N-1)} \sum_{\{\sigma_i\}} \mathrm{e}^{-\beta H_c[J,h]}, \quad (10.72)
$$

where the classical energy function is given by

$$-\beta H_c[J,h] = \sum_{l=1}^{\mathcal{N}} \sum_{j=1}^{N-1} (K_J^z\, \sigma_{j,l}\sigma_{j+1,l} + K_h^x\, \sigma_{j,l}\sigma_{j,l+1}), \qquad (10.73)$$

and corresponds to a classical Ising model with couplings $K_J^z = K^z(J/\mathcal{N})$, and $K_h^x = K^x(h/\mathcal{N})$. We have thus managed to relate the trace of e^{-H} of a quantum problem in d dimensions with the partition function of a classical statistical-mechanics problem in $d+1$ dimensions

$$Z(K_J^z, K_h^x) = \sum_{\{\sigma\}} e^{-\beta H_c[J,h]}. \qquad (10.74)$$

The quantum-classical mapping described above allows us to derive the self-duality of the two-dimensional classical Ising model. The quantum self-duality, which is basically a unitary mapping, i.e. $H[J,h] = UH[h,J]U^\dagger$, implies

$$\mathrm{Tr}\; e^{-H[J,h]} = \mathrm{Tr}\; Ue^{-H[h,J]}U^\dagger = \mathrm{Tr}\; e^{-H[h,J]}, \qquad (10.75)$$

because of the cyclic invariance of the trace operation and the fact that $U^\dagger U = \mathbb{1}$. This equality, eqn (10.75), in turn leads us trivially to the classical self-dual relation[15]

$$\frac{Z(K_J^z, K_h^x)}{A(K_J^z)^{\mathcal{N}(N-1)}} = \frac{Z(K_h^z, K_J^x)}{A(K_h^z)^{\mathcal{N}(N-1)}}. \qquad (10.76)$$

By identifying $K_z = K_J^z$, $K_x = K_h^x$, and $K_z^* = K_h^z$, $K_x^* = K_J^x$, and using eqn (10.71), one arrives at the relation for the dual couplings

$$e^{-2K_z^*} = \tanh K_x, \qquad e^{-2K_x^*} = \tanh K_z, \qquad (10.77)$$

or, more symmetrically,

$$\sinh 2K_x \sinh 2K_z^* = 1, \qquad \sinh 2K_z \sinh 2K_x^* = 1, \qquad (10.78)$$

which correspond to the dual relations for the $d=2$ anisotropic Ising model case. In the isotropic case, $K_z = K_x$, eqn (10.77) reduces to eqn (10.3),

$$e^{-2K_z^*} = \tanh K_z, \qquad (10.79)$$

and the quantum self-dual point, $h=J$, corresponds to $K_z^* = K_z$. We have in this way shown that quantum and classical dualities, which seem to have a different origin and physical interpretation, are mathematically related and represent two sides of the same entity.

---

[15] Compare to eqn (10.2) and note the lack of the factor 2. This is because of the use of different boundary conditions.

11
Numerical methods

This short chapter introduces a few typical numerical methods used in modern studies of phase transitions and critical phenomena in spin systems. The first section describes the dynamics of a generic system with discrete degrees of freedom following the master equation. This section serves as a theoretical basis for the method of Monte Carlo simulations explained in the second section. Another useful numerical technique is the transfer matrix method, described in the last section, and that is applied for numerically exact evaluation of the free energy and related physical quantities.

11.1   Master equation

Monte Carlo simulations are realized as the numerical implementation of stochastic dynamics, which is conveniently represented using the *master equation*. We first introduce the master equation and analyze the properties of its solutions in order to establish a theoretical basis for Monte Carlo simulations that reproduce equilibrium expectation values of physical quantities by stochastic dynamics.

We often use notations for the Ising model in the present chapter but in principle the discussions here apply to arbitrary systems with discrete degrees of freedom. The Ising model is a classical spin system for which there is no intrinsic dynamics, such as a Newton or Schrödinger equation of motion, which determine the time evolution of microscopic degrees of freedom. We, nevertheless, think it natural that each spin flips from time to time under the influence of thermal agitation from the environment. The master equation is a useful way to formulate this idea in terms of stochastic changes of spin configurations. This stochastic dynamics, built on concepts of probability theory, represents a fictitious dynamics that, as we will see, allow us to study the thermodynamic properties of the model.

Let us denote a spin configuration by an alphabet, e.g. $a = \{1, -1, -1, \cdots, 1\}$. We describe the state of a system using the probability that the system has a configuration a at time t, $P(a, t)$. The Ising model with N spins has the total number of configurations 2^N and hence we have a complete stochastic description of the system at time t if we know the 2^N values of $P(a, t)$ for all possible as. The master equation describes how this set of probabilities evolves with time.

Suppose that the configuration changes from a to b with the *transition probability* $w(a \to b)\Delta t$ (≥ 0) in a small time interval Δt.[1] Then, the probability that the system has the configuration a decreases by $w(a \to b)\Delta t \cdot P(a, t)$ because the

---

[1] The transition probability $w(a \to b)$ is a conditional probability.

system was in a with probability $P(a,t)$ and then has changed to b with probability $w(a \rightarrow b)\Delta t$. Similarly, the probability that the system is in configuration a would increase by the amount $w(b \rightarrow a)\Delta t \cdot P(b,t)$ if there is an influx of the configuration from $b(\neq a)$. Thus, the net change of the probability should be the balance of these two contributions,

$$P(a, t + \Delta t) - P(a, t) = - \sum_{b(\neq a)} w(a \rightarrow b)P(a,t)\Delta t + \sum_{b(\neq a)} w(b \rightarrow a)P(b,t)\Delta t.$$

(11.1)

This is the master equation.

Implicit in the above discussion is the concept of a *Markov process*, where the state of the system at the next time step $t + \Delta t$ is determined only by the present state at time t and is unaffected by the previous states at $t - \Delta t, t - 2\Delta t, \cdots$. This is a reasonable assumption, although it is usually hard to deduce rigorously from more fundamental rules, and we follow the convention to adopt this assumption here. Also, it is sometimes useful to take the continuous-time limit $\Delta t \rightarrow 0$ and write the master equation as a differential equation. In the present section we use the discrete representation with an application to Monte Carlo simulations in mind.

It is convenient to rewrite the master equation in a compact form as

$$P(a, t + \Delta t) = \Big(1 - \sum_{b(\neq a)} w(a \rightarrow b)\Delta t\Big) P(a,t) + \sum_{b(\neq a)} w(b \rightarrow a)\Delta t \cdot P(b,t)$$

$$\equiv \sum_{b} \mathcal{L}_{ab} P(b,t), \tag{11.2}$$

where

$$\mathcal{L}_{aa} = 1 - \sum_{b(\neq a)} w(a \rightarrow b)\Delta t, \quad \mathcal{L}_{ab} = w(b \rightarrow a)\Delta t \quad (b \neq a), \tag{11.3}$$

or, in a matrix-vector notation,

$$\boldsymbol{P}(t + \Delta t) = \mathcal{L}\boldsymbol{P}(t). \tag{11.4}$$

The ath component of vector $\boldsymbol{P}(t)$ is $P(a,t)$. Suppose that the *stochastic matrix* $\mathcal{L}$ has right eigenvectors and corresponding eigenvalues,

$$\mathcal{L}\boldsymbol{e}_\alpha = \lambda_\alpha \boldsymbol{e}_\alpha. \tag{11.5}$$

If the set of eigenvectors $\{\boldsymbol{e}_\alpha\}$ is complete, we can expand the probability $\boldsymbol{P}(t)$ as

$$\boldsymbol{P}(t) = \sum_\alpha c_\alpha(t)\boldsymbol{e}_\alpha, \tag{11.6}$$

and the time evolution (11.4) is expressed as

$$\boldsymbol{P}(t + \Delta t) = \sum_\alpha c_\alpha(t)\mathcal{L}\boldsymbol{e}_\alpha = \sum_\alpha c_\alpha(t)\lambda_\alpha \boldsymbol{e}_\alpha. \tag{11.7}$$

After n steps of time evolution from the initial state $t = 0$, the probability becomes

$$\boldsymbol{P}(n\Delta t) = \sum_{\alpha} c_{\alpha}(0)\lambda_{\alpha}^{n}\boldsymbol{e}_{\alpha}. \qquad (11.8)$$

The behavior of the probability as a function of time step n is therefore dictated by the eigenvalue spectrum $\{\lambda_{\alpha}\}$.

The requirement of the conservation of probability, $\sum_{a} P(a, t) = 1$, places a strong constraint on the eigenvalue spectrum. Since the sum of all components of the left-hand side of eqn (11.8) is unity, so is the right-hand side,

$$\sum_{\alpha} c_{\alpha}(0)\lambda_{\alpha}^{n} \sum_{a} e_{\alpha}(a) = 1, \qquad (11.9)$$

where $e_{\alpha}(a)$ is the ath component of $\boldsymbol{e}_{\alpha}$. This relation holds for arbitrary n. For the left-hand side of eqn (11.9) to be independent of n, the largest eigenvalue λ_0 should be unity, with the corresponding eigenvector satisfying $\sum_{a} e_0(a) = 1$, and all other eigenvectors must satisfy $\sum_{a} e_{\alpha}(a) = 0$ ($\alpha \neq 0$). It is further necessary that $c_0(0) = 1$ and $|\lambda_{\alpha}| < 1$ for $\alpha \neq 0$. This latter condition on the eigenvalues λ_{α} comes from the observation that the probability $P(a, n\Delta t)$ should not grow indefinitely with n. Since all matrix elements of $\mathcal{L}$ are positive semi-definite,[2] the Perron–Frobenius theorem[3] guarantees that the eigenvector for the leading eigenvalue λ_0 is non-degenerate, as long as the matrix $\mathcal{L}$ is irreducible (which means that any configuration can be reached from any other configuration after a finite number of steps). This irreducibility condition is satisfied in Monte Carlo simulations by standard choices of the transition probability, as will be explained in the next section.

We conclude that the system evolves toward a unique equilibrium distribution $\boldsymbol{e}_0$ as $n \to \infty$,

$$\boldsymbol{P}(n\Delta t) = \boldsymbol{e}_0 + c_1(0)\mathrm{e}^{-n|\log\lambda_1|}\boldsymbol{e}_1 + \cdots \qquad (11.10)$$

with the relaxation time $1/|\log\lambda_1|$, where $\lambda_1(< 1)$ is the second largest eigenvalue of $\mathcal{L}$. It is in this sense guaranteed that the stochastic dynamics under the master equation eventually realizes the equilibrium distribution $\boldsymbol{e}_0$.

Discussions in this section are very general and apply to any choices of the transition probability and matrix $\mathcal{L}$ as long as the latter is irreducible and positive semi-definite. Accordingly, the equilibrium distribution $\boldsymbol{e}_0$ may not necessarily be the Gibbs–Boltzmann distribution $\mathrm{e}^{-\beta H}/Z$. We study in the next section the conditions that the transition probability must satisfy for the Gibbs–Boltzmann distribution to be realized as the equilibrium distribution.

---

[2] The diagonal element $\mathcal{L}_{aa}$ in eqn (11.3) should be positive semi-definite because it is the probability that the system stays in the present configuration a.

[3] This theorem states that an irreducible square matrix with positive semi-definite elements has its largest (in magnitude) eigenvalue positive and the corresponding eigenvector is non-degenerate with all its components being positive.

11.2   Monte Carlo simulation

It is necessary to choose appropriate transition probabilities in Monte Carlo simulations so that the equilibrium distribution is of the Gibbs–Boltzmann form $P(a,t) = \mathrm{e}^{-\beta H(a)}/Z \equiv P_{\mathrm{eq}}(a)$. Suppose that an equilibrium has been achieved in the master equation (11.1) with $P(a,t) = P_{\mathrm{eq}}(a)$. Then, the left-hand side vanishes and consequently

$$\sum_{b(\neq a)} w(a \to b) P_{\mathrm{eq}}(a) = \sum_{b(\neq a)} w(b \to a) P_{\mathrm{eq}}(b). \tag{11.11}$$

This relation constrains the possible form of the transition probability. A sufficient condition for the above relation to hold is to equate both sides term by term,

$$w(a \to b) P_{\mathrm{eq}}(a) = w(b \to a) P_{\mathrm{eq}}(b). \tag{11.12}$$

This relation is called the *detailed balance condition*. If we use the Gibbs–Boltzmann distribution for $P_{\mathrm{eq}}(a)$ in the above equation, the ratio of the transition probabilities satisfies

$$\frac{w(a \to b)}{w(b \to a)} = \mathrm{e}^{-\beta(H(b)-H(a))}. \tag{11.13}$$

Common choices of the transition probability that satisfy the detailed balance condition are the *heat-bath method*

$$w(a \to b) = \frac{\mathrm{e}^{-\beta H(b)}}{\mathrm{e}^{-\beta H(a)} + \mathrm{e}^{-\beta H(b)}}, \tag{11.14}$$

and the *Metropolis method*

$$w(a \to b) = \mathrm{e}^{-\beta(H(b)-H(a))_+}, \tag{11.15}$$

where $(f)_+ = f$ if $f \geq 0$ and 0 otherwise.

> **EXERCISE 11.1** Confirm that the heat-bath and Metropolis methods satisfy the detailed balance condition.

We next have to determine what types of transitions are allowed, that is, what combinations of a and b would have $w(a \to b) > 0$. The process of a *single-spin flip* is often used in Monte Carlo simulations of the Ising model, in which only a single spin is flipped in a given time step. We choose site i and decide whether to flip S_i to $-S_i$ according to the probability $w(a \to b)$, where

$$a = \{S_1, S_2, \cdots, S_i, \cdots, S_N\}, \quad b = \{S_1, S_2, \cdots, -S_i, \cdots, S_N\}. \tag{11.16}$$

The transition probabilities for all other processes are considered vanishing. It is clear that any configuration $\{S_1, S_2, \cdots, S_N\}$ can be reached from any other configuration $\{S_1', S_2', \cdots, S_N'\}$ by successively flipping spins at sites where $S_i \neq S_i'$. Thus, this choice of the transition probability generates an irreducible matrix $\mathcal{L}$. For the single-spin flip process, the heat-bath and Metropolis methods are written as

$$w(S_i \rightarrow -S_i) = \frac{e^{-\beta H(-S_i)}}{e^{-\beta H(S_i)} + e^{-\beta H(-S_i)}} = \frac{e^{-\beta \Delta E_i}}{1 + e^{-\beta \Delta E_i}}, \tag{11.17}$$

and

$$w(S_i \rightarrow -S_i) = e^{-\beta (\Delta E_i)_+}, \tag{11.18}$$

respectively. Here, $\Delta E_i = H(-S_i) - H(S_i)$ and the dependence of H on spins other than S_i has been suppressed for simplicity of notation.

Equations (11.17) and (11.18) are particularly useful if we notice that ΔE_i can be written by the local spin configuration around site i. Suppose that the Hamiltonian has the following expression

$$H = H_i + H' = -\sum_j J_{ij} S_i S_j + H', \tag{11.19}$$

where H' is the part that does not include S_i, and $J_{ij} = J$ for a neighboring pair $\langle ij \rangle$ and 0 otherwise in the summation. Then, the change of the energy by a single-spin flip is

$$\Delta E_i = H(-S_i) - H(S_i) = 2 \sum_j J_{ij} S_i S_j. \tag{11.20}$$

This quantity is easily calculated numerically and can be inserted into eqn (11.17) or eqn (11.18) to evaluate the transition probability.

> **EXERCISE 11.2** Consider a simple two-spin system with $H = -JS_1S_2$. There are four possible configurations of spins, $a = \{1, 1\}, b = \{1, -1\}, c = \{-1, 1\}, d = \{-1, -1\}$. Write the transition probabilities between these four configurations using the heat-bath method under the single-spin flip process. Write, further, the matrix $\mathcal{L}$ and evaluate its right eigenvalues and eigenvectors. Confirm that the largest eigenvalue is unity with the corresponding eigenvector being the Gibbs–Boltzmann distribution. Observe that other eigenvectors have both positive and negative components and satisfy $\sum_a e_\alpha(a) = 0$ ($\alpha \neq 0$).

In Monte Carlo simulations one regards the calculation of the expectation value of a physical quantity $\widehat{\mathcal{O}}$ as an average over the configurations generated by the stochastic dynamics. Assume, for instance, that we want to evaluate the magnetization m of the Ising model in the Gibbs–Boltzmann ensemble

$$m = \langle \widehat{m} \rangle_{P_{\text{eq}}} = \frac{1}{N} \left\langle \sum_{i=1}^N S_i \right\rangle_{P_{\text{eq}}} = \sum_a P_{\text{eq}}(a) \left(\frac{1}{N} \sum_{i=1}^N S_i \right), \tag{11.21}$$

where the sum is performed over all 2^N configurations. Monte Carlo methods choose a subset of configurations according to the stochastics dynamics generated by the transition probabilities w. One then constructs an *estimator* of m by choosing a number of sampled configurations, e.g. M configurations,

$$m_E = \frac{1}{M} \sum_{\tilde{a}} \hat{m}(\tilde{a}), \tag{11.22}$$

where the M configurations $\tilde{a}$ are sampled according to P_{eq}, a probability measure on the space of configurations. The estimator m_E has the property that it becomes a more accurate estimate of m as M grows larger. That means that for a sufficiently large number of sampled configurations one can come arbitrarily close to the desired expectation value. Indeed, the *strong law of large numbers* guarantees that

$$\text{Prob} \left\{ \lim_{M \to \infty} \frac{1}{M} \sum_{\tilde{a}} \hat{\mathcal{O}}(\tilde{a}) = \langle \hat{\mathcal{O}} \rangle_{P_{eq}} \right\} = 1, \tag{11.23}$$

i.e. it converges with probability 1.

The sequence of statistically independent sampled configurations $\{\tilde{a}\}$ constitutes a *random walk*. The initial probability distribution $P(a, t = 0)$ and the transition probability $w(a \to b)$ characterize the random walk. If the equilibrium distribution exists, under the hypothesis of ergodicity,[4] then the random walk converges to P_{eq}, irrespective of the initial distribution $P(a, t = 0)$. Since an infinite random walk cannot be simulated in practice, the natural question that arises is: What is the error involved if the random walk is of length M? The answer to that question is provided by the central limit theorem, which states that for large M the estimator $\mathcal{O}_E = 1/M \sum_{\tilde{a}} \hat{\mathcal{O}}(\tilde{a})$ becomes normally distributed about $\langle \hat{\mathcal{O}} \rangle_{P_{eq}}$ with variance

$$\text{var} \left\{ \langle \hat{\mathcal{O}} \rangle_{P_{eq}} \right\} = \frac{1}{M} \left(\langle \hat{\mathcal{O}}^2 \rangle_{P_{eq}} - \langle \hat{\mathcal{O}} \rangle^2_{P_{eq}} \right) = \frac{\sigma^2}{M}, \tag{11.24}$$

if the M random variables $\tilde{a}$ are statistically independent. So, the central limit theorem allows us to assign numerical confidence limits to our best estimate of the value of a physical average

$$\langle \hat{\mathcal{O}} \rangle_{P_{eq}} \approx \mathcal{O}_E \pm \frac{\sigma}{\sqrt{M}}, \tag{11.25}$$

where the uncertainty in the estimate represents a main contribution to the statistical error.

In practice, the structure of the algorithm is as follows. Notice that we choose $\Delta t = 1$.

1. Initialize the spin configuration $\{S_i\}$, typically to a random configuration or a perfectly ferromagnetic state.

2. Choose site i for flip trial.

3. Calculate the energy change ΔE_i.

4. Generate a uniform random number r between 0 and 1 and compare it with $w(S_i \to -S_i)$. Execute the flip $S_i \to -S_i$ if $r < w(S_i \to -S_i)$ and keep S_i unchanged if $r > w(S_i \to -S_i)$. In this way, the transition takes place with the designated transition probability.

[4] Ergodicity is the requirement that our Markov process can reach any configuration of our system from an arbitrary configuration, if we run it long enough.

5. Calculate the physical quantities of interest for the present configuration $\{S_i\}$.

6. Repeat 2 to 5 until sufficient statistics has been collected.

Sample codes are listed in Appendix A.20.

There are a number of points to be noticed in the actual implementation of the algorithm. For example, the effects of the initial condition should be discarded because it takes some time, $\tau = 1/|\log \lambda_1|$, for the system to equilibrate, as discussed in the previous section. Measurements of physical quantities are also to be performed at some intervals of Monte Carlo steps, not at each single step, since consecutive spin configurations are correlated, which prevents us from collecting data drawn independently from the equilibrium distribution. If these conditions are satisfied, spin configurations thus generated are considered to be drawn independently from the Gibbs–Boltzmann distribution, and hence the simple averages of physical quantities as in eqn (11.22) are usually reliable approximations of the canonical average. Also, as is always the case in data analysis, statistical and systematic errors must be properly estimated, the former as in eqn (11.25) and a typical example of the latter being the finite-size effects. The method of finite-size scaling is useful for this purpose.

11.3 Numerical transfer matrix method

The numerical transfer matrix method is another popular technique to numerically evaluate physical quantities especially in two dimensions. According to the transfer matrix method described in Chapter 9, we introduce a 2×2 transfer matrix for the one-dimensional Ising model and diagonalize it to obtain the partition function in the thermodynamic limit using the largest eigenvalue. The two-dimensional case is similar in spirit but is much more complicated in practice because a large $2^L \times 2^L$ matrix, where L is the linear length of a finite-size square lattice, should be diagonalized. The numerical transfer matrix method is used instead to evaluate the partition function of a long strip of size $M \times 2^L$, where M can, in principle, be chosen as large as we wish.

Suppose that we have successfully evaluated the partition function numerically up to the lth row indicated by circles in Fig. 11.1, starting from the bottom row and tracing out all spin variables shown in black dots. The result is stored as 2^L numbers, $Z(S_1^{(l)}, S_2^{(l)}, \cdots, S_L^{(l)})$ for $\{S_1^{(l)} = \pm 1, S_2^{(l)} = \pm 1, \cdots, S_L^{(l)} = \pm 1\}$. We next evaluate the effects of the interaction between $S_1^{(l)}$ and $S_1^{(l+1)}$, the latter being the spin just above $S_1^{(l)}$ in the next row, as

$$\tilde{Z}(S_1^{(l+1)}, S_2^{(l)}, \cdots, S_L^{(l)}) = \sum_{S_1^{(l)}=\pm 1} e^{K S_1^{(l)} S_1^{(l+1)}} Z(S_1^{(l)}, S_2^{(l)}, \cdots, S_L^{(l)}). \tag{11.26}$$

Then, the interaction between $S_2^{(l)}$ and $S_2^{(l+1)}$ is taken into account as

$$\tilde{Z}(S_1^{(l+1)}, S_2^{(l+1)}, \cdots, S_L^{(l)}) = \sum_{S_2^{(l)}=\pm 1} e^{K S_2^{(l)} S_2^{(l+1)}} \tilde{Z}(S_1^{(l+1)}, S_2^{(l)}, \cdots, S_L^{(l)}). \tag{11.27}$$

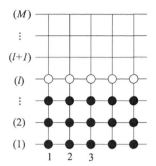

Fig. 11.1 The numerical transfer matrix method evaluates the partition function row by row for the two-dimensional lattice.

The interaction between $S_1^{(l+1)}$ and $S_2^{(l+1)}$ will be considered later. By repeating similar processes, we obtain the values of $\tilde{Z}(S_1^{(l+1)}, S_2^{(l+1)}, \cdots, S_L^{(l+1)})$. The transfer from the lth row to the $(l+1)$th row completes by adding horizontal interactions within the row, assuming periodic boundary conditions, as

$$Z(S_1^{(l+1)}, S_2^{(l+1)}, \cdots, S_L^{(l+1)}) = e^{K(S_1^{(l+1)} S_2^{(l+1)} + S_2^{(l+1)} S_3^{(l+1)} + \cdots + S_L^{(l+1)} S_1^{(l+1)})}$$

$$\times \tilde{Z}(S_1^{(l+1)}, S_2^{(l+1)}, \cdots, S_L^{(l+1)}). \tag{11.28}$$

It is straightforward to reach the final Mth row by repeating this procedure to have $Z(S_1^{(M)}, S_2^{(M)}, \cdots, S_L^{(M)})$. Assuming a free boundary at the final row, that is, there is no $(M+1)$th row, the total partition function is the sum of $Z(S_1^{(M)}, S_2^{(M)}, \cdots, S_L^{(M)})$ over all possible values of $S_i^{(M)} = \pm 1 \ (\forall i)$. The initial value is the Boltzmann factor for the horizontal interactions,

$$Z(S_1^{(1)}, \cdots, S_L^{(1)}) = e^{K(S_1^{(1)} S_2^{(1)} + S_2^{(1)} S_3^{(1)} + \cdots + S_L^{(1)} S_1^{(1)})}. \tag{11.29}$$

A clear advantage of this method is that we do not have to directly diagonalize a large matrix and the computational cost is linear in M. The memory and computational cost in terms of L is exponential, 2^L, which limits the actual size somewhere between $L = 10$ and $L = 20$.

Since the partition function and its logarithm can thus be obtained numerically, physical quantities are evaluated by numerical differentiations with respect to the temperature or the external field.

Although the simple two-dimensional Ising model has an exact solution and thus does not need to be analyzed by numerical methods, the method of numerical transfer matrix is useful for more general cases, such as random systems and non-Ising models.

For further reading

A comprehensive description of the history and citations to original papers on the modern theory of critical phenomena are found in

1. Domb, C. (1996). *The Critical Point*. Taylor and Francis, London.

Introductory accounts on many of the topics discussed in the present book are found in references 2 and 3 in the following list, whereas references 4 and 5 include very detailed descriptions of various aspects of phase transitions and critical phenomena.

2. Yeomans, Y. M. (1992). *Statistical Mechanics of Phase Transitions*. Oxford University Press, Oxford.
3. Cardy, J. (1996). *Scaling and Renormalization in Statistical Physics*. Cambridge University Press, Cambridge.
4. Goldenfeld, N. (1992). *Lectures on Phase Transitions and the Renormalization Group*. Westview Press, Oxford.
5. Binney, J. J., Dowrick, N. J., Fisher, A. J. and Newman, M. E. J. (1992). *The Theory of Critical Phenomena*. Oxford University Press, Oxford.

Reference 6 is relatively close to the present book in its scope and level. Reference 7 includes many pedagogical problems.

6. Herbut, I. (2007). *A Modern Approach to Critical Phenomena*. Cambridge University Press, Cambridge.
7. Kardar, M. (2007). *Statistical Physics of Fields*. Cambridge University Press, Cambridge.

For technical details of field-theoretical methods, the following books are appropriate,

8. Ma, S.-k. (2000). *Modern Theory of Critical Phenomena*. Westview Press, Oxford.
9. Zinn-Justin, J. (2007). *Phase Transitions and Renormalization Group*. Oxford University Press, Oxford.
10. Amit, D. J. and Martin-Mayor, V. (2005). *Field Theory, the Renormalization Group, and Critical Phenomena* (3rd edn). World Scientific, Singapore.

A very detailed rendition of conformal field theory can be found in

11. Di Francesco, P., Mathieu P. and Sénéchal D. (1997). *Conformal Field Theory*. Springer Verlag, New York.

For a comprehensive introduction to Monte Carlo simulations, see

12. Landau, D. P. and Binder K. (2000). *A Guide to Monte Carlo Simulations in Statistical Physics*. Cambridge University Press, Cambridge.

The series edited by Domb and Green, later by Domb and Lebowitz, is an extensive collection of reviews.

13. Domb, C. and Green, M. S. (ed.) (1972–1976). *Phase Transitions and Critical Phenomena*, vols 1–6. Academic Press, London.
Domb, C. and Lebowitz, J. L. (ed.) (1983–2001). *Phase Transitions and Critical Phenomena*, vols 7–20. Academic Press, London.

Appendix A

A.1 Saddle-point method

Suppose that a function $f(x)$ has a maximum at $x = x_0$ as illustrated in Fig. A.1. Then, the integral

$$I = \int_{-\infty}^{\infty} e^{Nf(x)}\, \mathrm{d}x \tag{A.1}$$

is evaluated asymptotically in the limit $N \to \infty$ as

$$I \approx e^{Nf(x_0)}, \tag{A.2}$$

which simply amounts to keeping the maximum value of the integrand. This is the result of applying the *saddle-point method* or the *method of steepest descents*.

The expansion of $f(x)$ around x_0 starts from the quadratic term since its first-order term vanishes at the maximum x_0,

$$f(x) \approx f(x_0) + \frac{1}{2}(x - x_0)^2 f''(x_0) \quad (f''(x_0) < 0). \tag{A.3}$$

Cubic and higher-order terms become non-negligible for larger $|x - x_0|$, where $f(x)$ is significantly smaller than $f(x_0)$. In the integrand of eqn (A.1), $f(x)$ is multiplied by N and is exponentiated, which leads to overwhelmingly smaller values of $e^{Nf(x)}$ for $x \neq x_0$ compared to $e^{Nf(x_0)}$ for large N. As a concrete example, if $\Delta f = f(x) - f(x_0) = -1$ and $N = 10$, then $r \equiv e^{Nf(x)}/e^{Nf(x_0)} = 4.5 \times 10^{-5}$. For the same difference of the functional values $\Delta f = -1$, $N = 100$ gives $r = 3.7 \times 10^{-44}$, and $r = 5.1 \times 10^{-435}$ for $N = 1000$. It should be clear that the leading contribution comes only from the immediate neighborhood of x_0 in the limit $N \to \infty$. We may therefore ignore the cubic and higher-order terms in the expansion of eqn (A.3).

The asymptotic expression can be evaluated by the Gaussian integral using eqn (A.3),

$$I \approx \int_{-\infty}^{\infty} \mathrm{d}x \, \exp\left(Nf(x_0) - \frac{N}{2}(x - x_0)^2 |f''(x_0)|\right)$$

$$= \exp\left(Nf(x_0) + \frac{1}{2}\log 2\pi - \frac{1}{2}\log\left(N|f''(x_0)|\right)\right). \tag{A.4}$$

Since the term $Nf(x_0)$ is much larger than the other terms in the exponent, we often keep only $e^{Nf(x_0)}$ as the asymptotic form. This is eqn (A.2).

The name 'saddle-point method' comes from the behavior of the function in the complex plane: The function $f(z)$, an analytical continuation of $f(x)$ to the complex plane, is maximum at $z = x_0$ along the real axis but is minimum along the path

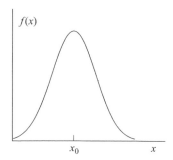

Fig. A.1 The function $f(x)$ is assumed to have a maximum at x_0.

parallel to the imaginary axis, as can be verified from eqn (A.3). See also Fig. A.2. This method is also called the *method of steepest descents* since the integration runs along the path where the change of the functional value is largest.

Let us comment on the higher-order terms for those readers who are curious about how the higher-order terms affect the result. For simplicity, choose $x_0 = 0$ and assume that $f(x)$ is an even function. We expand $f(x)$ around $x = 0$ and keep terms up to the fourth order. Then, the integral I to be evaluated has the following form,

$$I \approx e^{Nf(0)} \int_{-\infty}^{\infty} \exp\left(-\frac{aN}{2}x^2 + bNx^4\right) dx, \tag{A.5}$$

where b should be negative since the integral diverges otherwise. As $f(x)$ is maximum at $x = 0$, a should be positive. Set $x = t/\sqrt{aN}$ to have

$$I \approx \frac{e^{Nf(0)}}{\sqrt{aN}} \int_{-\infty}^{\infty} \exp\left(-\frac{t^2}{2} + \frac{bt^4}{a^2N}\right) dt. \tag{A.6}$$

A naive next step consists of ignoring the fourth-order term in the limit $N \to \infty$ as its coefficient has the factor N^{-1}. This corresponds to the saddle-point result (A.4). This argument applies to any higher-order terms not just the fourth-order one.

Let us evaluate the effects of corrections due to the fourth-order term in a more careful way. The fourth-order term in the exponent prevents us from performing the integral directly, so we expand e^{bt^4/a^2N} around $t = 0$. The nth-order term of the expansion is proportional to $t^{4n}/(n!N^n)$. Its integral

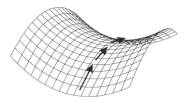

Fig. A.2 The path in the integral is chosen along the arrow such that the function becomes largest at the saddle point.

$$\int_{-\infty}^{\infty} \frac{t^{4n}}{n! N^n} \exp\left(-\frac{t^2}{2}\right) dt \tag{A.7}$$

is proportional to $(2n)!/(n! N^n)$ for large N. Since $(2n)! \gg n!$, the coefficient of the expansion is larger for higher-order terms and consequently the expansion does not converge. Thus, this is not a Taylor expansion but it is an *asymptotic expansion*.

An asymptotic expansion does not converge and is anomalous in this sense. It is, nevertheless, useful to terminate the expansion at an appropriate finite order and then take the limit of a very small parameter value N^{-1}, in which case the difference between the approximate value using the asymptotic expansion and the correct value can be reduced arbitrarily. In the present example of eqn (A.6), the correction of the first term $(n = 1)$ gives

$$\int_{-\infty}^{\infty} \exp\left(-\frac{t^2}{2}\right)\left(1 + \frac{b}{a^2 N} t^4\right) dt = \sqrt{2\pi}\left(1 + \frac{3b}{a^2 N}\right) \approx \sqrt{2\pi} \exp\left(\frac{3b}{a^2 N}\right). \tag{A.8}$$

Then, the asymptotic expansion of the integral to first order reads

$$I \approx \exp\left(N f(0) + \frac{1}{2}\log 2\pi - \frac{1}{2}\log(aN) + \frac{3b}{a^2 N}\right). \tag{A.9}$$

This value is close to the correct integral value for large N. For the quartic function $f(x) = -x^2/2 - x^4$, for instance, the direct integral value I_1 and the value I_2 of the formula (A.9), in their logarithmic form, are compared as $\log I_1 = -0.3860, \log I_2 = -0.5324$ for $N = 10$, $\log I_1 = -1.4099, \log I_2 = -1.4137$ for $N = 100$, and $\log I_1 = -2.5379, \log I_2 = -2.5379$ for $N = 1000$. This example shows the numerical accuracy of the asymptotic expansion for sufficiently large N.

A.2   Expressing the susceptibility in terms of correlation functions

The magnetic susceptibility, a thermodynamic quantity, can be written in terms of correlation functions. We show this fact for a model with Ising variables by starting from the definition of the magnetization of a system with Hamiltonian H_0 and external field h,[1]

$$M = \frac{\sum_{\{S_i\}} (S_1 + \cdots + S_N) e^{-\beta H_0 + \beta h \sum_i S_i}}{\sum_{\{S_i\}} e^{-\beta H_0 + \beta h \sum_i S_i}}. \tag{A.10}$$

The susceptibility is the h-derivative of M in the limit $h \to 0$. If we write χ for the susceptibility per spin, we have

$$N\chi = \lim_{h \to 0} \frac{\partial M}{\partial h} = \frac{\beta \sum_{\{S_i\}} (S_1 + \cdots + S_N)^2 e^{-\beta H_0}}{\sum_{\{S_i\}} e^{-\beta H_0}}$$

$$-\beta \left(\frac{\sum_{\{S_i\}} (S_1 + \cdots + S_N) e^{-\beta H_0}}{\sum_{\{S_i\}} e^{-\beta H_0}}\right)^2. \tag{A.11}$$

---

[1] Notice that the factor $\beta = 1/T$ is not absorbed in the definition of the Hamiltonian.

The first term on the right-hand side is a sum of correlation functions and the second term is the square of the spontaneous magnetization,[2]

$$N\chi = \beta \sum_{i,j=1}^{N} \langle S_i S_j \rangle - \beta M^2. \tag{A.12}$$

If the system is translationally invariant,

$$\langle S_i S_j \rangle = \langle S_0 S_r \rangle \quad (r = |i - j|), \quad \langle S_i \rangle = \frac{M}{N} = m \ (\forall i), \tag{A.13}$$

and we may rewrite the susceptibility as

$$\chi = \beta \sum_l \big(\langle S_0 S_l \rangle - \langle S_0 \rangle \langle S_l \rangle \big). \tag{A.14}$$

This equation shows that the susceptibility is the sum of connected correlation functions (defined by the quantity in the parentheses on the right-hand side) and connects the divergence of χ at the critical point with the two-point correlation function. This relation holds not only for the Ising model but also for other systems and is known as the *static susceptibility sum rule*. In the paramagnetic (disordered) phase, the second term on the right-hand side of eqn (A.14) vanishes because there is no spontaneous magnetization.

This sum rule is a consequence of the *linear response theory* and is a special example of the *fluctuation–dissipation theorem*. Consider an arbitrary system whose Hamiltonian H_0 is modified because of the presence of an external inhomogeneous field $B(\boldsymbol{r})$ (a magnetic field being an example) as

$$H = H_0 - \int d\boldsymbol{r} \, O(\boldsymbol{r}) B(\boldsymbol{r}), \tag{A.15}$$

where $O(\boldsymbol{r})$ is the system variable that linearly couples to the external field (the magnetization being an example). The free energy of the system is $F = -\beta^{-1} \log Z$ in terms of the partition function

$$Z = \mathrm{Tr} \exp \Big(-\beta H_0 + \beta \int d\boldsymbol{r} \, O(\boldsymbol{r}) B(\boldsymbol{r}) \Big). \tag{A.16}$$

Let us define a *generalized isothermal susceptibility*

$$\chi(\boldsymbol{r}, \boldsymbol{r}') = -\frac{\delta^2 F}{\delta B(\boldsymbol{r}) \delta B(\boldsymbol{r}')} \tag{A.17}$$

as the second-order functional derivative of the free energy with the result

$$\chi(\boldsymbol{r}, \boldsymbol{r}') = \frac{1}{\beta} \left(\frac{1}{Z} \frac{\delta^2 Z}{\delta B(\boldsymbol{r}) \delta B(\boldsymbol{r}')} - \frac{1}{Z} \frac{\delta Z}{\delta B(\boldsymbol{r})} \cdot \frac{1}{Z} \frac{\delta Z}{\delta B(\boldsymbol{r}')} \right) \tag{A.18}$$

$$= \beta \Big(\langle O(\boldsymbol{r}) O(\boldsymbol{r}') \rangle - \langle O(\boldsymbol{r}) \rangle \langle O(\boldsymbol{r}') \rangle \Big) \equiv \beta G(\boldsymbol{r}, \boldsymbol{r}'). \tag{A.19}$$

---

[2] Rigorously speaking, it is necessary to take the thermodynamic limit $N \to \infty$ first and then the limit $h \to 0$ to evaluate the spontaneous magnetization. See Section 5.6.

If the system is translationally invariant, $G(\boldsymbol{r}, \boldsymbol{r}') = G(\boldsymbol{r} - \boldsymbol{r}')$, and we have

$$\chi = \int \mathrm{d}\boldsymbol{r}\chi(\boldsymbol{r}) \equiv \beta \int \mathrm{d}\boldsymbol{r}\, G(\boldsymbol{r}). \qquad (A.20)$$

This is a very profound result that explains why a divergent χ is associated with the increase in the range of the two-point correlation function. In the case of a fluid, $O(\boldsymbol{r}) = \rho(\boldsymbol{r})$ may represent the particle density and $\chi = \kappa$ is the compressibility. Therefore, a divergent compressibility is correlated to the increase of both density fluctuations and the range of the density–density correlation function. When the correlation length ξ becomes comparable to the wavelength of light, the latter will be strongly scattered by density inhomogeneities and the phenomenon of *critical opalescence* emerges.

Equation (A.20) connects the response χ of the system to an external perturbation B with the fluctuations G in equilibrium. It thus has a similar physical content to the fluctuation–dissipation theorem explained in critical dynamics, eqn (2.115).

A.3   Rushbrooke's inequality

We prove here Rushbrooke's inequality

$$\alpha_- + 2\beta + \gamma_- \geq 2. \qquad (A.21)$$

This inequality is related to the scaling relation $\alpha + 2\beta + \gamma = 2$, the difference being, first that the equality is replaced by an inequality, and secondly that the critical exponents α and γ are restricted to those below the critical point. The present section uses only a thermodynamic stability condition in contrast to the scaling theory in Chapter 3.

The Helmholtz free energy F of a gas is a function of temperature T and volume V. Correspondingly, the Helmholtz free energy of a magnetic system is a function of T and m, the magnetization per spin. We define the heat capacity at fixed magnetization m by the second-order derivative of F with respect to T, corresponding to the heat capacity at fixed volume for a gas,

$$C_m = T\left(\frac{\partial S}{\partial T}\right)_m = -T\left(\frac{\partial^2 F}{\partial T^2}\right)_m. \qquad (A.22)$$

The entropy is also a function of temperature and magnetization, $S(T, m)$. We now consider the situation where m is determined as a function of the external field h and temperature T and replace m in $S(T, m)$ by $m(h, T)$ to have $S(T, m(h, T))$, which we express by the same symbol $S(T, h)$ for simplicity. The T-derivative of $S(T, h)$ corresponds to the heat capacity at constant pressure for a gas,

$$C_h = T\left(\frac{\partial S(T, h)}{\partial T}\right)_h \left(= -T\left(\frac{\partial^2 G}{\partial T^2}\right)_h\right), \qquad (A.23)$$

where G is the Gibbs free energy. Using the relation between $S(T, h)$ and $S(T, m)$, that is $S(T, h) = S(T, m(h, T))$, we have

$$\left(\frac{\partial S(T, h)}{\partial T}\right)_h = \left(\frac{\partial S(T, m)}{\partial T}\right)_m + \left(\frac{\partial S(T, m)}{\partial m}\right)_T \left(\frac{\partial m}{\partial T}\right)_h. \qquad (A.24)$$

Then, if we write χ for the susceptibility $\partial m/\partial h$, we find

$$\chi(C_h - C_m) = \left(\frac{\partial m}{\partial h}\right)_T \cdot T \left(\frac{\partial S}{\partial m}\right)_T \left(\frac{\partial m}{\partial T}\right)_h. \tag{A.25}$$

We use here the relation to be derived later,

$$\left(\frac{\partial m}{\partial h}\right)_T \left(\frac{\partial S}{\partial m}\right)_T = \left(\frac{\partial m}{\partial T}\right)_h, \tag{A.26}$$

to obtain

$$\chi(C_h - C_m) = T \left\{\left(\frac{\partial m}{\partial T}\right)_h\right\}^2. \tag{A.27}$$

Now, the internal energy E satisfies

$$C_m = T \left(\frac{\partial S}{\partial T}\right)_m = \left(\frac{\partial E}{\partial T}\right)_m, \tag{A.28}$$

and E is a monotonically increasing function of T due to the thermodynamic stability of a macroscopic system.[3] This implies $C_m \geq 0$, and hence from eqn (A.27),

$$\chi C_h \geq T \left\{\left(\frac{\partial m}{\partial T}\right)_h\right\}^2. \tag{A.29}$$

We take the limit $h \to 0$ and substitute the definitions of critical exponents below the critical point, $\chi \propto (T_c - T)^{-\gamma_-}, C_h \propto (T_c - T)^{-\alpha_-}, \partial m/\partial T \propto (T_c - T)^{\beta-1}$ to get Rushbrooke's inequality (A.21).

It remains to show the identity (A.26). First, the following relation is derived,

$$\left(\frac{\partial h}{\partial m}\right)_T \left(\frac{\partial m}{\partial T}\right)_h \left(\frac{\partial T}{\partial h}\right)_m = -1. \tag{A.30}$$

For this purpose it is useful to note the functional relation $f(h, m, T) = 0$ for some f (not to be confused with the free energy), which shows that two variables among h, m and T determine the value of the remaining one. An equivalent expression is $h(m, T)$. Let us thus differentiate $f(h(m, T), m, T) = 0$ with respect to m to have

$$\frac{\partial f}{\partial h}\left(\frac{\partial h}{\partial m}\right)_T + \frac{\partial f}{\partial m} = 0. \tag{A.31}$$

Similarly, from $f(h, m(h, T), T) = 0$,

$$\frac{\partial f}{\partial m}\left(\frac{\partial m}{\partial T}\right)_h + \frac{\partial f}{\partial T} = 0, \tag{A.32}$$

and from $f(h, m, T(m, h)) = 0$

$$\frac{\partial f}{\partial T}\left(\frac{\partial T}{\partial h}\right)_m + \frac{\partial f}{\partial h} = 0. \tag{A.33}$$

---

[3] Otherwise, the system heats up ($\Delta T > 0$) as its energy decreases ($\Delta E < 0$).

These three relations immediately lead to eqn (A.30). We now apply the Maxwell relation

$$\left(\frac{\partial S}{\partial m}\right)_T = -\left(\frac{\partial h}{\partial T}\right)_m, \tag{A.34}$$

derived by differentiation of

$$-S = \left(\frac{\partial F}{\partial T}\right)_m, \quad h = \left(\frac{\partial F}{\partial m}\right)_T, \tag{A.35}$$

to eqn (A.30) to prove (A.26).

A.4   Cumulants

The nth moment of a stochastic variable x with the probability density function $P(x)$ is defined as

$$\langle x^n \rangle = \int \mathrm{d}x \, x^n P(x). \tag{A.36}$$

The average of $\mathrm{e}^{\mathrm{i}xk}$ can be expanded in terms of the moments as

$$\langle \mathrm{e}^{\mathrm{i}kx} \rangle = \sum_{n=0}^{\infty} \frac{(\mathrm{i}k)^n}{n!} \langle x^n \rangle, \tag{A.37}$$

the exponentiation of which defines the cumulant $\langle x^n \rangle_c$,

$$\sum_{n=0}^{\infty} \frac{(\mathrm{i}k)^n}{n!} \langle x^n \rangle = \exp\left(\sum_{n=1}^{\infty} \frac{(\mathrm{i}k)^n}{n!} \langle x^n \rangle_c\right). \tag{A.38}$$

For example, the first cumulant is equal to the average (the first moment) and the second cumulant is the variance, as can be easily verified from the above definition,

$$\langle x \rangle_c = \langle x \rangle, \quad \langle x^2 \rangle_c = \langle x^2 \rangle - \langle x \rangle^2. \tag{A.39}$$

The Gaussian distribution has the special property that $\langle \mathrm{e}^{\mathrm{i}kx} \rangle$ can be evaluated explicitly to give

$$\langle \mathrm{e}^{\mathrm{i}kx} \rangle = \exp\left(\mathrm{i}k\langle x \rangle - \frac{k^2}{2} \langle x^2 \rangle_c\right). \tag{A.40}$$

This shows that all higher-order cumulants than the second-order one vanish. This is a very convenient property of cumulants in comparison with moments, the latter being non-vanishing to higher orders for the Gaussian distribution.

A.5   Renormalization group equations from the ϵ expansion

In this appendix we derive the differential renormalization group equation, eqns (4.56) and (4.57), from the ϵ expansion in a d-dimensional system.

The basic strategy is to start from the ϕ^4 model (4.40) and carry out the processes of coarse graining (integration of short-range degrees of freedom), rescaling of length and renormalization of the spin degrees of freedom (i.e. fields) by using a perturbative expansion in powers of the quartic term. As we will see below, a main mathematical tool is the Gaussian integration.

It is convenient to write the basic Hamiltonian (4.40) with $h = 0$,

$$H = \int d\mathbf{r} \left\{ (\nabla \phi(\mathbf{r}))^2 + t\phi(\mathbf{r})^2 + u\phi(\mathbf{r})^4 \right\}, \tag{A.41}$$

in the Fourier-transformed (wave-number) expression, often called the momentum space representation following the convention of field theory,

$$\phi(\mathbf{r}) = \frac{1}{(2\pi)^d} \int_0^\Lambda d\mathbf{q}\, e^{i\mathbf{q}\cdot\mathbf{r}} \phi(\mathbf{q}), \quad \phi(\mathbf{q}) = \int d\mathbf{r}\, e^{-i\mathbf{q}\cdot\mathbf{r}} \phi(\mathbf{r}), \tag{A.42}$$

as in eqn (2.75). Here, we have used the same symbol ϕ both for the original and Fourier-transformed variables to simplify the notation. The integral over momentum is cutoff at the absolute value of momentum $|\mathbf{q}_{\max}| = \Lambda$, so that we have approximated the volume of integration to a hypersphere of radius Λ. Singularities are expected to come from the long-wavelength (small-q) modes, and therefore the cutoff is supposed not to affect the results, which is indeed the case, as is shown in Section 4.2.2. The momentum cutoff Λ is proportional to the inverse of the shortest length scale, the lattice constant a, which will be gradually increased (i.e. coarse grained) as the renormalization group proceeds. This means that Λ will be gradually decreased by successive elimination of the large-q (short-length) degrees of freedom.

The Hamiltonian (A.41) is written as

$$H = \int_0^\Lambda \frac{d\mathbf{q}}{(2\pi)^d} (t + q^2) \phi(\mathbf{q}) \phi(-\mathbf{q})$$

$$+ \frac{u}{(2\pi)^{4d}} \int_0^\Lambda d\mathbf{q}_1 \cdots d\mathbf{q}_4 (2\pi)^d \delta(\mathbf{q}_1 + \cdots + \mathbf{q}_4) \phi(\mathbf{q}_1) \phi(\mathbf{q}_2) \phi(\mathbf{q}_3) \phi(\mathbf{q}_4), \tag{A.43}$$

as can be verified by using

$$\frac{1}{(2\pi)^d} \int d\mathbf{r}\, e^{i\mathbf{q}\cdot\mathbf{r}} = \delta(\mathbf{q}). \tag{A.44}$$

The quartic term in eqn (A.43) will be denoted as V in this appendix, i.e. $H = H_0 + V$. We will consider perturbations in powers of V.

A.5.1   Gaussian model: zeroth-order contribution

The zeroth-order contribution is the Gaussian model because the quartic term V is absent, i.e. $H = H_0$. Although we have already analyzed the Gaussian model in Section 4.2.1, it is instructive to formulate the momentum-space renormalization group for this model to illustrate the basic ideas used in higher-order calculations. As in the real-space renormalization group approach, one needs to perform two steps. The first consists of integrating out (partial trace) the large momenta (short-wavelength) degrees of freedom while the second step is concerned with the rescaling of (momentum) variables and spin fields.

Let us introduce separate notations for spin variables with $|\boldsymbol{q}|$ smaller and larger than Λ/b,

$$\phi(\boldsymbol{q}) = \begin{cases} S(\boldsymbol{q}) & 0 \leq |\boldsymbol{q}| < \Lambda/b \\ \sigma(\boldsymbol{q}) & \Lambda/b \leq |\boldsymbol{q}| \leq \Lambda \end{cases} \tag{A.45}$$

where b is the scaling factor to be chosen slightly larger than 1. Notice that the spatial rescaling by b corresponds to the momentum (wave number) rescaling by $1/b$ as they are dimensionally inverse to each other. The Gaussian Hamiltonian is separated into two parts, $H = H_S + H_\sigma$, where

$$H_S = \int_0^{\Lambda/b} \frac{\mathrm{d}\boldsymbol{q}}{(2\pi)^d} (t + q^2) S(\boldsymbol{q}) S(-\boldsymbol{q}), \quad H_\sigma = \int_{\Lambda/b}^{\Lambda} \frac{\mathrm{d}\boldsymbol{q}}{(2\pi)^d} (t + q^2) \sigma(\boldsymbol{q}) \sigma(-\boldsymbol{q}). \tag{A.46}$$

Correspondingly, the partition function is the product of two parts, $Z = Z_S Z_\sigma$, where

$$Z_S = \int \prod_{|\boldsymbol{q}| < \Lambda/b} \mathrm{d}S(\boldsymbol{q}) \, \mathrm{e}^{-H_S}, \quad Z_\sigma = \int \prod_{\Lambda/b \leq |\boldsymbol{q}| \leq \Lambda} \mathrm{d}\sigma(\boldsymbol{q}) \, \mathrm{e}^{-H_\sigma}. \tag{A.47}$$

If one wishes to be mathematically more rigorous, one should define the system over a finite volume Ω, and then perform the thermodynamic limit. In that case one needs to replace integrals over $\boldsymbol{q}$ by sums, as indicated in Section 2.9, and the measure of integration in eqn (A.47) should include a factor $\Omega^{-1/2}$ per mode $\boldsymbol{q}$. The final result is the same.

At the Gaussian level there is no coupling between S and σ variables. Following the spirit of coarse graining of the short length scales, we perform the integration only over the σ variables, i.e. a partial trace, and study what happens to the remaining part H_S. This process is equivalent to eliminating fluctuations at length scales $a < |\boldsymbol{r}| < ba$, i.e. to trace out the spin degrees of freedom within a region that corresponds to a block in the real-space language. The integral over σ is the Gaussian integral discussed in Section 2.9 and gives just a regular function of t and Λ, independent of S

$$Z_\sigma = \prod_{\Lambda/b \leq |\boldsymbol{q}| \leq \Lambda} \left(\frac{(2\pi)^{1+d/2}}{2} \sqrt{\frac{\pi}{t + q^2}} \right), \tag{A.48}$$

which contributes to the free energy but not to the critical phenomena. Thus, the part of H_S in Z is trivially kept intact by this process. This feature is characteristic of the Gaussian model.

The rest of the renormalization group calculation consists of rescaling the momentum space variables and renormalizing the spin degrees of freedom (rescaling of momenta and spin fields)[4]

$$\boldsymbol{q}_b \equiv b\boldsymbol{q}, \quad \phi(\boldsymbol{q}_b) \equiv \zeta(b)^{-1} S(\boldsymbol{q}). \tag{A.49}$$

The renormalized Hamiltonian for S now reads

$$H_S = \int_0^\Lambda \frac{\mathrm{d}\boldsymbol{q}_b}{(2\pi)^d} \, b^{-d}(t + b^{-2} q_b^2) \zeta^2 \phi(\boldsymbol{q}_b) \phi(-\boldsymbol{q}_b). \tag{A.50}$$

This has the same form as the original Gaussian Hamiltonian, eqn (A.43) with $u = 0$, if we choose

$$b^{-d-2}\zeta^2 = 1, \quad t_b \equiv b^{-d}\zeta^2 t. \tag{A.51}$$

The spin-renormalization factor is therefore fixed to $\zeta = b^{1+d/2}$, and the renormalized t satisfies

$$t_b = b^2 t, \tag{A.52}$$

in agreement with eqn (4.43).

A.5.2 First-order contribution

We next consider the effect of the quartic term V perturbatively. For this purpose we again separate the partition function into S and σ parts,

$$Z = \int \prod_{|\boldsymbol{q}|<\Lambda/b} \mathrm{d}S(\boldsymbol{q}) \int \prod_{\Lambda/b\leq|\boldsymbol{q}|\leq\Lambda} \mathrm{d}\sigma(\boldsymbol{q}) \, \mathrm{e}^{-H_S - H_\sigma - V(S,\sigma)}$$

$$= \int \prod_{|\boldsymbol{q}|<\Lambda/b} \mathrm{d}S(\boldsymbol{q}) \, \mathrm{e}^{-H_S} \, Z_\sigma \, \langle \mathrm{e}^{-V(S,\sigma)} \rangle_\sigma, \tag{A.53}$$

where V depends explicitly on both S and σ variables, and

$$Z_\sigma = \int \prod_{\Lambda/b\leq|\boldsymbol{q}|\leq\Lambda} \mathrm{d}\sigma(\boldsymbol{q}) \mathrm{e}^{-H_\sigma} \tag{A.54}$$

$$\langle \mathrm{e}^{-V(S,\sigma)} \rangle_\sigma = \frac{1}{Z_\sigma} \int \prod_{\Lambda/b\leq|\boldsymbol{q}|\leq\Lambda} \mathrm{d}\sigma(\boldsymbol{q}) \, \mathrm{e}^{-H_\sigma} \, \mathrm{e}^{-V(S,\sigma)}. \tag{A.55}$$

The expectation value $\langle \mathrm{e}^{-V} \rangle_\sigma$ in eqn (A.53) is rewritten as a cumulant expansion,

$$Z = Z_\sigma \int \prod_{|\boldsymbol{q}|<\Lambda/b} \mathrm{d}S(\boldsymbol{q}) \, \mathrm{e}^{-H_S} \, \mathrm{e}^{-\langle V \rangle_\sigma + (\langle V^2 \rangle_\sigma - \langle V \rangle_\sigma^2)/2 + \mathcal{O}(V^3)}. \tag{A.56}$$

---

[4] Rescaling of the spin field in real space is different. See eqns (3.7) and (4.42). They are related by the Fourier transformation in the following way $c(b)^{-1} = b^d \zeta(b)^{-1}$.

To evaluate the various contributions, it is necessary to separate the S- and σ-variables in the Hamiltonian. We write the quartic term $V(S, \sigma)$ as

$$V(S, \sigma) = \frac{u}{(2\pi)^{4d}} \int_0^\Lambda d\boldsymbol{q}_1 \cdots d\boldsymbol{q}_4 \, (2\pi)^d \delta(\boldsymbol{q}_1 + \cdots + \boldsymbol{q}_4) \phi(\boldsymbol{q}_1) \phi(\boldsymbol{q}_2) \phi(\boldsymbol{q}_3) \phi(\boldsymbol{q}_4)$$

$$= \frac{u}{(2\pi)^{4d}} \left\{ \int_0^{\Lambda/b} d\boldsymbol{q}_1 \cdots d\boldsymbol{q}_4 \, (2\pi)^d \delta(\boldsymbol{q}_1 + \cdots + \boldsymbol{q}_4) S(\boldsymbol{q}_1) S(\boldsymbol{q}_2) S(\boldsymbol{q}_3) S(\boldsymbol{q}_4) \right.$$

$$+ \int_0^{\Lambda/b} d\boldsymbol{q}_1 d\boldsymbol{q}_2 d\boldsymbol{q}_3 \int_{\Lambda/b}^\Lambda d\boldsymbol{q}_4 (2\pi)^d \delta(\boldsymbol{q}_1 + \cdots + \boldsymbol{q}_4) S(\boldsymbol{q}_1) S(\boldsymbol{q}_2) S(\boldsymbol{q}_3) \sigma(\boldsymbol{q}_4)$$

$$\cdots$$

$$\left. + \int_{\Lambda/b}^\Lambda d\boldsymbol{q}_1 \cdots d\boldsymbol{q}_4 \, (2\pi)^d \delta(\boldsymbol{q}_1 + \cdots + \boldsymbol{q}_4) \sigma(\boldsymbol{q}_1) \sigma(\boldsymbol{q}_2) \sigma(\boldsymbol{q}_3) \sigma(\boldsymbol{q}_4) \right\}$$

$$\equiv SSSS + 4SSS\sigma + 6SS\sigma\sigma + 4S\sigma\sigma\sigma + \sigma\sigma\sigma\sigma, \tag{A.57}$$

the latter expression having an obvious interpretation. For example, $4SSS\sigma$ stands for

$$\frac{u}{(2\pi)^{4d}} \left\{ \int_0^{\Lambda/b} d\boldsymbol{q}_1 d\boldsymbol{q}_2 d\boldsymbol{q}_3 \int_{\Lambda/b}^\Lambda d\boldsymbol{q}_4 (2\pi)^d \delta(\boldsymbol{q}_1 + \cdots + \boldsymbol{q}_4) S(\boldsymbol{q}_1) S(\boldsymbol{q}_2) S(\boldsymbol{q}_3) \sigma(\boldsymbol{q}_4) \right.$$

$$+ \int_0^{\Lambda/b} d\boldsymbol{q}_1 d\boldsymbol{q}_2 d\boldsymbol{q}_4 \int_{\Lambda/b}^\Lambda d\boldsymbol{q}_3 (2\pi)^d \delta(\boldsymbol{q}_1 + \cdots + \boldsymbol{q}_4) S(\boldsymbol{q}_1) S(\boldsymbol{q}_2) \sigma(\boldsymbol{q}_3) S(\boldsymbol{q}_4)$$

$$+ \int_0^{\Lambda/b} d\boldsymbol{q}_1 d\boldsymbol{q}_3 d\boldsymbol{q}_4 \int_{\Lambda/b}^\Lambda d\boldsymbol{q}_2 (2\pi)^d \delta(\boldsymbol{q}_1 + \cdots + \boldsymbol{q}_4) S(\boldsymbol{q}_1) \sigma(\boldsymbol{q}_2) S(\boldsymbol{q}_3) S(\boldsymbol{q}_4)$$

$$\left. + \int_0^{\Lambda/b} d\boldsymbol{q}_2 d\boldsymbol{q}_3 d\boldsymbol{q}_4 \int_{\Lambda/b}^\Lambda d\boldsymbol{q}_1 (2\pi)^d \delta(\boldsymbol{q}_1 + \cdots + \boldsymbol{q}_4) \sigma(\boldsymbol{q}_1) S(\boldsymbol{q}_2) S(\boldsymbol{q}_3) S(\boldsymbol{q}_4) \right\}. \tag{A.58}$$

Similar abbreviations have been used for the other terms in eqn (A.57). Our goal is to determine the renormalized Hamiltonian in terms of the long-wavelength variables S. We will proceed in a way similar to the block-spin method of Section 4.1.2. Therefore, the integral in the partition function is to be carried out only over the σ-variables, as in the zeroth-order (Gaussian) case.

Our task in this section is to evaluate the first-order contribution, i.e. $\langle V \rangle_\sigma$,

$$\langle V \rangle_\sigma = SSSS + 4SSS \langle \sigma \rangle_\sigma + 6SS \langle \sigma\sigma \rangle_\sigma + 4S \langle \sigma\sigma\sigma \rangle_\sigma + \langle \sigma\sigma\sigma\sigma \rangle_\sigma, \tag{A.59}$$

where the long-wavelength variables S have been kept intact, according to the definition of $\langle \cdots \rangle_\sigma$. The first term $SSSS$ is the unchanged quartic term for $|\boldsymbol{q}| < \Lambda/b$ and will be rescaled later. The odd-order terms $\langle \sigma \rangle_\sigma$ and $\langle \sigma\sigma\sigma \rangle_\sigma$ vanish identically as H_σ is even in σ in the definition of the integral, eqn (A.55). The final term in eqn (A.59), $\langle \sigma\sigma\sigma\sigma \rangle_\sigma$, gives an additive constant and plays no role in the calculation of parameter changes under a renormalization group transformation (but contributes to the regular part of the free energy). Thus, only the term $6SS\langle\sigma\sigma\rangle_\sigma$ yields a non-trivial correction of $\mathcal{O}(V)$ to the renormalized Hamiltonian.

Let us write $SS\langle\sigma\sigma\rangle_\sigma$ explicitly as

$$
u\int_0^{\Lambda/b}\frac{\mathrm{d}\boldsymbol{q}_1\mathrm{d}\boldsymbol{q}_2}{(2\pi)^{2d}}\,S(\boldsymbol{q}_1)S(\boldsymbol{q}_2)\int_{\Lambda/b}^{\Lambda}\frac{\mathrm{d}\boldsymbol{q}_3\mathrm{d}\boldsymbol{q}_4}{(2\pi)^{2d}}\,(2\pi)^d\delta(\boldsymbol{q}_1+\cdots+\boldsymbol{q}_4)\langle\sigma(\boldsymbol{q}_3)\sigma(\boldsymbol{q}_4)\rangle_\sigma
$$

$$
= u\int_0^{\Lambda/b}\frac{\mathrm{d}\boldsymbol{q}_1\mathrm{d}\boldsymbol{q}_2}{(2\pi)^{2d}}\,S(\boldsymbol{q}_1)S(\boldsymbol{q}_2)(2\pi)^d\delta(\boldsymbol{q}_1+\boldsymbol{q}_2)\int_{\Lambda/b}^{\Lambda}\frac{\mathrm{d}\boldsymbol{q}}{(2\pi)^d}\frac{1}{2(t+q^2)}, \tag{A.60}
$$

where we have used

$$
\langle\sigma(\boldsymbol{q}_3)\sigma(\boldsymbol{q}_4)\rangle_\sigma = (2\pi)^d\delta(\boldsymbol{q}_3+\boldsymbol{q}_4)\frac{1}{2(t+q_3^2)}, \tag{A.61}
$$

as in eqn (2.84).[5] The last integral in eqn (A.60) will be denoted as $I_1/2$ hereafter,

$$
I_1 = \int_{\Lambda/b}^{\Lambda}\frac{\mathrm{d}\boldsymbol{q}}{(2\pi)^d}\frac{1}{t+q^2}. \tag{A.62}
$$

We now rescale $\boldsymbol{q}$ and renormalize S as

$$
\boldsymbol{q}_{b,i} \equiv b\boldsymbol{q}_i, \quad \phi(\boldsymbol{q}_{b,i}) \equiv \zeta(b)^{-1}S(\boldsymbol{q}_i) \quad (i=1,2). \tag{A.63}
$$

Then, eqn (A.60) becomes

$$
u\zeta^2 b^{-2d}\int_0^{\Lambda}\frac{\mathrm{d}\boldsymbol{q}_{b,1}}{(2\pi)^d}\,\phi(\boldsymbol{q}_{b,1})\phi(-\boldsymbol{q}_{b,1})b^d\cdot\frac{I_1}{2}, \tag{A.64}
$$

where we have used

$$
\delta(\boldsymbol{q}_1+\boldsymbol{q}_2) = b^d\delta(\boldsymbol{q}_{b,1}+\boldsymbol{q}_{b,2}). \tag{A.65}
$$

This result (A.64) represents a correction to the SS term in the Hamiltonian, the q-independent part, as

$$
t_b \equiv \zeta^2 b^{-d}t + u\zeta^2 b^{-d}\cdot 6\cdot\frac{I_1}{2} = \zeta^2 b^{-d}(t+3uI_1), \tag{A.66}
$$

where the first term on the right-hand side is the Gaussian contribution, eqn (A.51). The factor 6 comes from $6SS\langle\sigma\sigma\rangle_\sigma$. For $\zeta = b^{1+d/2}$ as required from the invariance of the q-dependent part of the quadratic term, we find

$$
t_b = b^2(t+3uI_1). \tag{A.67}
$$

---

[5] Notice that the factor of temperature T in eqn (2.84) is absorbed in the definition of the Hamiltonian in this section.

The remaining $SSSS$ term in eqn (A.59) needs just rescaling of $\boldsymbol{q}$ and renormalization of S,

$$
\begin{aligned}
SSSS &= \frac{u}{(2\pi)^{4d}} \int_0^{\Lambda/b} d\boldsymbol{q}_1 \cdots d\boldsymbol{q}_4 \, (2\pi)^d \delta(\boldsymbol{q}_1 + \cdots + \boldsymbol{q}_4) S(\boldsymbol{q}_1) \cdots S(\boldsymbol{q}_4) \\
&= \frac{u\zeta^4 b^{-4d+d}}{(2\pi)^{4d}} \\
&\quad \cdot \int_0^{\Lambda} d\boldsymbol{q}_{b,1} \cdots d\boldsymbol{q}_{b,4} \, (2\pi)^d \delta(\boldsymbol{q}_{b,1} + \cdots + \boldsymbol{q}_{b,4}) \phi(\boldsymbol{q}_{b,1}) \cdots \phi(\boldsymbol{q}_{b,4}). \quad \text{(A.68)}
\end{aligned}
$$

Thus, the renormalized value of u is

$$
u_b \equiv \zeta^4 b^{-3d} u = b^{4-d} u, \quad \text{(A.69)}
$$

as has been found already in eqn (4.43).

A.5.3   Second-order contribution

The renormalization group equation for t, eqn (A.67), has a non-trivial correction term $3uI_1$ that may possibly yield to a non-Gaussian fixed point, $t^* \neq 0, u^* \neq 0$. Equation (A.69) for u, in contrast, has yet no correction to the simple Gaussian evaluation, eqn (4.43), and the only fixed point is $u^* = 0$. Therefore, we need to calculate the next order correction only for the coefficient u of the quartic term.

To find non-trivial contributions to the second-order correction, $\langle V^2 \rangle_\sigma - \langle V \rangle_\sigma^2$, see eqn (A.56), it is useful to write $\langle V^2 \rangle_\sigma$ and $\langle V \rangle_\sigma^2$ explicitly as

$$
\begin{aligned}
\langle V^2 \rangle_\sigma &= \langle (S_1 S_2 S_3 S_4 + 4 S_1 S_2 S_3 \sigma_4 + 6 S_1 S_2 \sigma_3 \sigma_4 + 4 S_1 \sigma_2 \sigma_3 \sigma_4 + \sigma_1 \sigma_2 \sigma_3 \sigma_4) \\
&\quad \cdot (S_5 S_6 S_7 S_8 + 4 S_5 S_6 S_7 \sigma_8 + 6 S_5 S_6 \sigma_7 \sigma_8 + 4 S_5 \sigma_6 \sigma_7 \sigma_8 + \sigma_5 \sigma_6 \sigma_7 \sigma_8) \rangle_\sigma \quad \text{(A.70)} \\
\langle V \rangle_\sigma^2 &= \langle S_1 S_2 S_3 S_4 + 4 S_1 S_2 S_3 \sigma_4 + 6 S_1 S_2 \sigma_3 \sigma_4 + 4 S_1 \sigma_2 \sigma_3 \sigma_4 + \sigma_1 \sigma_2 \sigma_3 \sigma_4 \rangle_\sigma \\
&\quad \cdot \langle S_5 S_6 S_7 S_8 + 4 S_5 S_6 S_7 \sigma_8 + 6 S_5 S_6 \sigma_7 \sigma_8 + 4 S_5 \sigma_6 \sigma_7 \sigma_8 + \sigma_5 \sigma_6 \sigma_7 \sigma_8 \rangle_\sigma . \quad \text{(A.71)}
\end{aligned}
$$

From the argument in the preceding paragraph, we look for $SSSS$-type terms in eqn (A.70), i.e. quartic in S, since these are the terms contributing to the renormalized u. A generic term such as $S_1 S_2 S_3 S_4 \sigma_5 \sigma_6 \sigma_7 \sigma_8$ means

$$
\begin{aligned}
S_1 S_2 S_3 S_4 \sigma_4 \sigma_5 \sigma_6 \sigma_7 &= u^2 \int_0^{\Lambda/b} \frac{d\boldsymbol{q}_1 d\boldsymbol{q}_2 d\boldsymbol{q}_3 d\boldsymbol{q}_4}{(2\pi)^{4d}} S(\boldsymbol{q}_1) S(\boldsymbol{q}_2) S(\boldsymbol{q}_3) S(\boldsymbol{q}_4) \\
&\quad \cdot \int_{\Lambda/b}^{\Lambda} \frac{d\boldsymbol{q}_5 d\boldsymbol{q}_6 d\boldsymbol{q}_7 d\boldsymbol{q}_8}{(2\pi)^{4d}} \sigma(\boldsymbol{q}_5) \sigma(\boldsymbol{q}_6) \sigma(\boldsymbol{q}_7) \sigma(\boldsymbol{q}_8) \\
&\quad \cdot (2\pi)^{2d} \delta(\boldsymbol{q}_1 + \boldsymbol{q}_2 + \boldsymbol{q}_3 + \boldsymbol{q}_4) \delta(\boldsymbol{q}_5 + \boldsymbol{q}_6 + \boldsymbol{q}_7 + \boldsymbol{q}_8). \quad \text{(A.72)}
\end{aligned}
$$

Thus, the following classes of second-order corrections to the quartic term result:

(i) $S_1 S_2 S_3 S_4 \langle \sigma_5 \sigma_6 \sigma_7 \sigma_8 \rangle_\sigma$.
    In $\langle V^2 \rangle_\sigma - \langle V \rangle_\sigma^2$, this term cancels with the same term in eqn (A.71).

(ii) $S_1 S_2 S_3 \langle \sigma_4 \sigma_6 \sigma_7 \sigma_8 \rangle_\sigma S_5$.

This term vanishes for the following reason. We first notice that $\langle \sigma_4 \sigma_6 \sigma_7 \sigma_8 \rangle_\sigma = \langle \sigma(q_4) \sigma(q_6) \sigma(q_7) \sigma(q_8) \rangle_\sigma$ vanishes unless q_4, q_6, q_7 and q_8 are paired as $(q_4 = -q_6, q_7 = -q_8)$, $(q_4 = -q_7, q_6 = -q_8)$, or $(q_4 = -q_8, q_6 = -q_7)$ due to the quadratic structure of H_σ in eqn (A.46).[6] For example, if q_4 is independent of q_6, q_7 and q_8, then the integral over $\sigma(q_4)$ will be decoupled from the others,

$$\langle \sigma(q_4) \sigma(q_6) \sigma(q_7) \sigma(q_8) \rangle_\sigma = \langle \sigma(q_4) \rangle_\sigma \langle \sigma(q_6) \sigma(q_7) \sigma(q_8) \rangle_\sigma, \qquad (A.73)$$

which vanishes as H_σ is even in σ. In other words

$$\langle \sigma_4 \sigma_6 \sigma_7 \sigma_8 \rangle_\sigma = \langle \sigma_4 \sigma_6 \rangle_\sigma \langle \sigma_7 \sigma_8 \rangle_\sigma + \langle \sigma_4 \sigma_7 \rangle_\sigma \langle \sigma_6 \sigma_8 \rangle_\sigma + \langle \sigma_4 \sigma_8 \rangle_\sigma \langle \sigma_6 \sigma_7 \rangle_\sigma. \qquad (A.74)$$

The constraint $q_4 + q_6 = 0, q_7 + q_8 = 0$ is incompatible with the other constraint $q_1 + q_2 + q_3 + q_4 = 0, q_5 + q_6 + q_7 + q_8 = 0$, which comes from expressions such as eqn (A.58). The reason is that $q_7 + q_8 = 0$ and $q_5 + q_6 + q_7 + q_8 = 0$ mean $q_5 + q_6 = 0$, which together with $q_4 + q_6 = 0$ leads to $q_4 = q_5$. This is impossible since $|q_4| \geq \Lambda/b$ (q_4 is in σ) and $|q_5| < \Lambda/b$ (q_5 is in S). Similar analyses apply to the combinations $(q_4 = -q_7, q_6 = -q_8)$ and $(q_4 = -q_8, q_6 = -q_7)$.

(iii) $S_1 S_2 \langle \sigma_3 \sigma_4 \sigma_7 \sigma_8 \rangle_\sigma S_5 S_6$.

This gives a non-trivial contribution. As noticed above, the four-body expectation value breaks up into three terms,

$$\langle \sigma_3 \sigma_4 \sigma_7 \sigma_8 \rangle_\sigma = \langle \sigma_3 \sigma_4 \rangle_\sigma \langle \sigma_7 \sigma_8 \rangle_\sigma + \langle \sigma_3 \sigma_7 \rangle_\sigma \langle \sigma_4 \sigma_8 \rangle_\sigma + \langle \sigma_3 \sigma_8 \rangle_\sigma \langle \sigma_4 \sigma_7 \rangle_\sigma. \qquad (A.75)$$

The first term on the right-hand side represents $S_1 S_2 \langle \sigma_3 \sigma_4 \rangle_\sigma S_5 S_6 \langle \sigma_7 \sigma_8 \rangle_\sigma$, which cancels with the same term in $\langle V \rangle_\sigma^2$ of eqn (A.71). Thus, only the last two terms in eqn (A.75) contribute, both of which give the same value. We then evaluate the second term and double its contribution,

$$2 S_1 S_2 S_5 S_6 \langle \sigma_3 \sigma_7 \rangle_\sigma \langle \sigma_4 \sigma_8 \rangle_\sigma = 2u^2 \int_0^{\Lambda/b} \frac{dq_1 dq_2 dq_5 dq_6}{(2\pi)^{4d}} S(q_1) S(q_2) S(q_5) S(q_6)$$

$$\cdot \int_{\Lambda/b}^{\Lambda} \frac{dq_3 dq_7}{(2\pi)^{2d}} \langle \sigma(q_3) \sigma(q_7) \rangle_\sigma \int_{\Lambda/b}^{\Lambda} \frac{dq_4 dq_8}{(2\pi)^{2d}} \langle \sigma(q_4) \sigma(q_8) \rangle_\sigma$$

$$\cdot (2\pi)^{2d} \delta(q_1 + q_2 + q_3 + q_4) \delta(q_5 + q_6 + q_7 + q_8). \qquad (A.76)$$

---

[6] The evaluation of these correlation functions constitute an application of the general Wick's theorem, which can be straightforwardly understood as a consequence of simple Gaussian integration. See also Section 6.6.

The factor after the integral over q_3 and q_7 (second and third lines in eqn (A.76)) is evaluated as

$$\int_{\Lambda/b}^{\Lambda} \frac{dq_3 dq_7}{(2\pi)^{2d}} (2\pi)^d \delta(q_3 + q_7) \frac{1}{2(t + q_3^2)} \int_{\Lambda/b}^{\Lambda} \frac{dq_4 dq_8}{(2\pi)^{2d}} (2\pi)^d \delta(q_4 + q_8) \frac{1}{2(t + q_4^2)}$$

$$\cdot (2\pi)^{2d} \delta(q_1 + q_2 + q_3 + q_4) \delta(q_5 + q_6 + q_7 + q_8)$$

$$= \int_{\Lambda/b}^{\Lambda} \frac{dq_3}{(2\pi)^d} \frac{1}{2(t + q_3^2)} \int_{\Lambda/b}^{\Lambda} \frac{dq_4}{(2\pi)^d} \frac{1}{2(t + q_4^2)}$$

$$\cdot (2\pi)^{2d} \delta(q_1 + q_2 + q_3 + q_4) \delta(q_5 + q_6 - q_3 - q_4)$$

$$= \int_{\Lambda/b}^{\Lambda} \frac{dq_3}{(2\pi)^d} \frac{1}{2(t + q_3^2)} \frac{1}{2(t + (q_3 - q_5 - q_6)^2)} (2\pi)^d \delta(q_1 + q_2 + q_5 + q_6)$$

$$\equiv \frac{I_2 + \mathcal{O}(q_5, q_6)}{4} (2\pi)^d \delta(q_1 + q_2 + q_5 + q_6). \tag{A.77}$$

Here, the integral over q_3, with q_5 and q_6 set to 0, has been defined as I_2,

$$I_2 = \int_{\Lambda/b}^{\Lambda} \frac{dq}{(2\pi)^d} \frac{1}{(t + q^2)^2}. \tag{A.78}$$

Then, the present contribution (A.76) reads

$$\frac{2u^2}{(2\pi)^{4d}} \int_0^{\Lambda/b} dq_1 dq_2 dq_5 dq_6 S(q_1) S(q_2) S(q_5) S(q_6)$$

$$\cdot (2\pi)^d \delta(q_1 + q_2 + q_5 + q_6) \cdot \frac{I_2 + \mathcal{O}(q_5, q_6)}{4}. \tag{A.79}$$

The correction term $\mathcal{O}(q_5, q_6)$ will be neglected hereafter since it corresponds to spatial derivatives of S, and thus is not relevant to the renormalization group evaluation of the simple quartic term. Rescaling of q and renormalization of S change eqn (A.79) into

$$2u^2 b^{-4d+d} \zeta^4 \cdot \frac{I_2}{4} \cdot \frac{1}{(2\pi)^{4d}} \int_0^{\Lambda} dq_{b,1} dq_{b,2} dq_{b,5} dq_{b,6}$$

$$\cdot (2\pi)^d \delta(q_{b,1} + q_{b,2} + q_{b,5} + q_{b,6}) \phi(q_{b,1}) \phi(q_{b,2}) \phi(q_{b,5}) \phi(q_{b,6}). \tag{A.80}$$

This shows that the $\mathcal{O}(V^2)$ correction to the coefficient u of the quartic term is, with $\zeta = b^{1+d/2}$ taken into account,

$$\frac{1}{2} u^2 b^{4-d} I_2 \times 6 \times 6 \times \frac{1}{2} = 9u^2 b^{4-d} I_2, \tag{A.81}$$

where 6×6 comes from the corresponding numerical coefficients in eqn (A.70), and the final $1/2$ is due to the factor in front of $\langle V^2 \rangle_\sigma - \langle V \rangle_\sigma^2$ in the exponent of eqn (A.56).

(iv) $S_1 \langle \sigma_2 \sigma_3 \sigma_4 \sigma_8 \rangle_\sigma S_5 S_6 S_7$.

This vanishes for the same reason as in (ii).

(v) $\langle \sigma_1 \sigma_2 \sigma_3 \sigma_4 \rangle_\sigma S_5 S_6 S_7 S_8$.

This also vanishes for the same reason as in (i).

We have exhausted all possibilities and have the following renormalization group equations for t and u according to eqns (A.67), (A.69) and (A.81),

$$t_b = b^2(t + 3uI_1) \tag{A.82}$$

$$u_b = b^{4-d}u(1 - 9uI_2), \tag{A.83}$$

where the minus sign of $9uI_2$ is due to the difference in the sign in front of $\langle V \rangle_\sigma$ and $(\langle V^2 \rangle_\sigma - \langle V \rangle_\sigma^2)/2$ in the exponent of eqn (A.56).

A.5.4 Differential form of the renormalization group equation

The integrals I_1 and I_2 are simplified in the limit of b close to 1, or $1 - 1/b \equiv l \ll 1$ ($b \approx 1 + l$), corresponding to an infinitesimal renormalization group transformation. In that limit we can approximate the integrals in eqns (A.62) and (A.78) as ($n = 1, 2$)

$$I_n = \int_{\Lambda(1-l)}^{\Lambda} dq \, \mathcal{I}_n(q) \approx \mathcal{I}_n(\Lambda)\Lambda l, \tag{A.84}$$

with

$$\mathcal{I}_n(q) = \frac{S}{(2\pi)^d} q^{d-1} \frac{1}{(t+q^2)^n}, \tag{A.85}$$

where S is the surface area of the unit sphere in the d-dimensional q space (not to be confused with the spin field).[7] Then,

$$I_1 = \frac{\Lambda^d l S}{(2\pi)^d} \frac{1}{t + \Lambda^2} \equiv \frac{cl}{t + \Lambda^2} \tag{A.86}$$

$$I_2 = \frac{cl}{(t + \Lambda^2)^2}. \tag{A.87}$$

Therefore, for small l, eqns (A.82) and (A.83) reduce to

$$\Delta t = t_b - t \approx (1 + 2l)\left(t + \frac{3cl}{t + \Lambda^2}u\right) - t \approx 2lt + \frac{3cl}{t + \Lambda^2}u, \tag{A.88}$$

$$\Delta u = u_b - u \approx (1 + \epsilon l)u\left(1 - \frac{9cl}{(t + \Lambda)^2}u\right) - u \approx \epsilon lu - \frac{9cl}{(t + \Lambda)^2}u^2, \tag{A.89}$$

---

[7] The surface area of a unit sphere S can be computed in the following way

$$\int_{-\infty}^{\infty} dx_1 \cdots dx_d \, e^{-(x_1^2 + \cdots + x_d^2)} = \pi^{d/2} = S \int_0^{\infty} dx \, x^{d-1} e^{-x^2} = \frac{S}{2} \Gamma(d/2),$$

so that $S = 2\pi^{d/2}/\Gamma(d/2)$, where $\Gamma(x)$ is the gamma function.

where $\epsilon = 4 - d$. We finally have the desired differential equations with $\mathrm{d}b = l$

$$\frac{\mathrm{d}t}{\mathrm{d}b} = 2t + \frac{3c}{t + \Lambda^2} u \tag{A.90}$$

$$\frac{\mathrm{d}u}{\mathrm{d}b} = \epsilon u - \frac{9c}{(t + \Lambda^2)^2} u^2. \tag{A.91}$$

A.6 Symmetry and Noether's theorem

Symmetry plays important roles in physics and is often related to conservation laws. For instance, invariance of the Lagrangian under time displacement implies the conservation of energy. *Noether's theorem* is a formal mathematical statement about the consequences of a field theory having continuous symmetries: To every continuous group of transformations that leave the action invariant corresponds a conserved charge. We assume in the present appendix that the reader has some basic knowledge of the concepts and notation of geometry or general relativity such as covariant and contravariant tensor quantities and an implicit summation over repeated indices, known as Einstein summation.[8]

A.6.1 Principle of stationary action

Consider a classical field theory whose action is given by

$$S[\Phi] = \int_\Gamma \mathrm{d}^d r \, \mathcal{L}(\Phi, \partial_\mu \Phi), \tag{A.92}$$

with the Lagrangian density $\mathcal{L}$ containing only up to first-order derivatives of the fields ϕ_i, collectively denoted by $\Phi(\mathbf{r}) = \{\phi_1(\mathbf{r}), \cdots\}$. This ensures that the equations of motion for the fields, the *field equations*, are second-order differential equations in the time variable, as will be shown later. Notice that the Lagrangian density does not depend explicitly on the coordinates r^μ, where $\mu = 1, \ldots, d$.

To write down the field equations we need to apply a variational principle, by considering variations of the action $\delta S[\Phi] \approx S[\Phi + \delta\Phi] - S[\Phi]$, such that variations of the field $\delta\Phi$ are zero on the boundary $\partial\Gamma$ of the volume Γ. The explicit form of the variation is

$$\delta S[\Phi] = \int_\Gamma \mathrm{d}^d r \, \delta\mathcal{L}(\Phi, \partial_\mu \Phi) = \int_\Gamma \mathrm{d}^d r \, \left(\frac{\partial \mathcal{L}}{\partial \phi_i} \delta\phi_i + \frac{\partial \mathcal{L}}{\partial(\partial_\mu \phi_i)} \partial_\mu \delta\phi_i \right). \tag{A.93}$$

Integrating the second term by parts and remembering that $\delta\phi_i = 0$ on the boundary $\partial\Gamma$, we obtain

$$\delta S[\Phi] = \int_\Gamma \mathrm{d}^d r \, \left(\frac{\partial \mathcal{L}}{\partial \phi_i} - \partial_\mu \left(\frac{\partial \mathcal{L}}{\partial(\partial_\mu \phi_i)} \right) \right) \delta\phi_i. \tag{A.94}$$

---

[8] There exists a version of the theorem for classical discrete systems but we are going to discuss only the field-theory formulation.

From the stationarity condition that $\delta S[\Phi] = 0$ for arbitrary $\delta \Phi$ (satisfying the vanishing boundary condition), this expression leads to the following equation of motion of the field, the *Euler–Lagrange equations*, one for each field ϕ_i,

$$\frac{\partial \mathcal{L}}{\partial \phi_i} - \partial_\mu \left(\frac{\partial \mathcal{L}}{\partial (\partial_\mu \phi_i)} \right) = 0. \tag{A.95}$$

This is a set of partial differential equations for the fields. When one of the variables is time, this equation is a second-order differential equation of time t when the Lagrangian density is quadratic in $\partial_t \phi_i$, as is usually assumed.

A.6.2 Symmetries and conserved charges

A sufficient condition for a transformation to be called a *symmetry transformation* of the theory is that it either preserves the functional form of the Lagrangian density or it only changes the Lagrangian density by the addition of a divergence. Then, under the mapping $\Phi(\boldsymbol{r}) \to \Phi'(\boldsymbol{r}')$,

$$\mathcal{L}'(\Phi'(\boldsymbol{r}'), \partial'_\mu \Phi'(\boldsymbol{r}')) = \mathcal{L}(\Phi'(\boldsymbol{r}'), \partial'_\mu \Phi'(\boldsymbol{r}')) + \partial'_\mu \Theta^\mu(\boldsymbol{r}'). \tag{A.96}$$

This preserves the Euler–Lagrange equations of motion since the last divergent term is rewritten as a surface contribution after integration using Gauss theorem and hence does not affect the variational calculations leading to eqn (A.94).

Suppose that one performs a transformation that affects both the coordinates and the fields, i.e. $\boldsymbol{r} \to \boldsymbol{r}'$ $(r'^\mu = \Lambda^\mu_\nu r^\nu)$ and $\Phi(\boldsymbol{r}) \to \Phi'(\boldsymbol{r}') = \mathcal{F}(\Phi(\boldsymbol{r}))$. Then, the action changes to

$$S'[\Phi'] = \int_{\Gamma'} \mathrm{d}^d \boldsymbol{r}' \, \mathcal{L}(\Phi', \partial'_\mu \Phi') = \int_\Gamma \mathrm{d}^d \boldsymbol{r} \, \mathfrak{J}(\boldsymbol{r}) \, \mathcal{L}(\mathcal{F}(\Phi), \boldsymbol{\Lambda}^{-1} \cdot \partial \mathcal{F}(\Phi)), \tag{A.97}$$

where $\mathfrak{J}(\boldsymbol{r})$ is the Jacobian of the transformation.

As a simple example, consider a scale transformation $\boldsymbol{r} \to b^{-1} \boldsymbol{r}$, $\Phi(\boldsymbol{r}) \to \Phi'(\boldsymbol{r}') = b^\Delta \Phi(\boldsymbol{r})$, with Δ the scaling dimension of the field and with the Jacobian $\mathfrak{J}(\boldsymbol{r}) = b^{-d}$. Then, the transformed action is

$$S'[\Phi'] = b^{-d} \int_\Gamma \mathrm{d}^d \boldsymbol{r} \, \mathcal{L}(b^\Delta \Phi, b^{(1+\Delta)} \partial_\mu \Phi). \tag{A.98}$$

If the Lagrangian density corresponds to a real scalar field $\phi(\boldsymbol{r})$

$$\mathcal{L}(\phi, \partial_\mu \phi) = \partial_\mu \phi \, \partial^\mu \phi + \lambda \phi^n, \tag{A.99}$$

with λ a coupling constant and n an integer, the action is scale invariant $S[\Phi] = S'[\Phi']$ only if $\Delta = d/2 - 1$, and $n = d/\Delta = 2d/(d-2)$.

We now reflect on the effect of infinitesimal coordinate transformations on the action of the form

$$r'^\mu = r^\mu + \varepsilon^a \frac{\delta r^\mu}{\delta \varepsilon^a} \ , \quad \Phi'(\boldsymbol{r}') = \Phi(\boldsymbol{r}) + \varepsilon^a \frac{\delta \mathcal{F}}{\delta \varepsilon^a}(\boldsymbol{r}), \tag{A.100}$$

where $\{\varepsilon^a\}$ is a set of infinitesimal, coordinate-dependent, parameters characterizing the continuous mapping. For example, a simple uniform translation has $r'^\mu = r^\mu + \epsilon^\mu$

with a constant vector ϵ^μ. To first order in ε^a, the Jacobian matrix and its inverse transform as

$$\frac{\partial r'^\nu}{\partial r^\mu} = \delta_\mu^\nu + \partial_\mu\left(\varepsilon^a \frac{\delta r^\nu}{\delta\varepsilon^a}\right), \quad \frac{\partial r^\nu}{\partial r'^\mu} = \delta_\mu^\nu - \partial_\mu\left(\varepsilon^a \frac{\delta r^\nu}{\delta\varepsilon^a}\right), \quad \mathfrak{J}(\mathbf{r}) = 1 + \partial_\mu\left(\varepsilon^a \frac{\delta r^\mu}{\delta\varepsilon^a}\right). \text{(A.101)}$$

Then, the transformed action becomes

$$S'[\Phi'] = \int_\Gamma \mathrm{d}^d \mathbf{r}\, \mathfrak{J}(\mathbf{r})\, \mathcal{L}\left(\Phi + \varepsilon^a \frac{\delta\mathcal{F}}{\delta\varepsilon^a}, \left(\delta_\mu^\nu - \partial_\mu\left(\varepsilon^a \frac{\delta r^\nu}{\delta\varepsilon^a}\right)\right)\partial_\nu\left(\Phi + \varepsilon^a \frac{\delta\mathcal{F}}{\delta\varepsilon^a}\right)\right). \quad \text{(A.102)}$$

Let us evaluate $\delta S = S' - S$, which does not vanish in general. For this purpose, we expand the Lagrangian in the integrand of eqn (A.102) as

$$\mathcal{L}\left(\Phi + \varepsilon^a \frac{\delta\mathcal{F}}{\delta\varepsilon^a}, \partial_\mu\Phi + \partial_\mu\left(\varepsilon^a \frac{\delta\mathcal{F}}{\delta\varepsilon^a}\right) - \partial_\mu\left(\varepsilon^a \frac{\delta r^\nu}{\delta\varepsilon^a}\right)\partial_\nu\Phi\right)$$

$$= \mathcal{L}(\Phi, \partial_\mu\Phi) + \varepsilon^a \frac{\delta\mathcal{F}}{\delta\varepsilon^a} \frac{\partial\mathcal{L}}{\partial\Phi}$$

$$+ \partial_\mu\left(\varepsilon^a \frac{\delta\mathcal{F}}{\delta\varepsilon^a}\right)\frac{\partial\mathcal{L}}{\partial(\partial_\mu\Phi)} - \partial_\mu\left(\varepsilon^a \frac{\delta r^\nu}{\delta\varepsilon^a}\right)\partial_\nu\Phi \frac{\partial\mathcal{L}}{\partial(\partial_\mu\Phi)}. \quad \text{(A.103)}$$

Here, since Φ represents a collection of fields $\{\phi_i\}$, the above expression assumes a summation over the field index; for instance $\partial\mathcal{L}/(\partial(\partial_\mu\Phi))\partial_\nu\Phi$ means $\partial\mathcal{L}/(\partial(\partial_\mu\phi_j))\partial_\nu\phi_j$. With the Jacobian contribution of eqn (A.101) taken into account, the change in the action is

$$\delta S = \int_\Gamma \mathrm{d}^d \mathbf{r}\, \left(\varepsilon^a \frac{\delta\mathcal{F}}{\delta\varepsilon^a}\frac{\partial\mathcal{L}}{\partial\Phi} + \partial_\mu\left(\varepsilon^a \frac{\delta\mathcal{F}}{\delta\varepsilon^a}\right)\frac{\partial\mathcal{L}}{\partial(\partial_\mu\Phi)}\right.$$

$$\left. - \partial_\mu\left(\varepsilon^a \frac{\delta r^\nu}{\delta\varepsilon^a}\right)\partial_\nu\Phi \frac{\partial\mathcal{L}}{\partial(\partial_\mu\Phi)} + \partial_\mu\left(\varepsilon^a \frac{\delta r^\mu}{\delta\varepsilon^a}\right)\mathcal{L}\right). \quad \text{(A.104)}$$

Notice that the action is invariant, $\delta S = 0$, if ε^a is a constant independent of the coordinates, i.e. a global symmetry. This means that for a coordinate-dependent ε^a, δS involves only the first-order derivative of ε^a with respect to the coordinates. We therefore drop terms involving ε^a without derivatives. Equation (A.104) therefore becomes

$$\delta S = -\int_\Gamma \mathrm{d}^d \mathbf{r}\, J_a^\mu\, \partial_\mu\varepsilon^a, \quad \text{(A.105)}$$

with

$$J_a^\mu = \left(\frac{\partial\mathcal{L}}{\partial(\partial_\mu\Phi)}\partial_\nu\Phi - \delta_\nu^\mu\mathcal{L}\right)\frac{\delta r^\nu}{\delta\varepsilon^a} - \frac{\partial\mathcal{L}}{\partial(\partial_\mu\Phi)}\frac{\delta\mathcal{F}}{\delta\varepsilon^a}, \quad \text{(A.106)}$$

where J_a^μ is called the *canonical current* associated with the transformation. Integration by parts yields

$$\delta S = \int_\Gamma \mathrm{d}^d \mathbf{r}\, (\partial_\mu J_a^\mu)\, \varepsilon^a. \quad \text{(A.107)}$$

If $\delta S = 0$ for arbitrary small ε^a, the divergence of the current should vanish,

$$\partial_\mu J_a^\mu = 0. \quad \text{(A.108)}$$

This is a conservation law because, if the zeroth coordinate is time and the other coordinates are spatial, this equation can be written as

$$\frac{\partial \rho}{\partial t} + \nabla \cdot \boldsymbol{J} = 0, \tag{A.109}$$

where ρ is the zeroth component of the current vector, J^0, and $\boldsymbol{J} = (J^1, J^2, J^3, \ldots, J^d)$. The subscript a plays no role in this example. This is the well-known charge–current conservation. The emergence of a conservation law, eqn (A.108), as a consequence of symmetry (or invariance $\delta S = 0$) is the statement of Noether's theorem.

We point out that there is some freedom in the definition of the current. One may add to the canonical current the divergence of an antisymmetric tensor, $B_a^{\nu\mu} = -B_a^{\mu\nu}$, $J_a^\mu \to J_a^\mu + \partial_\nu B_a^{\nu\mu}$, and still eqn (A.108) is satisfied.

In classical field theories, time t plays the role of a coordinate and is typically associated with the zeroth component of r^μ, i.e. $r^0 = t$. One can then define a *charge* associated with the current J_a^0 in the following way

$$Q_a = \int_V \mathrm{d}^{d-1}\boldsymbol{r} \, J_a^0, \tag{A.110}$$

with the integral performed over the spatial coordinates in volume V. Now, consider its time derivative, using eqn (A.108),

$$\partial_0 Q_a = \int_V \mathrm{d}^{d-1}\boldsymbol{r} \, \partial_0 J_a^0 = -\int_V \mathrm{d}^{d-1}\boldsymbol{r} \, \partial_i J_a^i = -\int_{\partial V} \mathrm{d}S^i \, J_a^i. \tag{A.111}$$

This vanishes, $\partial_0 Q_a = 0$, if the current becomes zero sufficiently rapidly towards the surface. Therefore, under these conditions, Noether's theorem also implies the (time) conservation of a charge associated with a conserved current.

A.6.3   Energy–momentum tensor

The *energy–momentum tensor* or the *stress tensor* is the conserved current associated with translational invariance and, therefore, Noether's theorem can be used to derive it. Its preponderant role in conformal field theory compels us to look at its properties in more detail.

For a translationally invariant system, the Lagrangian density must be invariant under the transformation $r'^\mu = r^\mu + \epsilon^\mu$. The corresponding conserved canonical current is, according to eqn (A.106),

$$T_\nu^\mu = \frac{\partial \mathcal{L}}{\partial(\partial_\mu \Phi)} \partial_\nu \Phi - \delta_\nu^\mu \mathcal{L}, \tag{A.112}$$

which satisfies $\partial_\mu T_\nu^\mu = 0$, see eqn (A.108). The four-momentum is the conserved charge, using eqn (A.110),

$$P^\nu = \int_V \mathrm{d}^{d-1}\boldsymbol{r} \, T^{0\nu}, \tag{A.113}$$

where the doubly contravariant energy–momentum tensor has been used,

$$T^{\mu\nu} = g^{\rho\nu} T^{\mu}_{\rho} = \frac{\partial \mathcal{L}}{\partial(\partial_{\mu}\Phi)} \partial^{\nu}\Phi - g^{\mu\nu}\mathcal{L}. \tag{A.114}$$

The energy density corresponds to the component $T^{0}_{0} = T^{00}$ of the tensor and the total energy is P^{0}, which defines the Hamiltonian. For a real scalar field $\phi(\boldsymbol{r})$ with Lagrangian (in the Euclidean space-time),

$$\mathcal{L}(\phi, \partial_{\mu}\phi) = \frac{1}{2}(\partial_{\mu}\phi\,\partial^{\mu}\phi + m^{2}\phi^{2}), \tag{A.115}$$

the doubly covariant energy–momentum tensor $(T_{\mu\nu} = g_{\mu\rho} T^{\rho}_{\nu})$ is

$$T_{\mu\nu} = \partial_{\mu}\phi\,\partial_{\nu}\phi - g_{\mu\nu}\mathcal{L}, \tag{A.116}$$

and turns out to be symmetric $T_{\mu\nu} = T_{\nu\mu}$. For an arbitrary field theory, though, it is not necessarily true that the conserved canonical current produces a symmetric tensor. We have seen in the previous section that there is some freedom in the definition of the conserved current. By adding the divergence of an antisymmetric tensor, $B^{\rho\mu\nu} = -B^{\mu\rho\nu}$, to the conserved canonical current, one can generate another conserved current $T^{\mu\nu} \to T^{\mu\nu} + \partial_{\rho}B^{\rho\mu\nu}$, since $\partial_{\mu}\partial_{\rho}B^{\rho\mu\nu} = 0$. The important fact is that this addition does not alter the conservation law. In field theories with rotation (Euclidean) or Lorentz (Minkowski) invariance, $T^{\mu\nu}$ can be made symmetric. See also Appendix A.10.[9]

A.6.4   Generators of symmetry transformations

It is instructive to show the relation between the canonical current and the generators of infinitesimal transformations. These generators $\{g_a\}$ are defined as

$$\Phi'(\boldsymbol{r}) = (1 - i\varepsilon^{a} g_a)\Phi(\boldsymbol{r}). \tag{A.117}$$

If ε^{a} is coordinate independent, the transformation defines a global symmetry. From eqn (A.100), we may rewrite the fields in terms of the transformed coordinate to lowest order in ε^{a}

$$\Phi'(\boldsymbol{r}') = \Phi(\boldsymbol{r}) + \varepsilon^{a}\frac{\delta\mathcal{F}}{\delta\varepsilon^{a}}(\boldsymbol{r}) = \Phi(\boldsymbol{r}') - \varepsilon^{a}\frac{\delta r^{\mu}}{\delta\varepsilon^{a}}\partial'_{\mu}\Phi(\boldsymbol{r}') + \varepsilon^{a}\frac{\delta\mathcal{F}}{\delta\varepsilon^{a}}(\boldsymbol{r}'), \tag{A.118}$$

and thus identify

$$ig_a\Phi(\boldsymbol{r}) = \frac{\delta r^{\mu}}{\delta\varepsilon^{a}}\partial_{\mu}\Phi(\boldsymbol{r}) - \frac{\delta\mathcal{F}}{\delta\varepsilon^{a}}(\boldsymbol{r}). \tag{A.119}$$

The conserved canonical current of eqn (A.106) in terms of the generators can be written as

$$J^{\mu}_{a} = i\frac{\partial\mathcal{L}}{\partial(\partial_{\mu}\Phi)}g_a\Phi - \delta^{\mu}_{\nu}\frac{\delta r^{\nu}}{\delta\varepsilon^{a}}\mathcal{L}. \tag{A.120}$$

---

[9] One can define a symmetric energy-momentum tensor in an alternative way, as in eqn (6.35). In other words, it can be defined as the functional derivative of the action with respect to the metric.

As an example, consider the case of an infinitesimal translation by a vector $\epsilon^\mu = \varepsilon^\mu$, i.e. $r'^\mu = r^\mu + \epsilon^\mu$. Then, $\delta\mathcal{F}/\delta\epsilon^\mu = 0$, and $\delta r^\mu/\delta\epsilon^\nu = \delta^\mu_\nu$, which leads us to the well-known generator of translations

$$g_a = P_\nu = -\mathrm{i}\partial_\nu. \tag{A.121}$$

Another example of great interest for conformal transformations is the case of an infinitesimal homogeneous Lorentz transformation in Euclidean space-time

$$r'^\mu = r^\mu + \varepsilon^\mu_\nu r^\nu, \tag{A.122}$$

where $\varepsilon^\mu_\nu = \omega^{\mu\rho}g_{\rho\nu} = \omega_{\rho\nu}g^{\rho\mu}$, with an antisymmetric tensor $\omega^{\mu\nu} = -\omega^{\nu\mu}$ and the flat space metric $g_{\mu\nu} = \mathrm{diag}(1,1,\cdots,1)$. This implies, using $g^{\mu\rho}g_{\rho\nu} = \delta^\mu_\nu$ and $g_{\mu\nu} = g_{\nu\mu}$,

$$\frac{\delta r^\mu}{\delta\omega_{\rho\nu}} = \frac{1}{2}(g^{\rho\mu}r^\nu - g^{\nu\mu}r^\rho). \tag{A.123}$$

Similarly, the effect of the Lorentz transformation on the field Φ is

$$\mathcal{F}(\Phi) = (1 - \frac{\mathrm{i}}{2}\omega_{\rho\nu}S^{\rho\nu})\Phi, \tag{A.124}$$

with a Hermitian matrix $S^{\rho\nu}$. It follows that

$$\frac{\delta\mathcal{F}}{\delta\omega_{\rho\nu}} = -\frac{\mathrm{i}}{2}S^{\rho\nu}\Phi. \tag{A.125}$$

By inserting eqns (A.123) and (A.125) into eqn (A.119), we find the generators of the Lorentz transformation as

$$2g_a = L^{\rho\nu} = \mathrm{i}(r^\rho\partial^\nu - r^\nu\partial^\rho) + S^{\rho\nu}. \tag{A.126}$$

A.6.5 Goldstone theorem

The Goldstone theorem states that there exists a massless (zero-energy) mode if a continuous global symmetry is spontaneously broken. To formulate it more precisely, consider a Lagrangian density

$$\mathcal{L}(\Phi, \partial_\mu\Phi) = \frac{1}{2}\partial_\mu\Phi\partial^\mu\Phi + V(\Phi), \tag{A.127}$$

where $\Phi(\boldsymbol{r}) = (\phi_1(\boldsymbol{r}), \cdots, \phi_n(\boldsymbol{r}))$ is an n-component vector field. We may define the mass matrix as

$$(M^2)_{ij} = \frac{\partial^2 V}{\partial\phi_i\partial\phi_j} \tag{A.128}$$

because, for a simple single-component Gaussian field theory,

$$\mathcal{L}(\phi, \partial_\mu\phi) = \frac{1}{2}\partial_\mu\phi\,\partial^\mu\phi + \frac{1}{2}m^2\phi^2, \tag{A.129}$$

the parameter m is called the mass in field theory. Suppose that, under a global symmetry operation $\mathfrak{g}_\alpha$ (see Appendix A.7), the field $\bar{\Phi}$ (representing a broken symmetry state) changes into another state $\mathfrak{g}_\alpha \bar{\Phi} = e^{-i\varepsilon^a g_a} \bar{\Phi}(\neq \bar{\Phi})$. Here, g_a is the generator of an infinitesimal transformation,

$$\Phi'(\boldsymbol{r}) = (1 - i\varepsilon^a g_a)\Phi(\boldsymbol{r}). \tag{A.130}$$

Since the system has the symmetry under $\mathfrak{g}$, the potential V is invariant under an infinitesimal transformation,

$$\frac{\partial V(\mathfrak{g}_\alpha \Phi)}{\partial \varepsilon^a}\bigg|_{\varepsilon^a = 0} = \frac{\partial V}{\partial \phi_j} \frac{\partial (\mathfrak{g}_\alpha \Phi)_j}{\partial \varepsilon^a}\bigg|_{\varepsilon^a = 0} = -i\frac{\partial V}{\partial \phi_j}(g_a \Phi)_j = 0 \tag{A.131}$$

for generic Φ. Then, by differentiating this last equation with respect to ϕ_i, we obtain

$$\frac{\partial}{\partial \phi_i}\left(\frac{\partial V}{\partial \phi_j}(g_a \Phi)_j\right) = \frac{\partial^2 V}{\partial \phi_i \partial \phi_j}(g_a \Phi)_j + \frac{\partial V}{\partial \phi_j}\frac{\partial (g_a \Phi)_j}{\partial \phi_i} = 0. \tag{A.132}$$

If we restrict ourselves to the state of broken symmetry, $\Phi = \bar{\Phi}$, the derivative of the potential vanishes by definition, $\partial V/\partial \phi_j|_{\Phi=\bar{\Phi}} = 0$, and therefore the mass matrix has a massless mode,

$$(M^2)_{ij}(g_a \bar{\Phi})_j = 0. \tag{A.133}$$

A.7 Basics of group theory and Lie algebras

A minimum amount of basic knowledge, mainly the definitions and notations, is summarized in this appendix for the group theory and Lie algebra.

A.7.1 Group and its representations

Symmetries may be classified as external or space-time (e.g. the Lorentz group, the group of Lorentz transformation) and internal. The latter refers to the set of transformations that leave the Hamiltonian or action invariant and act on the fields but not on the coordinates; these are the symmetries of the physical laws. The set of transformations forms a *group* $\mathcal{G}$ and is written as

$$\mathcal{G} = \{\mathfrak{g}_\alpha\}, \tag{A.134}$$

with group elements $\mathfrak{g}_\alpha$: A group is a non-empty set of elements, which is closed under an associative product $(\mathfrak{g}_\alpha \cdot \mathfrak{g}_\beta) \cdot \mathfrak{g}_\gamma = \mathfrak{g}_\alpha \cdot (\mathfrak{g}_\beta \cdot \mathfrak{g}_\gamma)$, contains an identity, $\mathfrak{e}$, and all of its elements are invertible, i.e. $\mathfrak{g}_\alpha \cdot \mathfrak{g}_\alpha^{-1} = \mathfrak{e}$. The number of elements of the group defines its *order*, which may be finite, denumerable infinite (discrete), or non-denumerable infinite (continuous). In general, groups of symmetries in physics are either finite or non-denumerable infinite (known as Lie groups, see below). Besides, a group $\mathcal{G}$ may be *Abelian*, in which all its group elements commute $\mathfrak{g}_\alpha \cdot \mathfrak{g}_{\alpha'} = \mathfrak{g}_{\alpha'} \cdot \mathfrak{g}_\alpha, \forall \alpha, \alpha'$, or non-Abelian. A group representing symmetries of a physical system may be local (also called gauge), meaning that the symmetry applies to subsystems of the original physical system, or global, where all the degrees of freedom are involved in the transformation. Invariant physical observables, O, are those physical quantities that remain invariant under the symmetry group $\mathcal{G}$.

Table A.1 shows representative examples of physical models displaying different kinds of symmetries. For instance, the Heisenberg Hamiltonian without external field

Table A.1 Examples of models displaying different kinds of symmetries. The group (or subgroup) of symmetries involved is written in parentheses. BCS stands for the Bardeen-Cooper–Schrieffer model of superconductivity.

Symmetry	Discrete	Continuous
	Ising [$\mathbb{Z}_2$]	XY [$SO(2)$]
Global		
	p-clock [$\mathbb{Z}_p$]	Heisenberg [$SO(3)$]
	Z_2 gauge [$\mathbb{Z}_2$]	$U(1)$ gauge [$U(1)$]
Local		
		BCS [hidden $SU(2)$]

is invariant under all the geometric symmetries of the lattice as well as under the group $SO(3)$ that applies to the spin space (the Special Orthogonal group in three dimensions, which is non-Abelian). This means that, if we rotate *all* spins through an arbitrary angle about a fixed axis in three dimensions, the Hamiltonian remains invariant because the interaction is written as the inner product of two spins, $\boldsymbol{S}_i \cdot \boldsymbol{S}_j$. Similarly, if one considers the ϕ^4 field theory model of eqn (5.23) and performs the $\mathbb{Z}_2$ mapping $\phi(\boldsymbol{r}) \to -\phi(\boldsymbol{r})$ at all $\boldsymbol{r}$, the Hamiltonian (or action) remains invariant. The Abelian group $\mathbb{Z}_p$ is composed of p elements, each of which corresponds to the rotation in a two-dimensional space by an angle that is a multiple of $2\pi/p$. The p-clock model has the same expression as the XY model

$$H = -J \sum_{\langle ij \rangle} \cos(\phi_i - \phi_j), \tag{A.135}$$

the difference being in the values of the angle variable; ϕ_i is an integer multiple of $2\pi/p$. This Hamiltonian clearly has the global symmetry of $\mathbb{Z}_p$, $\phi_i \to \phi_i - 2\pi k/p$ ($\forall i$) with an integer k. The XY model is recovered in the limit $p \to \infty$. Some accounts are given on the lattice gauge theory listed in Table A.1 in Section 7.7.

With each group element $\mathfrak{g}_\alpha$ there is an associated matrix that will be denoted as $\mathcal{O}_\alpha = \mathcal{O}(\mathfrak{g}_\alpha)$ satisfying the same group relations. The set $\{\mathcal{O}_\alpha\}$ forms a *representation* of the group $\mathcal{G}$. A representation is a homomorphic mapping of the group $\mathcal{G}$ onto a set of matrices $\mathcal{O}$ such that: $\mathcal{O}(\mathfrak{e}) = \mathbb{1}$ and $\mathcal{O}_\alpha \mathcal{O}_\beta = \mathcal{O}(\mathfrak{g}_\alpha \mathfrak{g}_\beta)$. The dimension of the representation, $\dim(\mathcal{O})$, is the dimension of the (vector) space on which it acts. In this book, by a representation we simply mean a non-singular $\dim(\mathcal{O}) \times \dim(\mathcal{O})$ matrix representation. A representation is *irreducible* if its invariant subspaces under the action of all the elements of the group are only the empty space and the full space. A completely *reducible* representation can be written as a direct sum of irreducible representations, known as irreps.

Consider the Hamiltonian of a quantum system, H, which commutes with $\mathcal{O}_\alpha$, i.e. $[H, \mathcal{O}_\alpha] = 0$. The set of eigenstates $\{|\Psi_n\rangle\}$ with the same eigenvalue E_n form an invariant subspace

$$HO_\alpha|\Psi_n\rangle = O_\alpha H|\Psi_n\rangle = E_n O_\alpha|\Psi_n\rangle. \tag{A.136}$$

This implies that $|\Psi'_n\rangle = O_\alpha|\Psi_n\rangle$ is also an eigenstate with the same eigenvalue. When the dimension of this invariant subspace is larger than one, the energy eigenvalue E_n is *degenerate*. The dimension of the degenerate subspace is equal to the dimension of the representation of $\mathcal{G}$ associated with the eigenstate $|\Psi_n\rangle$. If the group $\mathcal{G}$ is Abelian, all the irreps are one-dimensional and there is no degeneracy induced by $\mathcal{G}$. It is important to emphasize that symmetry does not always imply degeneracy.

A.7.2   Lie algebra

An interesting case, because of the physical consequences, is realized by the presence of continuous symmetries, such as $SO(3)$. We have already seen that the set of continuous transformations forms a group, known as the *Lie group*. Lie groups play a fundamental role in physics. In the Lie group, a notion of continuity or *closeness* should be defined such that a finite transformation of the group can be generated by a series of infinitesimal ones. A simple example is the group of spatial translations, for which a finite amount of translation is achieved by the accumulation of infinitesimal translations.

For a one-parameter continuous group, a representation of its elements can be written as

$$O_\mu(\theta) = e^{i\theta X_\mu}, \tag{A.137}$$

where θ is a continuous parameter and the X_μs are the *generators* of the Lie group.[10] The representations of the group elements are defined such that $\theta = 0$ represents the identity operator $\mathbb{1}$ and an infinitesimal transformation $\delta\theta$ is expressed as

$$O_\mu(\delta\theta) = \mathbb{1} + i\,\delta\theta\,X_\mu. \tag{A.138}$$

The generators form a *Lie algebra*: A real/complex Lie algebra $\mathcal{L}$ is a linear space over real/complex numbers, where a *Lie product* $[\,,\,]$ is defined that satisfies the following rules, $(\mathfrak{a}, \mathfrak{b}, \mathfrak{c} \in \mathcal{L}$ and $\alpha, \beta \in \mathbb{R}$ or $\mathbb{C})$:

$$[\alpha\mathfrak{a} + \beta\mathfrak{b}, \mathfrak{c}] = \alpha[\mathfrak{a}, \mathfrak{c}] + \beta[\mathfrak{b}, \mathfrak{c}]$$

$$[\mathfrak{a}, \mathfrak{b}] = -[\mathfrak{b}, \mathfrak{a}]$$

$$0 = [\mathfrak{a}, [\mathfrak{b}, \mathfrak{c}]] + [\mathfrak{b}, [\mathfrak{c}, \mathfrak{a}]] + [\mathfrak{c}, [\mathfrak{a}, \mathfrak{b}]]. \tag{A.139}$$

An example of a Lie product is the commutator $[\mathfrak{a}, \mathfrak{b}] = \mathfrak{a}\mathfrak{b} - \mathfrak{b}\mathfrak{a}$. The Poisson bracket used in classical mechanics also satisfies the above condition. The Lie algebra is required to be closed by the Lie product, i.e. $[X_\mu, X_\nu] \in \mathcal{L}$,

$$[X_\mu, X_\nu] = iC_{\mu\nu}^\gamma\,X_\gamma, \tag{A.140}$$

where an implicit summation over the repeated index γ is assumed (Einstein summation). The coefficient $C_{\mu\nu}^\gamma$ is called the *structure constant* of the algebra. The relation of eqn (A.140) is also often called the Lie algebra in the physics literature.

---

[10] Precisely speaking, a generator is defined by the differentiation of an element of the Lie group near the identity $\mathfrak{e}$, not necessarily by using the representation of the element. We, nevertheless, refer to an element and its representation interchangeably in this book as long as it causes no confusion.

A.8   Basics of homotopy theory

We formulate the theory of homotopy described in Section 5.8 in a little more formal way. Let $\mathcal{X}$ and $\mathcal{Y}$ be two topological spaces and consider the set of maps $\mathcal{F} = \{f_i\}$ from closed curves in $\mathcal{X}$ to $\mathcal{Y}$. In the example of the XY model in two dimensions given in Section 5.8, $\mathcal{X}$ is $\mathbb{R}^2$ and $\mathcal{Y}$ is $\mathbb{S}^1$. Two maps f_0 and f_1 are *homotopic* if they can be continuously deformed into each other. For example, any loop Γ on the surface of a sphere is homotopic to a point since Γ can be continuously deformed to a point, as shown in Fig. 5.6. On the other hand, the surface of a doughnut has loops that can be shrunk to a point while other loops, e.g. the ones wrapping the main circumference, cannot be. This equivalence relation defines an equivalence class known as the *homotopy class*. Mathematically, let I be the interval $[0, 1]$ and x_0 a point in $\mathcal{X}$. Then, a loop is defined as a continuous map

$$\Gamma : I \ni x \rightarrow \Gamma(x) \in \mathcal{X}, \tag{A.141}$$

with $\Gamma(0) = \Gamma(1) = x_0$. The *product* of two loops Γ_1, Γ_2 based at x_0 is another loop $\Gamma_1 \circ \Gamma_2 : I \rightarrow \mathcal{X}$ such that

$$\Gamma_1 \circ \Gamma_2(x) = \begin{cases} \Gamma_1(2x) & 0 \leq x \leq 1/2 \\ \Gamma_2(2x - 1) & 1/2 \leq x \leq 1 \end{cases}, \tag{A.142}$$

while the *inverse* of a loop Γ is $\Gamma^{-1}(x) = \Gamma(1 - x)$ that corresponds to traversing the loop in the opposite direction. The product or composition of loops is associative, i.e. $\Gamma_1 \circ (\Gamma_2 \circ \Gamma_3) = (\Gamma_1 \circ \Gamma_2) \circ \Gamma_3$. The loops by themselves do not have a group structure, but the equivalence classes of loops form a group (see Appendix A.7).

Let us start by defining the equivalence relation 'homotopic to'. Two loops Γ_1 and Γ_2 based at x_0 are homotopic, symbolically written $\Gamma_1 \sim \Gamma_2$, if there is a continuous map Υ, called the *homotopy*,

$$\Upsilon : I \times I \ni (x, t) \rightarrow \Upsilon(x, t) \in \mathcal{X}, \tag{A.143}$$

such that $\Upsilon(x, 0) = \Gamma_1(x), \Upsilon(x, 1) = \Gamma_2(x), \forall x \in I$, and $\Upsilon(0, t) = \Upsilon(1, t) = x_0, \forall t \in I$. As anticipated, the relation $\sim$ defines an equivalence relation with the class of loops homotopic to a representative loop Γ denoted as $[\Gamma]$. The product or composition of homotopy classes is defined by $[\Gamma_1] \circ [\Gamma_2] = [\Gamma_1 \circ \Gamma_2]$. The *first homotopy group* or the *fundamental group*, $\pi_1(\mathcal{X}, x_0)$, is the set of homotopy classes of loops based at $x_0 \in \mathcal{X}$. If the topological space $\mathcal{X}$ is arcwise connected,[11] then two groups based at different points x_0 and x_1 are isomorphic, $\pi_1(\mathcal{X}, x_0) \cong \pi_1(\mathcal{X}, x_1)$, which means that we can simply write $\pi_1(\mathcal{X})$ for the fundamental group. In our example of the XY model, the fundamental group is $\pi_1(\mathbb{S}^1) = \mathbb{Z}$, the group of integers under addition.

The first homotopy group classifies classes of loops in a given topological space $\mathcal{X}$. One may sometimes wish to assign other groups to $\mathcal{X}$. Indeed, it is possible to define homotopy classes of n-dimensional ($n \geq 1$) spheres $\mathbb{S}^n$ in $\mathcal{X}$ with the property that they realize *higher homotopy groups*, $\pi_n(\mathcal{X})$. As before, if $\mathcal{X}$ is arcwise connected, one does not need to specify the base point. The n-loop based at $x_0 \in \mathcal{X}$ is a continuous map

---

[11] $\mathcal{X}$ is arcwise connected if, for any x_0 and $x_1 \in \mathcal{X}$, there exists a path joining these two points.

$$\Gamma : I^n \ni x^n \to \Gamma(x^n) \in \mathcal{X}, \tag{A.144}$$

and the homotopy is defined in terms of the unit n-cube interval

$$\Upsilon : I^n \times I \ni (x^n, t) \to \Upsilon(x^n, t) \in \mathcal{X}. \tag{A.145}$$

Although the fundamental group may be non-Abelian, higher homotopy groups ($n > 1$) share the property of being always Abelian.

We would like to emphasize that homotopy groups classify maps, or specifically in our XY model example, classify spin configurations. Homeomorphisms between the topological spaces $\mathcal{X}$ and $\mathcal{Y}$ categorize those spaces into equivalent classes. The fact that homotopy groups can be defined gives a real intrinsic value to homotopy theory since that same group structure provides the laws for the combination of defects and the rules for their characterization and classification.

A.9 Restrictions on the type of conformal mappings

We show in this appendix that translation, rotation, dilatation and the special conformal transformation exhaust the list of possible conformal transformations for $d \geq 3$. Also shown is that the Cauchy–Riemann equations emerge for $d = 2$, which means that holomorphic transformations are allowed as conformal mappings in two dimensions in addition to the above-mentioned transformations. It is assumed also in the present and next appendices that the reader is familiar with the notations of geometry or general relativity. We consider an infinitesimal transformation of the Euclidean metric, $g_{\mu\nu} = \delta_{\mu\nu}$, where $\delta_{\mu\nu}$ is Kronecker's symbol, and consequently we do not distinguish between covariant and contravariant quantities, i.e. upper and lower indices.

Let us consider an infinitesimal coordinate transformation $r'^\mu = r^\mu + \epsilon^\mu(\mathbf{r})$. This induces local variations of the metric as ($\partial_\mu = \partial/\partial r^\mu$)

$$g'_{\mu\nu} = g_{\mu\nu} - (\partial_\mu \epsilon_\nu + \partial_\nu \epsilon_\mu) = g_{\mu\nu} + \delta g_{\mu\nu} \tag{A.146}$$

because of the definition

$$g'_{\mu\nu}(\mathbf{r}') = \frac{\partial r_\kappa}{\partial r'^\mu} \frac{\partial r_\lambda}{\partial r'^\nu} g_{\kappa\lambda}(\mathbf{r}), \tag{A.147}$$

and its consequence for an infinitesimal transformation $r'_\mu = r_\mu + \epsilon_\mu$

$$\frac{\partial r_\kappa}{\partial r'^\mu} = \delta_{\kappa\mu} - \partial_\mu \epsilon_\kappa. \tag{A.148}$$

According to eqn (A.146), the requirement of local angle preservation for conformal mappings

$$g'_{\mu\nu}(\mathbf{r}') = \Omega(\mathbf{r}) g_{\mu\nu}(\mathbf{r}) \tag{A.149}$$

implies

$$\partial_\mu \epsilon_\nu + \partial_\nu \epsilon_\mu = f(\mathbf{r}) g_{\mu\nu} \tag{A.150}$$

for some $f(\mathbf{r})$. Summing it over $\mu = \nu$, we find

$$2\partial^\mu \epsilon_\mu = f(\mathbf{r}) d, \tag{A.151}$$

so that eqn (A.150) becomes

$$\partial_\mu \epsilon_\nu + \partial_\nu \epsilon_\mu = \frac{2}{d} g_{\mu\nu} (\partial^\kappa \epsilon_\kappa). \qquad (A.152)$$

Now, let us apply ∂_κ to eqn (A.150) as

$$\partial_\kappa \partial_\mu \epsilon_\nu + \partial_\kappa \partial_\nu \epsilon_\mu = g_{\mu\nu} \partial_\kappa f, \qquad (A.153)$$

and write two equivalent equations obtained by the changes of indices $\mu \leftrightarrow \kappa$ and $\nu \leftrightarrow \kappa$,

$$\partial_\mu \partial_\kappa \epsilon_\nu + \partial_\mu \partial_\nu \epsilon_\kappa = g_{\kappa\nu} \partial_\mu f \qquad (A.154)$$

$$\partial_\nu \partial_\mu \epsilon_\kappa + \partial_\nu \partial_\kappa \epsilon_\mu = g_{\mu\kappa} \partial_\nu f. \qquad (A.155)$$

If we sum eqns (A.153) and (A.154) and subtract eqn (A.155) from the result, we reach

$$2\partial_\mu \partial_\kappa \epsilon_\nu = g_{\mu\nu} \partial_\kappa f + g_{\kappa\nu} \partial_\mu f - g_{\mu\kappa} \partial_\nu f. \qquad (A.156)$$

Summation of this equation over $\mu = \kappa$ results in

$$2\partial^2 \epsilon_\nu = (2 - d)\partial_\nu f, \qquad (A.157)$$

or, by a further differentiation,

$$2\partial^2 \partial_\mu \epsilon_\nu = (2 - d)\partial_\mu \partial_\nu f. \qquad (A.158)$$

Notice that this equation holds if we exchange μ and ν since the right-hand side is symmetric with respect to μ and ν. An application of ∂^2 to eqn (A.150) yields

$$\partial^2 (\partial_\mu \epsilon_\nu + \partial_\nu \epsilon_\mu) = g_{\mu\nu} \partial^2 f, \qquad (A.159)$$

the left-hand side of which coincides with the left-hand side of eqn (A.158) because of the symmetry of this latter equation under the exchange of μ and ν. We then arrive at

$$(2 - d)\partial_\mu \partial_\nu f = g_{\mu\nu} \partial^2 f. \qquad (A.160)$$

Summation of both sides over $\mu = \nu$ yields

$$(2 - d)\partial^2 f = d \, \partial^2 f. \qquad (A.161)$$

If $d \neq 1, 2$, we conclude $\partial^2 f = 0$ and thus $\partial_\mu \partial_\nu f = 0$ according to eqn (A.160). Consequently, f is at most linear in coordinates, and ϵ_μ is therefore at most quadratic according to eqn (A.151). As discussed in Sections 6.2 and 6.3, infinitesimal transformations of at most quadratic order correspond to translation, rotation, dilatation and the special conformal transformation. This is what we planned to show.

In two dimensions, $d = 2$, we have the additional result that eqn (A.152) yields the Cauchy–Riemann equations. Let us write the right-hand side of eqn (A.152) explicitly as

$$g_{\mu\nu} (\partial^\kappa \epsilon_\kappa) = \delta_{\mu\nu} (\partial^1 \epsilon_1 + \partial^2 \epsilon_2). \qquad (A.162)$$

Then, the diagonal case $\mu = \nu = 1$ of eqn (A.152) is

$$2\partial^1 \epsilon_1 = \partial^1 \epsilon_1 + \partial^2 \epsilon_2, \qquad (A.163)$$

and the off-diagonal case $\mu = 1, \nu = 2$ is

$$\partial^1 \epsilon_2 + \partial^2 \epsilon_1 = 0. \tag{A.164}$$

These are the Cauchy–Riemann equations, $\partial^1 \epsilon_1 = \partial^2 \epsilon_2$, $\partial^1 \epsilon_2 = -\partial^2 \epsilon_1$. We thus conclude that any holomorphic function represents a conformal mapping in two dimensions.

A.10   Properties of the energy–momentum tensor

In this appendix we derive eqn (6.36) and discuss a few important properties of the energy–momentum tensor in two dimensions.

Since a coordinate transformation $z \to z + \epsilon(z)$ does not change the value of a correlation function $\langle X_n \rangle$ and since the partition function does not change by the same transformation, the numerator of the definition

$$\langle X_n \rangle = \frac{\displaystyle\int \mathcal{D}\phi \, X_n \mathrm{e}^{-S}}{\displaystyle\int \mathcal{D}\phi \, \mathrm{e}^{-S}} \tag{A.165}$$

also remains invariant,

$$\int \mathcal{D}\phi \, (\delta X_n) \, \mathrm{e}^{-S} - \int \mathcal{D}\phi \, X_n (\delta S) \, \mathrm{e}^{-S} = 0, \tag{A.166}$$

or more explicitly,

$$\sum_i \langle \phi_1(z_1, \bar{z}_1) \cdots \delta_{\epsilon\bar\epsilon} \phi_i(z_i, \bar{z}_i) \cdots \phi_n(z_n, \bar{z}_n) \rangle$$

$$+ \frac{1}{2\pi} \int_{\bar{D}} \mathrm{d}^2 \boldsymbol{r} \, \partial^\mu \epsilon^\nu \langle T_{\mu\nu}(\boldsymbol{r}) \phi_1(z_1, \bar{z}_1) \cdots \phi_n(z_n, \bar{z}_n) \rangle = 0, \tag{A.167}$$

where $\bar{D}$ is the region outside D defined in Section 6.4. We rewrite a part of the second term by integration by parts as

$$\frac{1}{2\pi} \int_{\bar{D}} \mathrm{d}^2 \boldsymbol{r} \, \partial^\mu \epsilon^\nu T_{\mu\nu}(\boldsymbol{r}) = \frac{1}{2\pi} \oint_C \mathrm{d}r_\lambda \, \omega^{\lambda\mu} \epsilon^\nu T_{\mu\nu}(\boldsymbol{r}) - \frac{1}{2\pi} \int_{\bar{D}} \mathrm{d}^2 \boldsymbol{r} \, \epsilon^\nu \partial^\mu T_{\mu\nu}(\boldsymbol{r}), \quad \text{(A.168)}$$

where the first term on the right-hand side is the surface (indeed, line) contribution and $\omega^{\lambda\mu}$ is an antisymmetric tensor $\omega^{12} = -\omega^{21} = 1$, $\omega^{11} = \omega^{22} = 0$. Since ϵ^ν can be chosen arbitrarily in $\bar{D}$ and the final result should not depend upon this choice, we conclude from the second term that the energy–momentum tensor satisfies

$$\partial^\mu T_{\mu\nu} = 0. \tag{A.169}$$

The integrand of the first term on the right-hand side of eqn (A.168) can be rewritten in terms of complex variables

$$w = r^1 + \mathrm{i}\, r^2, \ \bar{w} = r^1 - \mathrm{i}\, r^2, \ \epsilon(w) = \epsilon^1 + \mathrm{i}\, \epsilon^2, \ \bar{\epsilon}(\bar{w}) = \epsilon^1 - \mathrm{i}\, \epsilon^2, \tag{A.170}$$

and using the following properties of the energy–momentum tensor, to be proved later,

$$T_{12} = T_{21}, \; T_{11} + T_{22} = 0, \tag{A.171}$$

it results

$$\mathrm{d}r_\lambda\, \omega^{\lambda\mu} \epsilon^\nu T_{\mu\nu} = \mathrm{i}\,\mathrm{d}w\, \epsilon(w)T(w) - \mathrm{i}\,\mathrm{d}\bar{w}\, \bar{\epsilon}(\bar{w})\bar{T}(\bar{w}), \tag{A.172}$$

where we have used the definition of $T(w)$ and $\bar{T}(\bar{w})$ in eqns (6.37) and (6.38). We insert this equation into eqn (A.168) and then into eqn (A.167) and use eqn (6.22) for $\delta_{\epsilon\bar{\epsilon}}\phi_i$ to complete the derivation of eqn (6.36).

It remains to be shown that the energy–momentum tensor is symmetric and traceless, eqn (A.171). For a conformally invariant theory, the action is invariant under rotations

$$z + \epsilon(z) = \mathrm{e}^{\mathrm{i}\theta} z \approx (1 - \mathrm{i}\theta)z. \tag{A.173}$$

This means $\epsilon^1 = \theta r^2$, $\epsilon^2 = -\theta r^1$, which leads to $\partial^2 \epsilon^1 = \theta$, $\partial^1 \epsilon^2 = -\theta$ with other derivatives being zero. Then, the definition of the energy–momentum tensor, eqn (6.35), becomes

$$\delta S = -\frac{\theta}{2\pi} \int \mathrm{d}^2 \boldsymbol{r} \, (T_{21} - T_{12}) = 0 \tag{A.174}$$

for arbitrary (but small) θ, from which we conclude $T_{12} = T_{21}$. Analogously, invariance under dilatations, $z + \epsilon(z) = (1 + a)z$, leads to $\partial^1 \epsilon^1 = \partial^2 \epsilon^2 = a$ and other derivatives vanish. It then follows that $T_{11} + T_{12} = 0$ from a similar argument as above.

Strictly speaking, the left-hand side of the definition of T and $\bar{T}$ in eqns (6.37) and (6.38) should include both w and $\bar{w}$ in the arguments as $T(w, \bar{w})$ and $\bar{T}(w, \bar{w})$. We show that T actually depends only on the holomorphic variable and $\bar{T}$ only on the antiholomorphic variable, which justifies the notation of eqns (6.37) and (6.38), i.e.

$$\partial_{\bar{z}} T(z, \bar{z}) = 0, \; \partial_z \bar{T}(z, \bar{z}) = 0. \tag{A.175}$$

To show this result, let us take the derivative of $T(z, \bar{z})$ with respect to $\bar{z}$,

$$\partial_{\bar{z}} T(z, \bar{z}) = \frac{1}{2}(\partial_1 + \mathrm{i}\partial_2) \cdot \frac{1}{4}\left(T_{11} - T_{22} - 2\mathrm{i}T_{12}\right). \tag{A.176}$$

By using eqns (A.171) and (A.169), this expression is easily seen to be zero. Similarly for $\partial_z \bar{T}(z, \bar{z}) = 0$.

A.11 Energy–momentum tensor of the Gaussian theory

We show in this appendix that the energy–momentum tensor of the two-dimensional critical Gaussian theory may be chosen as $T(z) = -(\partial_z \phi(z))^2$, which is used in

Section 6.6. It is illuminating to use the Cartesian coordinates and write the Lagrangian density as

$$\mathcal{L} = \frac{1}{2}(\partial_1\phi)^2 + \frac{1}{2}(\partial_2\phi)^2. \tag{A.177}$$

According to eqn (A.116), the energy–momentum tensor has the components

$$T_{11} = (\partial_1\phi)^2 - \mathcal{L} = \frac{1}{2}(\partial_1\phi)^2 - \frac{1}{2}(\partial_2\phi)^2 \tag{A.178}$$

$$T_{22} = (\partial_2\phi)^2 - \mathcal{L} = -\frac{1}{2}(\partial_1\phi)^2 + \frac{1}{2}(\partial_2\phi)^2 \tag{A.179}$$

$$T_{12} = (\partial_1\phi)(\partial_2\phi) = T_{21}. \tag{A.180}$$

These results satisfy the generic properties of eqn (A.171), $T_{11} + T_{22} = 0$, $T_{12} = T_{21}$. The holomorphic component of the energy–momentum tensor is, using eqn (6.37),

$$T(z) = \frac{1}{4}(T_{11} - T_{22} - 2\mathrm{i}\,T_{12}) = \frac{1}{4}\left((\partial_1\phi)^2 - (\partial_2\phi)^2 - 2\mathrm{i}\,(\partial_1\phi)(\partial_2\phi)\right). \tag{A.181}$$

This is rewritten as $T(z) = (\partial_z\phi)^2$, as can be verified by the relation

$$\partial_z\phi = \frac{1}{2}(\partial_1\phi - \mathrm{i}\,\partial_2\phi). \tag{A.182}$$

It is customary to choose the energy–momentum tensor with the opposite sign, $T(z) = -(\partial_z\phi)^2$. Such a change of the sign is allowed as long as it does not affect the important properties of the energy–momentum tensor such as the symmetry $T_{12} = T_{21}$ and tracelessness $T_{11} + T_{22} = 0$.

A.12   Existence of spontaneous magnetization in the two-dimensional Ising model

This section proves the existence of spontaneous magnetization in the Ising model on the square lattice at sufficiently low temperatures by a sophisticated version of the Peierls argument discussed in Section 7.1.

Consider the ferromagnetic Ising model on the square lattice of finite size N. All spins on the boundary are in the up state ($S_i = +1$) as depicted in Fig. 7.1. No external field h is applied. This boundary condition breaks the global inversion ($\mathbb{Z}_2$) symmetry in a finite-size system. The effects of boundaries diminish as the system size increases, and the present boundary condition is considered to be equivalent to an infinitesimally small external field in the thermodynamic limit. The average magnetization per spin is

$$m = \frac{N_+ - N_-}{N} = 1 - 2\frac{\langle N_-\rangle}{N}, \tag{A.183}$$

where N_+ (N_-) is the total number of up (down) spins in a given configuration of spins and satisfies $N = N_+ + N_-$. Our goal is to obtain an upper bound to $\langle N_-\rangle$ as

$$\langle N_-\rangle \leq \frac{1-\alpha}{2}N, \tag{A.184}$$

where α is a positive constant independent of N. If eqn (A.184) is proved, it readily follows that

$$m \geq \alpha > 0, \qquad (A.185)$$

according to eqn (A.183). In other words, there is a non-vanishing spontaneous magnetization.

To prove eqn (A.184), we first fix the configuration of spins and draw a (vertical or horizontal) line segment between two neighboring spins if they are antiparallel ($+-$ or $-+$) as in Fig. 7.1. It is evident that such line segments form closed polygons (domain walls) because all spins on the boundary are up and hence no line segment exists between two spins that lie on the boundary. For a given polygon of length Γ, the total number of down ($-$) spins the polygon encloses, N_Γ, is at most $(\Gamma/4)^2$, $N_\Gamma \leq (\Gamma/4)^2$, since the area that the domain wall encloses is maximal when the polygon is a square, i.e. $N_\square = (\Gamma/4)^2$. Notice that the lattice spacing is the unit of length. Thus, for a given (fixed) configuration of spins, the total number of down spins is bounded as

$$N_- \leq \sum_{\Gamma \geq 4} \left(\frac{\Gamma}{4}\right)^2 \sum_{j=1}^{\nu(\Gamma)} X_\Gamma^j, \qquad (A.186)$$

where $X_\Gamma^j = 1$ if the jth polygon of length Γ is present in the spin configuration and $X_\Gamma^j = 0$ otherwise. The quantity $\nu(\Gamma)$ in eqn (A.186) is the total number of possible polygons of length Γ in the system. We provide a bound on $\nu(\Gamma)$ below. The sum over Γ starts at four because a closed polygon has at least four edges, and includes only even numbers since we are dealing with a square lattice. The average over spin configurations of eqn (A.186) is

$$\langle N_- \rangle \leq \sum_{\Gamma \geq 4} \left(\frac{\Gamma}{4}\right)^2 \sum_{j=1}^{\nu(\Gamma)} \left\langle X_\Gamma^j \right\rangle. \qquad (A.187)$$

The next step is to estimate an upper bound for $\langle X_\Gamma^j \rangle$. The thermal average of X_Γ^j is calculated as[12]

$$\left\langle X_\Gamma^j \right\rangle = \frac{\sum_{\{S_i\}} X_\Gamma^j \, \mathrm{e}^{-\beta H}}{\sum_{\{S_i\}} \mathrm{e}^{-\beta H}} = \frac{{\sum'}_{\{S_i\}} \, \mathrm{e}^{-\beta H}}{\sum_{\{S_i\}} \mathrm{e}^{-\beta H}}, \qquad (A.188)$$

where the sum $\sum'_{\{S_i\}}$ in the numerator is restricted to those spin configurations $\{C\}$ in which the jth polygon is realized ($X_\Gamma^j = 1$). An upper bound for $\left\langle X_\Gamma^j \right\rangle$, eqn (A.188), is obtained by restricting the sum over $\{S_i\}$ in the denominator to some special configurations of spins $\{C^*\}$. Especially convenient configurations to be left in the denominator are the following ones. For each configuration C of spins in the numerator of eqn (A.188) we associate another configuration C^* obtained from C by reversing all spins inside the jth polygon of length Γ. Since a pair of spins neighboring across a line segment are antiparallel to each other in C and parallel in C^*, we find a relation between the energies of the two configurations

---

[12] Notice that the factor $\beta = 1/T$ is not absorbed in the definition of the Hamiltonian.

$$H(C^*) = H(C) - 2 \sum_{\text{domain wall}} J = H(C) - 2J\Gamma. \qquad (A.189)$$

By restricting the sum over $\{S_i\}$ in the denominator of eqn (A.188) to $\{C^*\}$, we have an upper bound as

$$\left\langle X_\Gamma^j \right\rangle \le \frac{\sum_{\{C\}} e^{-\beta H}}{\sum_{\{C^*\}} e^{-\beta H}} = e^{-2\beta J\Gamma}, \qquad (A.190)$$

because all other factors coming from $H(C)$ in eqn (A.189) cancel each other between the numerator and denominator. Therefore, we have

$$\frac{\langle N_- \rangle}{N} \le \sum_{\Gamma \ge 4} \left(\frac{\Gamma}{4}\right)^2 \frac{\nu(\Gamma)}{N} e^{-2\beta J\Gamma}. \qquad (A.191)$$

We next provide an upper bound for the number of polygons of length Γ, i.e. $\nu(\Gamma)$. It is upper-bounded by $3^{\Gamma-1}N$ for the following reason. Consider drawing a closed polygon of length Γ by starting from a given site of the present finite square lattice system. There are at most N sites to start from, and we draw a line segment (a step in the polygon) in any one of them. Then, we have at most three choices of paths to go a step further, if we exclude the previous path. Thus, the total number of possible closed polygons is upper-bounded by $3^{\Gamma-1}N$, where the -1 in the exponent comes from the condition to reach the starting point at the final step, i.e. $\nu(\Gamma) \le 3^{\Gamma-1}N$. This is essentially the same discussion as in Section 7.1, except that we have taken into account the fact that the number $3^{\Gamma-1}N$ is a rigorous upper bound, not just an approximate estimate.

It is evident from eqn (A.191), with $\nu(\Gamma)$ replaced by $3^{\Gamma-1}N$, that $\langle N_- \rangle/N$ can be made arbitrarily small independently of N if $\beta = 1/T$ is large enough. Let us quantify this last statement. One can rewrite the right-hand side of the inequality (A.191) as

$$\sum_{\Gamma=4,6,8,\cdots} \frac{\Gamma^2}{48} q^\Gamma = \sum_{j=2,3,4,\cdots} \frac{j^2}{12} x^j, \qquad (A.192)$$

where $q = 3e^{-2\beta J}$, $x = q^2$, and evaluate this series in the following way

$$\sum_{j=2,3,4,\cdots}^{\mathcal{N}-1} j^2 x^j = x \left(\frac{d\mathcal{R}_0}{dx} + x\frac{d^2\mathcal{R}_0}{dx^2} - 1\right), \qquad (A.193)$$

where

$$\mathcal{R}_0 = \sum_{j=0,1,2,\cdots}^{\mathcal{N}-1} x^j = \frac{1 - x^{\mathcal{N}}}{1 - x}. \qquad (A.194)$$

If one assumes that $x < 1$, in the thermodynamic limit $(N, \mathcal{N}) \to \infty$, it results that

$$\sum_{j=2,3,4,\cdots}^{\infty} j^2 x^j = \frac{x^2}{(1 - x)^3}(4 - 3x + x^2). \qquad (A.195)$$

For sufficiently large β one can always fulfil the condition $q < 1$, and therefore the inequality is written as, denoting $\langle m_- \rangle = \lim_{N \to \infty} \langle N_- \rangle / N$,

$$\langle m_- \rangle \leq \frac{q^4}{12(1-q^2)^3}(4 - 3q^2 + q^4), \qquad (A.196)$$

and the right-hand side can be smaller than $1/2$, implying $m \geq \alpha > 0$. This completes the proof of eqn (A.184) and hence of eqn (A.185).

Throughout the above proof we treated the Ising system with a particular boundary condition, all spins up in the boundary, in the absence of an external field h. The spontaneous magnetization is actually defined as

$$\lim_{h \to +0} \lim_{N \to \infty} m(h), \qquad (A.197)$$

where the system has free or periodic boundary conditions. Intuitively, these two types of spontaneous magnetization are equivalent since both conditions (all spins up in the boundary, and the application of an infinitesimal external field h) select one out of the two degenerate states reflecting the $\mathbb{Z}_2$ symmetry. This equivalence has indeed been rigorously established.

A.13   Quantum version of the Mermin–Wagner theorem

In this appendix we prove the Mermin–Wagner theorem for the absence of spontaneous symmetry breaking for the quantum Heisenberg model in two and lower dimensions. For this purpose we first derive a few inequalities and then prove Bogoliubov's inequality. The latter inequality is used to prove the main theorem in the final section.

A.13.1   Quantum inequalities

Consider arbitrary quantum operators A and B, not necessarily Hermitian. The expectation value of the commutator

$$\left| \langle \Psi | [A^\dagger, B] | \Psi \rangle \right| = \left| \langle [A^\dagger, B] \rangle \right| = \left| \langle A^\dagger B - BA^\dagger \rangle \right| = \left| \langle A^\dagger B \rangle - \langle BA^\dagger \rangle \right| \quad (A.198)$$

over an arbitrary quantum state $|\Psi\rangle$ can certainly be bounded as

$$\left| \langle [A^\dagger, B] \rangle \right| \leq \left| \langle A^\dagger B \rangle \right| + \left| \langle BA^\dagger \rangle \right|, \qquad (A.199)$$

as a result of the triangle inequality. The expression $\langle A^\dagger B \rangle$ represents a scalar product. By using the Schwarz inequalities

$$\left| \langle A^\dagger B \rangle \right| \leq \sqrt{\langle A^\dagger A \rangle \langle B^\dagger B \rangle} \ , \ \left| \langle BA^\dagger \rangle \right| \leq \sqrt{\langle AA^\dagger \rangle \langle BB^\dagger \rangle} \ , \qquad (A.200)$$

eqn (A.199) can be written as

$$\left| \langle [A^\dagger, B] \rangle \right| \leq \sqrt{\langle A^\dagger A \rangle \langle B^\dagger B \rangle} + \sqrt{\langle AA^\dagger \rangle \langle BB^\dagger \rangle}. \qquad (A.201)$$

This implies, after noting, for example, that $(\sqrt{\langle A^\dagger A \rangle} - \sqrt{\langle B^\dagger B \rangle})^2 \geq 0$ or $\langle A^\dagger A \rangle + \langle B^\dagger B \rangle \geq 2\sqrt{\langle A^\dagger A \rangle \langle B^\dagger B \rangle}$,

$$2 \left| \langle [A^\dagger, B] \rangle \right| \leq \langle [A^\dagger, A]_+ \rangle + \langle [B^\dagger, B]_+ \rangle, \qquad (A.202)$$

with $[A^\dagger, A]_+ = A^\dagger A + A A^\dagger$, etc., representing anticommutators.

Another quite useful relation results after squaring inequality (A.201) and noticing that $\langle A^\dagger A\rangle\langle BB^\dagger\rangle + \langle AA^\dagger\rangle\langle B^\dagger B\rangle \geq 2\sqrt{\langle A^\dagger A\rangle\langle B^\dagger B\rangle\langle AA^\dagger\rangle\langle BB^\dagger\rangle}$,

$$\left|\langle[A^\dagger, B]\rangle\right|^2 \leq \langle[A^\dagger, A]_+\rangle\langle[B^\dagger, B]_+\rangle, \tag{A.203}$$

which represents a *generalized Heisenberg uncertainty relation* for arbitrary quantum operators. Notice that in the particular case where $A = A^\dagger$ and $B = B^\dagger$ are observables, i.e. Hermitian operators, the inequality above reduces to the standard Heisenberg uncertainty relation. The inequalities derived above are not used in the proof of the Mermin–Wagner theorem but are sometimes useful to discuss the properties of long-range order in quantum systems.

We have so far not taken into account the temperature. The standard generalization of the Mermin–Wagner theorem to quantum systems concerns finite temperatures. To this end we need to define a scalar product of two arbitrary operators A and B that involves the temperature $T = 1/\beta$. A standard scalar product of this kind, sometimes known as the *Duhamel two-point function*, is

$$(A, B)_\rho = \frac{1}{\beta}\int_0^\beta \mathrm{d}x \operatorname{Tr}\left[\rho A^\dagger(x)B\right], \tag{A.204}$$

where $\rho = \mathrm{e}^{-\beta H}/Z$ represents the density matrix of the canonical ensemble[13] and $A^\dagger(x) = \mathrm{e}^{xH}A^\dagger\mathrm{e}^{-xH}$. It is straightforward to prove that $(A, B)_\rho$ is a legitimate scalar product. In other words, it satisfies: (i) $(A, B+C)_\rho = (A, B)_\rho + (A, C)_\rho$, (ii) $(A, \lambda B)_\rho = \lambda(A, B)_\rho$ with λ a complex number, (iii) $(A, B)_\rho = (B, A)_\rho^*$ with $*$ meaning complex conjugation, and (iv) $(A, A)_\rho \geq 0$, and vanishes iff $A = 0$. One can then use the Schwarz inequality

$$|(A, B)_\rho|^2 \leq (A, A)_\rho (B, B)_\rho, \tag{A.205}$$

where the equality is satisfied whenever A and B are linearly dependent. This equation (A.205) constitutes the quantum version of the classical inequality (7.12).

A.13.2 Bogoliubov's inequality

Note that one can re-express eqn (A.204) in terms of the energy eigenvalues E_m and orthonormal eigenvectors, $H|m\rangle = E_m|m\rangle$, as

$$(A, B)_\rho = \frac{1}{\beta Z}\sum_{n,m}\langle n|A^\dagger|m\rangle\langle m|B|n\rangle\left(\frac{\mathrm{e}^{-\beta E_m} - \mathrm{e}^{-\beta E_n}}{E_n - E_m}\right), \tag{A.206}$$

where the last factor for the case $E_n = E_m$ is defined by the limit $E_n \to E_m$. This factor for general n and m is bounded as

$$0 < \left(\frac{\mathrm{e}^{-\beta E_m} - \mathrm{e}^{-\beta E_n}}{E_n - E_m}\right) \leq \frac{\beta}{2}\left(\mathrm{e}^{-\beta E_m} + \mathrm{e}^{-\beta E_n}\right), \tag{A.207}$$

---

[13] Notice that the factor $\beta = 1/T$ is not absorbed in the definition of the Hamiltonian since it is explicitly needed for integration.

because of the inequality

$$0 < \frac{e^{-v} - e^{-u}}{u - v} \le \frac{1}{2}(e^{-u} + e^{-v}), \tag{A.208}$$

the second inequality being a consequence of concavity of the exponential function. Equations (A.206) and (A.207) lead to the following inequality

$$(A, A)_\rho \le \frac{1}{2}\,\mathrm{Tr}\,\left[\rho[A^\dagger, A]_+\right] \equiv \frac{1}{2}\left\langle [A^\dagger, A]_+\right\rangle_\rho. \tag{A.209}$$

Similarly, one can generically define B in terms of another operator C as $B = [C^\dagger, H]$ and insert it into eqn (A.206) to obtain

$$\beta(A, B)_\rho = \left\langle [C^\dagger, A^\dagger]\right\rangle_\rho \;,\; \beta(B, B)_\rho = \left\langle [C^\dagger, [H, C]]\right\rangle_\rho. \tag{A.210}$$

The relation known as *Bogoliubov's inequality* is obtained by combining eqns (A.205) and (A.209)

$$\left|\left\langle [C^\dagger, A^\dagger]\right\rangle_\rho\right|^2 \le \frac{\beta}{2}\left\langle [A^\dagger, A]_+\right\rangle_\rho \left\langle [C^\dagger, [H, C]]\right\rangle_\rho. \tag{A.211}$$

A.13.3 Proof of the Mermin–Wagner theorem

The absence of spontaneous symmetry breaking can be proved for quantum spin systems with continuous symmetry in two and lower dimensions by using Bogoliubov's inequality. We show the example of the spin-1/2 ferromagnetic Heisenberg model on the square lattice with nearest-neighbor interactions.

The Hamiltonian

$$H = -J\sum_{\langle ij\rangle}(S_i^x S_j^x + S_i^y S_j^y + S_i^z S_j^z) - h\sum_i S_i^z \tag{A.212}$$

is first rewritten in terms of the raising and lowering operators $S_j^\pm = S_j^x \pm iS_j^y$ as

$$H = -J\sum_{\langle ij\rangle}\left(\frac{1}{2}S_i^+ S_j^- + \frac{1}{2}S_i^- S_j^+ + S_i^z S_j^z\right) - h\sum_i S_i^z. \tag{A.213}$$

It will be convenient to further rewrite it using the Fourier variables

$$S_j = \frac{1}{N}\sum_q e^{i q\cdot r_j} S_q, \quad S_q = \sum_j e^{-i q\cdot r_j} S_j \tag{A.214}$$

as

$$H = -\frac{2J}{N}\sum_q \gamma_q \left(\frac{1}{2}S_q^+ S_{-q}^- + \frac{1}{2}S_q^- S_{-q}^+ + S_q^z S_{-q}^z\right) - hS_0, \tag{A.215}$$

where

$$\gamma_q = \frac{1}{4}\sum_\delta e^{i q\cdot\delta} = \frac{1}{2}\left(\cos q_x + \cos q_y\right), \tag{A.216}$$

with $\boldsymbol{\delta}$ being the vector to neighboring sites on the square lattice, as in eqn (7.25). We have suppressed vector notations for subscripts for simplicity ($\boldsymbol{q} \to q$).

The following commutation relations will be used in the evaluation of various terms in Bogoliubov's inequality (A.211),

$$[S_q^\pm, S_{q'}^z] = \mp S_{q+q'}^\pm, \quad [S_q^+, S_{q'}^-] = 2S_{q+q'}^z, \tag{A.217}$$

which are consequences of the relations in the real space,

$$[S_j^\pm, S_j^z] = \mp S_j^\pm, \quad [S_j^+, S_j^-] = 2S_j^z. \tag{A.218}$$

It is the crux of the proof to choose the following operators as A and C in Bogoliubov's inequality,

$$A^\dagger = C = S_{-q}^-, \quad A = C^\dagger = S_q^+. \tag{A.219}$$

Then, the left-hand side of eqn (A.211) is

$$|\langle [C^\dagger, A^\dagger] \rangle|^2 = |\langle [S_q^+, S_{-q}^-] \rangle|^2 = |2\langle S_0^z \rangle|^2 = 4N^2 m_z^2, \tag{A.220}$$

where m_z is the magnetization per site along the z-axis. The subscript ρ for $\langle \cdots \rangle_\rho$ has been omitted for simplicity.

Next, the first factor on the right-hand side of eqn (A.211) is bounded as

$$\langle [A^\dagger, A]_+ \rangle = \langle [S_{-q}^-, S_q^+]_+ \rangle = 2\langle S_{-q}^x S_q^x + S_{-q}^y S_q^y \rangle$$
$$\le 2\langle S_{-q}^x S_q^x + S_{-q}^y S_q^y + S_{-q}^z S_q^z \rangle = 2\langle \boldsymbol{S}_{-q} \cdot \boldsymbol{S}_q \rangle. \tag{A.221}$$

The summation of both sides over $\boldsymbol{q}$ yields

$$\sum_q \langle [A^\dagger, A]_+ \rangle \le 2 \sum_q \langle \boldsymbol{S}_{-q} \cdot \boldsymbol{S}_q \rangle = N \sum_j \langle \boldsymbol{S}_j^2 \rangle = N^2 S(S+1) \tag{A.222}$$

with $S = 1/2$.

The commutation relation of H and C on the right-hand side of eqn (A.211) is calculated as, using (A.217),

$$[H, C] = -\frac{2J}{N} \sum_{q'} (\gamma_{q'} - \gamma_{q'-q})(S_{q'-q}^z S_{-q'}^- + S_{-q'}^- S_{q'-q}^z) + hS_{-q}^-. \tag{A.223}$$

Then we find, again using eqn (A.217),

$$\langle [C^\dagger, [H, C]] \rangle = \frac{J}{2N} \sum_{q'} \sum_{\delta} (1 - e^{i\boldsymbol{q}\cdot\boldsymbol{\delta}}) e^{i\boldsymbol{q}'\cdot\boldsymbol{\delta}} \left\langle 4S_{q'}^z S_{-q'}^z + S_{-q'}^+ S_{q'}^- + S_{-q'}^- S_{q'}^+ \right\rangle + 2hNm_z. \tag{A.224}$$

The expectation value of spin operators in the above equation is upper-bounded by $4\langle \boldsymbol{S}_{q'} \cdot \boldsymbol{S}_{-q'} \rangle$ as before, and we therefore obtain

$$|\langle [C^\dagger, [H, C]] \rangle| \le \frac{2J}{N} \sum_{q'} \sum_{\delta} |1 - e^{i\boldsymbol{q}\cdot\boldsymbol{\delta}}| \langle \boldsymbol{S}_{q'} \cdot \boldsymbol{S}_{-q'} \rangle + 2hNm_z \le JNq^2 S(S+1) + 2hNm_z. \tag{A.225}$$

Bogoliubov's inequality (A.211), with both sides divided by $\langle [C^\dagger, [H, C]] \rangle$ (> 0) and summed over q, reads

$$\frac{1}{2} \sum_q \langle [A^\dagger, A]_+ \rangle \geq \sum_q \frac{T \left| \langle [C^\dagger, A^\dagger] \rangle \right|^2}{\langle [C^\dagger, [H, C]] \rangle}. \tag{A.226}$$

If we insert eqns (A.220), (A.222) and (A.225) into this inequality, we obtain

$$\frac{N^2 S(S+1)}{2} \geq \frac{4T N^2 m_z^2}{N} \sum_q \frac{1}{J q^2 S(S+1) + 2h m_z}. \tag{A.227}$$

This is essentially equivalent to eqn (7.27) for the classical case and hence we can conclude $m_z \to 0$ as $h \to +0$ after $N \to \infty$. The case of $d < 2$ can be discussed similarly.

A.14   Replica symmetric solution of the SK model

This section derives the replica symmetric solution of the SK model, eqn (8.29), for the Ising spin glass with random infinite-range interactions.

A.14.1   Replica average of the partition function

Suppose that the interaction with quenched randomness in the Hamiltonian

$$H = -\sum_{i<j} J_{ij} S_i S_j - h \sum_i S_i \tag{A.228}$$

obeys the following distribution function,

$$P(J_{ij}) = \frac{1}{J} \sqrt{\frac{N}{2\pi}} \exp\left\{ -\frac{N}{2J^2} \left(J_{ij} - \frac{J_0}{N} \right)^2 \right\}. \tag{A.229}$$

We use the replica method to evaluate the configurational average of the free energy. The first step is to take the configurational average of the partition function raised to the power n,[14]

$$[Z^n] = \int \left(\prod_{i<j} dJ_{ij} P(J_{ij}) \right) \operatorname{Tr} \exp \left(\beta \sum_{i<j} J_{ij} \sum_{\alpha=1}^n S_i^\alpha S_j^\alpha + \beta h \sum_{i=1}^N \sum_{\alpha=1}^n S_i^\alpha \right), \tag{A.230}$$

---

[14] Notice that the factor $\beta = 1/T$ is not absorbed in the definition of the Hamiltonian, eqn (A.228).

where Tr stands for the sum over all spin variables and α is the replica index. The integral over J_{ij} can be carried out independently for each (ij) using eqn (A.229) to give the result

$$\mathrm{Tr}\exp\left\{\frac{1}{N}\sum_{i<j}\left(\frac{1}{2}\beta^2J^2\sum_{\alpha,\beta}S_i^\alpha S_j^\alpha S_i^\beta S_j^\beta+\beta J_0\sum_\alpha S_i^\alpha S_j^\alpha\right)+\beta h\sum_i\sum_\alpha S_i^\alpha\right\}$$

(A.231)

up to a trivial constant. We rearrange the sum over $i<j$ in the exponent as

$$[Z^n]=\mathrm{e}^{N\beta^2J^2n/4}\mathrm{Tr}\exp\left\{\frac{\beta^2J^2}{2N}\sum_{\alpha<\beta}\left(\sum_i S_i^\alpha S_i^\beta\right)^2\right.$$

$$\left.+\frac{\beta J_0}{2N}\sum_\alpha\left(\sum_i S_i^\alpha\right)^2+\beta h\sum_i\sum_\alpha S_i^\alpha\right\}$$

(A.232)

for sufficiently large N.

A.14.2 Reduction to a single-body problem

The trace over S_i^α in eqn (A.232) would be able to be taken independently at each i if the quadratic forms in the exponent were linear. Thus, we apply the Gaussian integral formula to the exponential of $\left(\sum_i S_i^\alpha S_i^\beta\right)^2$ with integral variable $q_{\alpha\beta}$ and to the exponential of $\left(\sum_i S_i^\alpha\right)^2$ with integral variable m_α to derive

$$[Z^n]=\mathrm{e}^{N\beta^2J^2n/4}\int\prod_{\alpha<\beta}\mathrm{d}q_{\alpha\beta}\int\prod_\alpha\mathrm{d}m_\alpha\exp\left(-\frac{N\beta^2J^2}{2}\sum_{\alpha<\beta}q_{\alpha\beta}^2-\frac{N\beta J_0}{2}\sum_\alpha m_\alpha^2\right)$$

$$\cdot\,\mathrm{Tr}\exp\left(\beta^2J^2\sum_{\alpha<\beta}q_{\alpha\beta}\sum_i S_i^\alpha S_i^\beta+\beta\sum_\alpha(J_0m_\alpha+h)\sum_i S_i^\alpha\right).$$

(A.233)

If we denote Tr also for the sum $\sum_{S_i^\alpha}$ of a single site, the expression after Tr in the above formula is

$$\left\{\mathrm{Tr}\exp\left(\beta^2J^2\sum_{\alpha<\beta}q_{\alpha\beta}S^\alpha S^\beta+\beta\sum_\alpha(J_0m_\alpha+h)S^\alpha\right)\right\}^N\equiv\exp\left(N\log\mathrm{Tr}\mathrm{e}^L\right),$$

(A.234)

where

$$L=\beta^2J^2\sum_{\alpha<\beta}q_{\alpha\beta}S^\alpha S^\beta+\beta\sum_\alpha(J_0m_\alpha+h)S^\alpha.$$

(A.235)

We now have

$$[Z^n] = e^{N\beta^2 J^2 n/4} \int \prod_{\alpha<\beta} dq_{\alpha\beta} \int \prod_{\alpha} dm_{\alpha}$$

$$\cdot \exp\left(-\frac{N\beta^2 J^2}{2}\sum_{\alpha<\beta} q_{\alpha\beta}^2 - \frac{N\beta J_0}{2}\sum_{\alpha} m_{\alpha}^2 + N \log \mathrm{Tr}e^L\right). \quad (A.236)$$

A.14.3   Saddle-point evaluation

The exponent of the integrand of the above equation is proportional to N, and we can use the saddle-point method to evaluate the integral. In the limit $N \to \infty$,

$$[Z^n] \approx \exp\left(-\frac{N\beta^2 J^2}{2}\sum_{\alpha<\beta} q_{\alpha\beta}^2 - \frac{N\beta J_0}{2}\sum_{\alpha} m_{\alpha}^2 + N \log \mathrm{Tr}e^L + \frac{N}{4}\beta^2 J^2 n\right)$$

$$\approx 1 + Nn\left\{-\frac{\beta^2 J^2}{4n}\sum_{\alpha\neq\beta} q_{\alpha\beta}^2 - \frac{\beta J_0}{2n}\sum_{\alpha} m_{\alpha}^2 + \frac{1}{n}\log \mathrm{Tr}e^L + \frac{1}{4}\beta^2 J^2\right\}.$$

Here, we have taken the limit $n \to 0$ with N kept large but finite. It is necessary to insert into $q_{\alpha\beta}$ and m_{α} the saddle-point values to extremize the expression inside $\{\cdots\}$.

The free energy is now written as, according to the replica method,

$$-\beta f = \lim_{n\to 0} \frac{[Z^n] - 1}{nN} = \lim_{n\to 0}\left\{-\frac{\beta^2 J^2}{4n}\sum_{\alpha\neq\beta} q_{\alpha\beta}^2\right.$$

$$\left. -\frac{\beta J_0}{2n}\sum_{\alpha} m_{\alpha}^2 + \frac{1}{4}\beta^2 J^2 + \frac{1}{n}\log \mathrm{Tr}e^L\right\}. \quad (A.237)$$

The saddle-point condition that the free energy is extremum with respect to the variable $q_{\alpha\beta}(\alpha \neq \beta)$ yields

$$q_{\alpha\beta} = \frac{1}{\beta^2 J^2}\frac{\partial}{\partial q_{\alpha\beta}}\log \mathrm{Tr}e^L = \frac{\mathrm{Tr}S^{\alpha}S^{\beta}e^L}{\mathrm{Tr}e^L} = \langle S^{\alpha}S^{\beta}\rangle_L, \quad (A.238)$$

where $\langle\cdots\rangle_L$ is the average with respect to the weight e^L. The extremum condition with respect to m_{α} can also be written as

$$m_{\alpha} = \frac{1}{\beta J_0}\frac{\partial}{\partial m_{\alpha}}\log \mathrm{Tr}e^L = \frac{\mathrm{Tr}S^{\alpha}e^L}{\mathrm{Tr}e^L} = \langle S^{\alpha}\rangle_L. \quad (A.239)$$

A.14.4   Replica symmetric solution

Further progress in the evaluation of eqn (A.237) is possible only if we know the explicit dependence of $q_{\alpha\beta}$ and m_{α} on the replica index α, β. A naive guess is that the physics should not depend on these indices because replicas have been introduced as a mathematical trick to take the configurational average. This idea suggests to assume *replica symmetry*, or independence of the parameters on the replica indices, $q_{\alpha\beta} = q$ ($\alpha \neq \beta$) and $m_\alpha = m$.

If we accept this replica symmetry, the free energy (A.237) reduces to, before the limit $n \to 0$,

$$-\beta f = \frac{\beta^2 J^2}{4n} \left\{ -n(n-1)q^2 \right\} - \frac{\beta J_0}{2n} nm^2 + \frac{1}{n} \log \text{Tr} e^L + \frac{1}{4}\beta^2 J^2. \qquad (A.240)$$

The third term on the right-hand side can be evaluated using the definition of L (A.235) and the Gaussian integral as

$$\log \text{Tr} e^L = \log \text{Tr} \sqrt{\frac{\beta^2 J^2 q}{2\pi}} \int dz$$

$$\cdot \exp \left(-\frac{\beta^2 J^2 q}{2} z^2 + \beta^2 J^2 q z \sum_\alpha S^\alpha - \frac{n}{2}\beta^2 J^2 q + \beta(J_0 m + h) \sum_\alpha S^\alpha \right)$$

$$= \log \int Dz \, \exp \left(n \log 2 \cosh(\beta J \sqrt{q} z + \beta J_0 m + \beta h) - \frac{n}{2}\beta^2 J^2 q \right)$$

$$= \log \left(1 + n \int Dz \log 2 \cosh \beta \tilde{H}(z) - \frac{n}{2}\beta^2 J^2 q + \mathcal{O}(n^2) \right). \qquad (A.241)$$

Here, $Dz = dz \, \exp(-z^2/2)/\sqrt{2\pi}$ and $\tilde{H}(z) = J\sqrt{q} z + J_0 m + h$. We insert eqn (A.241) into (A.240) and take the limit $n \to 0$ to find

$$-\beta f = \frac{\beta^2 J^2}{4}(1-q)^2 - \frac{1}{2}\beta J_0 m^2 + \int Dz \log 2 \cosh \beta \tilde{H}(z). \qquad (A.242)$$

This is the replica symmetric solution of the free energy of the SK model.

A.14.5 Order parameters

The integral variables $q_{\alpha\beta}$ and m_α introduced artificially for the Gaussian integrals are indeed order parameters as in the ferromagnetic model of Section 2.5. To confirm this fact, we notice that eqn (A.238) can be written as follows,

$$q_{\alpha\beta} = [\langle S_i^\alpha S_i^\beta \rangle] = \left[\frac{\text{Tr} S_i^\alpha S_i^\beta e^{-\beta \sum_\gamma H_\gamma}}{\text{Tr} e^{-\beta \sum_\gamma H_\gamma}} \right], \qquad (A.243)$$

where H_γ is the γth replica Hamiltonian,

$$H_\gamma = -\sum_{i<j} J_{ij} S_i^\gamma S_j^\gamma - h \sum_i S_i^\gamma. \qquad (A.244)$$

We can show the equivalence of eqns (A.238) and (A.243) almost in the same way as in the previous sections. First, notice that the denominator of eqn (A.243), Z^n, approaches unity in the limit $n \to 0$ and can thus be ignored. The numerator is expressed by inserting $S_i^\alpha S_i^\beta$ after the Tr symbol in the expression of $[Z^n]$. If we follow the calculation of Section A.14.1 with this fact in mind, we obtain the following in place of eqn (A.234),

$$\left(\mathrm{Tr}\,\mathrm{e}^L\right)^{N-1} \cdot \mathrm{Tr}\left(S^\alpha S^\beta \mathrm{e}^L\right). \tag{A.245}$$

The term $\log \mathrm{Tr}\,\mathrm{e}^L$ is proportional to n as eqn (A.237) suggests, and consequently $\mathrm{Tr}\,\mathrm{e}^L$ should approach unity as $n \to 0$. Hence, eqn (A.245) reduces to $\mathrm{Tr}(S^\alpha S^\beta \mathrm{e}^L)$ for $n \to 0$. Noting that the denominator approaches unity in eqn (A.238), we conclude that eqn (A.245) coincides with eqn (A.238). This completes the proof that eqn (A.243) and eqn (A.238) agree with each other. We find similarly that

$$m_\alpha = \left[\langle S_i^\alpha \rangle\right]. \tag{A.246}$$

Equation (A.246) shows that m is the conventional ferromagnetic order parameter. The other quantity $q_{\alpha\beta}$ represents the spin glass order parameter. To understand the latter interpretation, we notice that the trace operations for replicas other than α and β appear in exactly the same way in the denominator and numerator in eqn (A.243), which causes cancellation of these common factors to give

$$q_{\alpha\beta} = \left[\frac{\mathrm{Tr} S_i^\alpha \mathrm{e}^{-\beta H_\alpha}}{\mathrm{Tr}\,\mathrm{e}^{-\beta H_\alpha}} \frac{\mathrm{Tr} S_i^\beta \mathrm{e}^{-\beta H_\beta}}{\mathrm{Tr}\,\mathrm{e}^{-\beta H_\beta}}\right] = \left[\langle S_i^\alpha \rangle \langle S_i^\beta \rangle\right] = \left[\langle S_i \rangle^2\right] \equiv q. \tag{A.247}$$

In the high-temperature paramagnetic phase, $\langle S_i \rangle$ vanishes at any site i and hence $m = q = 0$. In the ferromagnetic phase, most spins align in the same direction (to be taken to be positive for example), $\langle S_i \rangle > 0$, and we have $m > 0$ and $q > 0$.

The spin glass phase has randomly frozen spins at most sites, in which $\langle S_i \rangle$ takes a site-dependent sign and absolute value. $\langle S_i \rangle$ changes randomly from site to site but this spatially random pattern does not change with time, and is frozen in this sense.

The randomly frozen spin state changes if the configuration of interactions $\{J_{ij}\}$ changes because the environment of each spin changes drastically. This suggests the interpretation that the average of $\langle S_i \rangle$ over the distribution of $\{J_{ij}\}$ is equivalent to the average over the possibilities of $\langle S_i \rangle > 0$ and $\langle S_i \rangle < 0$, and we may well have $m = \left[\langle S_i \rangle\right] = 0$. On the other hand, q is the configurational average of a positive quantity and does not vanish, implying the possibility of a phase characterized by $m = 0$, $q > 0$. This is the spin glass phase and q is the spin glass order parameter.

A.15   Integral for the partition function of the n-vector model

The integral used in the calculation of the partition function of the one-dimensional n-vector model of Section 9.2

$$G(K) = \int_{-\infty}^{\infty} \mathrm{d}x_1 \mathrm{d}x_2 \cdots \mathrm{d}x_n\, \delta(x_1^2 + x_2^2 + \cdots + x_n^2 - 1)\, \mathrm{e}^{Kx_1} \tag{A.248}$$

is evaluated in this section. The Fourier representation of the Dirac delta function yields

$$G(K) = \frac{1}{2\pi} \int_{-\infty}^{\infty} \mathrm{d}x_1 \mathrm{d}x_2 \cdots \mathrm{d}x_n \int_{-\infty+i\epsilon}^{\infty+i\epsilon} \mathrm{e}^{iu(x_1^2+\cdots+x_n^2-1)+Kx_1}\, \mathrm{d}u. \tag{A.249}$$

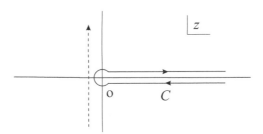

Fig. A.3 The path is changed from the dashed line to the full line to have the integral representation of the modified Bessel function of the first kind.

The integral over each x_i can be performed as a Gaussian integral.[15] The result is

$$G(K) = a \cdot \int_{-\infty+i\epsilon}^{\infty+i\epsilon} \exp\left(-iu + \frac{iK^2}{4u}\right) u^{-n/2} \, du, \tag{A.250}$$

where a trivial constant has been written as a. A change of integral variable as $u = -iKt/2$ gives, with b another trivial constant,

$$G(K) = b \cdot \left(\frac{K}{2}\right)^{1-n/2} \int_{-i\infty-\epsilon}^{i\infty-\epsilon} \exp\left(-\frac{K}{2}(t + t^{-1})\right) (-t)^{-n/2} \, dt. \tag{A.251}$$

The integration path starts from $-i\infty$ and extends to $i\infty$, always staying to the left of the imaginary axis. We deform the path so that it runs under the real axis, goes around the origin clockwise and returns to ∞ above the real axis, as shown in Fig. A.3. Then, the integral coincides with an integral representation of the modified Bessel function of the first kind, up to a constant, to give

$$G(K) = c \cdot \left(\frac{K}{2}\right)^{1-n/2} I_{n/2-1}(K). \tag{A.252}$$

A.16   Multiple Gaussian integral and lattice Green function

We show how to integrate the exponential of a quadratic form of N variables,

$$F_C(\boldsymbol{q}) = \int_{-\infty}^{\infty} e^{-\frac{1}{2}{}^t\boldsymbol{x}\cdot C\boldsymbol{x}+i{}^t\boldsymbol{x}\cdot\boldsymbol{q}} \, d\boldsymbol{x}, \tag{A.253}$$

where C is a real, positive definite symmetric matrix and ${}^t\boldsymbol{x} = (x_1, x_2, \cdots, x_N)$ is the transpose of vector $\boldsymbol{x}$. Let us write U for the orthogonal matrix that diagonalizes C and write $U^{-1}CU = D$ (diagonal matrix) and $U^{-1}\boldsymbol{x} = \boldsymbol{y}, U^{-1}\boldsymbol{q} = \boldsymbol{r}$. Then,

---

[15] We have introduced a small imaginary part $i\epsilon$ in u such that iu has a negative real part in order to guarantee the convergence of the integral. The conventional Gaussian integral formula applies in spite of the complex coefficient in front of x_i^2 by analytic continuation.

$$F_C(\boldsymbol{q}) = \int_{-\infty}^{\infty} e^{-\frac{1}{2}{}^t\boldsymbol{x} U U^{-1} C U U^{-1}\boldsymbol{x} + i^t\boldsymbol{x} U \cdot U^{-1}\boldsymbol{q}} \, \mathrm{d}\boldsymbol{x}$$

$$= \int_{-\infty}^{\infty} e^{-\frac{1}{2}{}^t\boldsymbol{y} \cdot D\boldsymbol{y} + i^t\boldsymbol{y} \cdot \boldsymbol{r}} \, \mathrm{d}\boldsymbol{y}. \tag{A.254}$$

Since D is diagonal, the final integral can be easily carried out as a single-variable Gaussian integral for each component of $\boldsymbol{y}$ separately. The result is an exponential of a quadratic form of $\boldsymbol{r}$. We write $d_1, d_2, \cdots, d_N$ for the diagonal elements of D (namely the eigenvalues of C) and express the result of the Gaussian integral in terms of the components of $\boldsymbol{q}$ using $U^{-1}\boldsymbol{q} = \boldsymbol{r}$,

$$F_C(\boldsymbol{q}) = \frac{(2\pi)^{N/2}}{(\det C)^{1/2}} e^{-\frac{1}{2}\sum_{n,l} q_n q_l G_{nl}}, \tag{A.255}$$

where $\det C = \prod_l d_l$, and G_{nl} is a matrix defined by

$$G_{nl} = \sum_m \frac{U_{nm}(U^{-1})_{ml}}{d_m}. \tag{A.256}$$

It turns out that G is the inverse matrix of C. To confirm it, we notice $C = UDU^{-1}$ to have

$$(GC)_{ij} = \sum_l G_{il} C_{lj} = \sum_l \sum_m \frac{U_{im}(U^{-1})_{ml}}{d_m} \sum_n U_{ln} d_n (U^{-1})_{nj}$$

$$= \sum_{m,n} \frac{U_{im}}{d_m} d_n (U^{-1})_{nj} \sum_l (U^{-1})_{ml} U_{ln}$$

$$= \sum_m U_{im}(U^{-1})_{mj} = \delta_{i,j}. \tag{A.257}$$

To summarize, the multiple Gaussian integral (A.253) is evaluated as

$$F_C(\boldsymbol{q}) = \frac{(2\pi)^{N/2}}{(\det C)^{1/2}} e^{-\frac{1}{2}\sum_{n,l} q_n q_l (C^{-1})_{nl}}. \tag{A.258}$$

Let us now apply this result to eqn (10.46). We first scale the variable from ϕ_i to $\sqrt{K}\phi_i$ to remove the coefficient K. This leaves $\sqrt{K}$ in the linear term. Comparison with eqn (A.253) reveals that the matrix C has the following non-vanishing elements in the present problem, with all the rest vanishing,

$$C_{nn} = 4, \quad C_{n,n+\delta} = -1. \tag{A.259}$$

The first equation is for diagonal elements and δ in the second equation is a vector connecting neighboring sites. We need the inverse $C^{-1} = G$ to apply the result (A.258). This goal is achieved by Fourier transformation because the system is translationally invariant and consequently C_{nm} depends only on the difference of indices $G_{nm} = G(n-m)$. By setting

$$C_{nm} = \frac{1}{(2\pi)^2} \int_{-\pi}^{\pi} \tilde{C}(k) e^{ik \cdot (n-m)} \, dk \qquad (A.260)$$

$$G_{nm} = \frac{1}{(2\pi)^2} \int_{-\pi}^{\pi} \tilde{G}(k) e^{ik \cdot (n-m)} \, dk, \qquad (A.261)$$

we have $\tilde{G}(k) = \tilde{C}(k)^{-1}$.[16] The reason is that the following quantity

$$(CG)_{nm} = \sum_{l} C_{nl} G_{lm}$$

$$= \frac{1}{(2\pi)^4} \int_{-\pi}^{\pi} \tilde{C}(k_1) \tilde{G}(k_2) \sum_{l} e^{ik_1 \cdot (n-l) + ik_2 \cdot (l-m)} \, dk_1 dk_2$$

$$= \frac{1}{(2\pi)^2} \int_{-\pi}^{\pi} \tilde{C}(k) \tilde{G}(k) e^{ik \cdot (n-m)} \, dk \qquad (A.262)$$

is equal to $\delta_{n,m}$, which means $\tilde{C}(k) \tilde{G}(k) = 1$. Thus, we have G_{nm} from eqn (A.261) if we find $\tilde{C}(k)$. $\tilde{C}(k)$ can be evaluated as the inverse Fourier transformation of C_{nm}. We set the lattice constant to unity and have, from eqn (A.259),

$$\tilde{C}(k) = \sum_{n} e^{-ik \cdot n} C_{l,l+n}$$

$$= 4 - (e^{-ik_x} + e^{ik_x} + e^{-ik_y} + e^{ik_y})$$

$$= 4 - 2 \cos k_x - 2 \cos k_y. \qquad (A.263)$$

Hence,

$$G_{nm} = G(n-m) = \frac{1}{2(2\pi)^2} \int_{-\pi}^{\pi} \frac{e^{ik \cdot (n-m)}}{2 - \cos k_x - \cos k_y} \, dk. \qquad (A.264)$$

This matrix G of eqn (A.264), the inverse of C of eqn (A.259), is the lattice Green function.

Let us return to the evaluation of eqn (10.46). By setting $q_j = 2\pi\sqrt{K} n_j$ in eqn (A.253) in consideration of eqn (10.46) after scaling of the integral variable, we have the integral result from eqn (A.258)

$$Z = e^{-2\pi^2 K \sum_{j,l} n_j n_l G(j-l)}. \qquad (A.265)$$

The trivial multiplicative factor has been dropped. This is eqn (10.47).

It should be noted here that $G(0)$ diverges because of the contribution coming from $|k| \to 0$ (the short-wavelength or large-system-size limit) according to eqn (A.264). Then, the term $j = l$ diverges in eqn (A.265) and consequently we have a senseless

---

[16] k and n, m, δ are actually two-dimensional vectors $\boldsymbol{k}, \boldsymbol{n}, \boldsymbol{m}, \boldsymbol{\delta}$ and $k \cdot (n-m)$ stands for the inner product $\boldsymbol{k} \cdot (\boldsymbol{n} - \boldsymbol{m})$. We do not use the vector notation here for simplicity of notation. The integral $\int dk$ is also the two-dimensional integral $\int dk_x dk_y$.

result $Z \to 0$. The neutrality condition $\sum_j n_j = 0$ helps us avoid this difficulty because this condition allows us to add $\left(\sum_j n_j\right)^2 G(0)(= 0)$ to the exponent,[17]

$$Z = e^{2\pi^2 K \sum_{j,l} n_j n_l \{G(0) - G(j-l)\}}. \tag{A.266}$$

Hence,

$$G(0) - G(r) = \frac{1}{2(2\pi)^2} \int_{-\pi}^{\pi} \frac{1 - e^{ik \cdot r}}{2 - \cos k_x - \cos k_y} \, dk_x dk_y \tag{A.267}$$

does not diverge as $k \to 0$.

The remaining task is the evaluation of $G(0) - G(r)$ as $r \to \infty$. We notice to this end that, as $k \cdot r$ in the exponent of the numerator of eqn (A.267) approaches 0, the numerator almost vanishes since $e^{ik \cdot r} \approx 1$ and thus does not contribute to the integral. This implies that the significant contribution comes from the range where $|k|$ is larger than c/r with c some constant. We thus write the integral in polar coordinates as

$$G(0) - G(r) \approx \frac{1}{2(2\pi)^2} \int_{c/r}^{\pi} \frac{1}{k^2/2} k dk \int_0^{2\pi} d\theta_k \approx \frac{1}{2\pi} \log r + \text{const.} \tag{A.268}$$

We have used here the fact that $e^{ik \cdot r}$ does not contribute to the integral for large r because $e^{ik \cdot r}$ moves rapidly on the unit circle in the complex plane as k changes and hence $1 - e^{ik \cdot r} \approx 1$. Equation (A.266) in that limit is finally

$$Z = e^{\pi K \sum_{j \neq l} n_j n_l \log |j-l|}, \tag{A.269}$$

where the term $j = l$ is excluded due to the condition $|j - l| \gg 1$. This is eqn (10.50).

A.17 Jordan–Wigner transformation

Spin-1/2 operators on a one-dimensional chain can be expressed in terms of Fermionic operators on the same chain as follows,

$$S_j^+ = (1 - 2n_1)(1 - 2n_2) \cdots (1 - 2n_{j-1}) a_j^\dagger \tag{A.270}$$

$$S_j^- = (1 - 2n_1)(1 - 2n_2) \cdots (1 - 2n_{j-1}) a_j \tag{A.271}$$

$$S_j^z = a_j^\dagger a_j - \frac{1}{2}, \tag{A.272}$$

where $n_j = a_j^\dagger a_j$ and $S_j^\pm = S_j^x \pm iS_j^y$. This is the Jordan–Wigner transformation. To prove the validity of this transformation, it is sufficient to show that the inverse transformation from spin to Fermion

$$a_j^\dagger = (-2)^{j-1} S_1^z S_2^z \cdots S_{j-1}^z S_j^+ \tag{A.273}$$

$$a_j = (-2)^{j-1} S_1^z S_2^z \cdots S_{j-1}^z S_j^- \tag{A.274}$$

---

[17] We first keep the system size finite so that $G(0)$ stays finite with $\sum_j n_j = 0$ and take the thermodynamic limit in the end. In this way $\left(\sum_j n_j\right)^2 G(0)$ always stays 0.

satisfies Fermionic commutation relations. The equivalence of eqns (A.273) and (A.274) to eqns (A.270)–(A.272) can be verified by insertion of the latter to the right-hand sides of the former.

Let us first recall a few properties of spin-1/2 operators. All components commute with each other at different sites, that is, if $j \neq l$,

$$\left[S_j^a, S_l^b\right] = 0 \quad (a, b = x, y, z). \tag{A.275}$$

On the same site, different components anticommute,

$$\left[S_j^x, S_j^y\right]_+ = \left[S_j^y, S_j^z\right]_+ = \left[S_j^z, S_j^x\right]_+ = 0, \tag{A.276}$$

where $[A, B]_+ = AB + BA$ represents the anticommutator. The spin size S is 1/2 and consequently the square of any component is 1/4, e.g. $(S_j^z)^2 = 1/4$.

These properties lead to anticommutation relations for the set $\{a_j, a_j^\dagger\}$ at different sites. For example,

$$\left[a_j, a_{j+1}^\dagger\right]_+ = \left[S_j^-, (-2)S_j^z S_{j+1}^+\right]_+ = -2\left[S_j^-, S_j^z\right]_+ S_{j+1}^+ = 0. \tag{A.277}$$

Similarly, it is straightforward to show $\left[a_j, a_l^\dagger\right]_+ = \left[a_j, a_l\right]_+ = \left[a_j^\dagger, a_l^\dagger\right]_+ = 0$ for arbitrary $j \neq l$. On the same site, proper anticommutation relations can be verified, for example, as

$$\left[a_j^\dagger, a_j\right]_+ = \left[S_j^+, S_j^-\right]_+ = \left[S_j^x + iS_j^y, S_j^x - iS_j^y\right]_+ = 2(S_j^x)^2 + 2(S_j^y)^2 = 1. \tag{A.278}$$

Other relations $\left[a_j, a_j\right]_+ = \left[a_j^\dagger, a_j^\dagger\right]_+ = 0$ are similarly easily checked. These results are sufficient to confirm that $\{a_j, a_j^\dagger\}$ are Fermionic operators.

Attention is needed to the boundary condition. If we impose a periodic boundary condition to the spin variable in a given Hamiltonian, products such as $S_N^x S_1^x$ and $S_N^y S_1^y$ may appear as in the Hamiltonian of the XY model, eqn (9.74). To rewrite these terms using the Fermionic operators, we notice

$$S_N^+ S_1^- = (1 - 2n_1) \cdots (1 - 2n_{N-1}) a_N^\dagger a_1$$

$$= (1 - 2n_1) \cdots (1 - 2n_{N-1})(1 - 2n_N)(1 - 2n_N) a_N^\dagger a_1$$

$$= -(-1)^\mathcal{U} a_N^\dagger a_1, \tag{A.279}$$

where the parity operator is given by

$$(-1)^\mathcal{U} = (1 - 2n_1) \cdots (1 - 2n_{N-1})(1 - 2n_N), \tag{A.280}$$

or equivalently

$$\mathcal{U} = \sum_{j=1}^N n_j. \tag{A.281}$$

Similarly, we find

$$S_N^- S_1^+ = (-1)^\mathcal{U} a_N a_1^\dagger. \tag{A.282}$$

It is appropriate to impose an antiperiodic boundary condition $a_{N+1} = -a_1$, $a_{N+1}^\dagger = -a_1^\dagger$ for $\mathcal{U}$ even and a periodic boundary condition $a_{N+1} = a_1$, $a_{N+1}^\dagger = a_1^\dagger$ for $\mathcal{U}$ odd because, then,

$$
\begin{aligned}
S_N^x S_1^x + S_N^y S_1^y &= \frac{1}{2}(S_N^+ S_1^- + S_N^- S_1^+) \\
&= -\frac{(-1)^\mathcal{U}}{2}(a_N^\dagger a_1 - a_N a_1^\dagger) \\
&= \frac{1}{2}(a_N^\dagger a_{N+1} + a_{N+1}^\dagger a_N).
\end{aligned}
\tag{A.283}
$$

This is of the same form as the other terms with $j = 1, \cdots, N-1$ in eqn (9.79). The difference in the boundary condition according to the parity of $\mathcal{U}$ affects the wave numbers as in eqns (9.103) and (9.104) in the context of Majorana fields.

A.18 Proof of Theorem 9.1

To prove Theorem 9.1 we first introduce the following theorem.

> THEOREM A.1 [Existence of the thermodynamic limit] *Consider the Ising model with uniform interactions. The free energy per spin*
>
> $$
> f = \frac{F}{N} = -\frac{T}{N}\log Z
> \tag{A.284}
> $$
>
> *converges to a limit as $N \to \infty$ if the number of sites on the surface of the system is sufficiently smaller than the total number of sites N.*

This theorem is valid irrespective of the sign of the interactions. It also holds for a model with many-body interactions. For simplicity, however, we present here the proof for the more restricted case of ferromagnetic, two-body, nearest-neighbor interactions only,

$$
H = -J \sum_{\langle ij \rangle} S_i S_j - h \sum_i S_i \quad (J, h > 0).
\tag{A.285}
$$

It will also be assumed that boundary conditions are free, although Theorem A.1 can in fact be proved for periodic boundaries as well. The thermodynamic limit will be taken by multiplications of system size by four or other appropriate integers.

Proof of Theorem A.1 under restricted conditions. It is useful to consider a slightly more general Hamiltonian than eqn (A.285),

$$
H = -\sum_{\langle ij \rangle} J_{ij} S_i S_j - h \sum_i S_i,
\tag{A.286}
$$

where J_{ij} is an adjustable parameter in the range $0 \le J_{ij} \le J$. The free energy corresponding to eqn (A.286) is a monotone decreasing function of an arbitrary

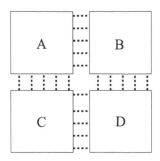

Fig. A.4 A system composed of four subsystems and the bonds connecting them.

interaction J_{kl},

$$-\frac{\partial F}{\beta \partial J_{kl}} = \langle S_k S_l \rangle \geq 0. \tag{A.287}$$

The last inequality, called the first *Griffiths inequality*, can be verified by the expansion of the Boltzmann factor,[18]

$$Z \cdot \langle S_k S_l \rangle = \sum_{\{S_i\}} S_k S_l \, e^{\beta \sum_{\langle ij \rangle} J_{ij} S_i S_j + \beta h \sum_i S_i}$$

$$= \sum_{\{S_i\}} S_k S_l \sum_{n=0}^{\infty} \frac{\beta^n}{n!} \left(\sum_{\langle ij \rangle} J_{ij} S_i S_j + h \sum_i S_i \right)^n \geq 0. \tag{A.288}$$

The final inequality holds because any term obtained by the expansion of the nth power is a product of J, h and S and the summation over S gives either 0 or a positive value.

Now, suppose that the system is composed of several identical subsystems connected by interactions at interfacial bonds as depicted in Fig. A.4. For example, if there are four subsystems,

$$H = H_A + H_B + H_C + H_D + H_{\text{int}}, \tag{A.289}$$

where interactions within H_A, H_B, H_C and H_D are uniform, whereas interfacial bonds are kept arbitrary as in eqn (A.286). If F represents the free energy of the total system with uniform interactions also at interfacial bonds, the monotonicity of the free energy, eqn (A.287), implies

$$F \leq F_A + F_B + F_C + F_D = 4F_A \tag{A.290}$$

because F is obtained from $F_A + F_B + F_C + F_D$ by increasing the values of interfacial bonds from 0 to J. The final equality comes from the fact that all subsystems are identical. Then, the free energy per spin is monotone decreasing with system size,

---

[18] Notice that the factor $\beta = 1/T$ is not absorbed in the definition of the Hamiltonian since it is formally needed for a Taylor expansion.

$$\frac{F}{N} \le \frac{F_A}{N_A}, \tag{A.291}$$

where $N_A = N/4$. Since the free energy per spin is bounded from below, as will be shown as Lemma A.2, the free energy per spin is concluded to converge to a limit in the thermodynamic limit. Although we have illustrated the idea for the case with four subsystems, the argument clearly holds for more general cases. ∎

Lemma A.2 *The free energy per spin of the system of eqn (A.285) is uniformly bounded from below for real values of the parameters.*

Proof. It is helpful to use the notation of the lattice gas. The translation $n_i = (1 - S_i)/2 \; (= 0, 1)$ applied to eqn (A.285) yields[19]

$$H = -4J \sum_{\langle ij \rangle} n_i n_j + (2Jc + 2h) \sum_i n_i \tag{A.292}$$

up to a trivial additive constant, which we ignore for simplicity, and c is the coordination number. The partition function is

$$Z = \sum_{\{n_i\}} e^{4K \sum_{\langle ij \rangle} n_i n_j - 2\beta(Jc+h) \sum_i n_i} = \sum_{k=0}^{N} y^k Q_k(K), \tag{A.293}$$

where $y = e^{-2\beta(Jc+h)}$ and

$$Q_k(K) = {\sum_{\{n_i\}}}' e^{4K \sum_{\langle ij \rangle} n_i n_j}. \tag{A.294}$$

The summation with a prime runs under the constraint $\sum_i n_i = k$. Since there exist k non-vanishing n_is and each of them has at most c non-vanishing neighboring n_js, the sum in the above exponent is bounded as

$$\sum_{\langle ij \rangle} n_i n_j \le ck. \tag{A.295}$$

Taking into account the number of combinations to choose k non-vanishing n_is out of N of them, we find

$$Q_k(K) \le \binom{N}{k} e^{4Kck}. \tag{A.296}$$

We insert this bound into eqn (A.293) to have

$$Z \le \sum_{k=0}^{N} y^k \binom{N}{k} e^{4Kck} = (1 + y\, e^{4Kc})^N. \tag{A.297}$$

Hence, the free energy per spin is uniformly bounded from below as

$$\frac{F}{N} \ge -T \log(1 + y\, e^{4Kc}). \quad ∎ \tag{A.298}$$

[19] The correspondence rule $n_i = (1 - S_i)/2$ is different from $n_i = (1 + S_i)/2$ in Section 1.5 just by the exchange of up and down states of all the Ising spins.

Proof of Theorem 9.1. We use the lattice gas representation, eqn (A.293). The difference between $z = e^{-2\beta h}$ and $y = e^{-2\beta(Jc+h)}$ can be adjusted by a simple shift of the real axis in the complex-h plane. The free energy per spin for general complex y

$$\frac{F}{N} = -\frac{T}{N} \log \left(\sum_{k=0}^{N} y^k Q_k(K) \right) \tag{A.299}$$

is analytic in the region R according to the assumption $Z \neq 0$ of the theorem. The same quantity is also uniformly bounded from above inside a circle with its center somewhere on the real axis in R,

$$\left| \frac{F}{N} \right| \leq \frac{T}{N} \log \left(\sum_{k=0}^{N} |y|^k Q_k(K) \right)$$

$$\leq \frac{T}{N} \log \left(\sum_{k=0}^{N} r^k Q_k(K) \right)$$

$$\leq T \log(1 + re^{4Kc}). \tag{A.300}$$

Here, r is the largest absolute value of y in the circle under consideration, and we have used the bound (A.296) for $Q_k(K)$. In consideration of Theorem A.1, we conclude that F/N converges uniformly to a limit within the above-mentioned circle due to Vitali's theorem stated below. The region of validity can be extended to the whole region R by repeating the discussion with the center of the circle shifted elsewhere inside the previous circle. ∎

THEOREM A.3[Vitali's theorem] *Consider a region D and a series of points $A \subset D$ accumulating to a point in D. If a series of regular functions defined in D and uniformly bounded in D converge to a limit on A, then the series of functions converge uniformly in D.*

A.19 Poisson summation formula

The Poisson summation formula replaces a sum of functional values over integers by an integral,

$$\sum_{l=-\infty}^{\infty} f(l) = \sum_{n=-\infty}^{\infty} \int_{-\infty}^{\infty} e^{2\pi i \phi n} f(\phi) \, d\phi. \tag{A.301}$$

To prove this relation it is sufficient to show that the sum over n on the right-hand side leaves only integer values of ϕ in the integral, i.e.

$$\sum_{n=-\infty}^{\infty} e^{2\pi i \phi n} = \sum_{l=-\infty}^{\infty} \delta(\phi - l). \tag{A.302}$$

For this purpose we first make use of the Fourier series of a periodic function $g(x)$ with period 1,

$$g(x) = \sum_{m=-\infty}^{\infty} e^{2\pi i m x} \, \tilde{g}(m).$$ (A.303)

The Fourier coefficient has the following expression with c an arbitrary real number,

$$\tilde{g}(m) = \int_{c}^{c+1} g(y) e^{-2\pi i m y} \, dy.$$ (A.304)

Insertion of this formula into eqn (A.303) yields

$$g(x) = \sum_{m=-\infty}^{\infty} \int_{c}^{c+1} g(y) e^{2\pi i m (x-y)} \, dy = \int_{c}^{c+1} g(y) \sum_{m=-\infty}^{\infty} e^{2\pi i m (x-y)} \, dy.$$ (A.305)

A necessary and sufficient condition for the above equation to hold for arbitrary x and c is that the part of summation in the integrand is a Dirac delta function that is non-vanishing only for integer $x - y$,

$$\sum_{m=-\infty}^{\infty} e^{2\pi i m (x-y)} = \sum_{l=-\infty}^{\infty} \delta(x - y - l).$$ (A.306)

This is eqn (A.302).

A.20 Sample codes for Monte Carlo simulation of the Ising model

Here are the sample codes of Monte Carlo simulations of the two-dimensional Ising model on the square lattice. The codes in this section are written as illustrations of the theory explained in Section 11.2. It is necessary to optimize the codes when the reader uses them for practical purposes.

A.20.1 Code in Fortran

Note that 'ran(iran)' is a random number generator between 0 and 1 and should be prepared appropriately.

```
!--------------------------------------------------------------
! Monte Carlo simulation of the two-dimensional Ising model
! Metropolis method.  Ferromagnetic interaction, J=1.
!--------------------------------------------------------------
        integer,parameter::L=40          !Linear size
        integer,parameter::mcs=100000    !Total MC steps (per spin)
        integer,parameter::discard=500   !# of steps to remove initial
                                          effects
        integer,parameter::measure=10    !Measurement interval
        integer::mcprocess,i,j           !Variables to control loops
        integer::i1,i2                   !Site index for flip trial
```

```fortran
      integer::spin(L,L)              !Spin configuration
      integer::ip(L),im(L)            !Table of right and left neighbors
      integer::iran                   !Random number seed
      real,parameter::T=2.0           !Temperature
      real::delta_E,energy            !Energy change and energy
      real::ecurrent                  !Local energy
      real::field                     !Local field
      real::denominator               !Measurement normalizer
!-------------------Initialization---------------------
! Table of nearest-neighbor sites for periodic boundaries
      do i=1,L
          ip(i)=i+1                             ! Right (upper) neighbor
          im(i)=i-1                             ! Left (lower) neighbor
      end do
      ip(L)=1                                   ! Right (up) of L is 1.
      im(1)=L                                   ! Left (bottom) of 1 is L.
!   Initial configuration.  All up.
      spin=1
!   Random number initialization
      iran=991963
!-------------------- Main loop ------------------------
      energy=0.0
 Main_loop: do mcprocess=1,mcs*L*L
!---------------------------------------------
          i1=ran(iran)*L+1 !Randomly choose a site for flip trial.
                           x coordinate.
          i2=ran(iran)*L+1 !Same for y coordinate.
          field=spin(ip(i1),i2)+spin(i1,ip(i2))+spin(im(i1),i2)
                           +spin(i1,im(i2))
                           !Sum of spin states around spin(i1,i2)
          delta_E=field*spin(i1,i2)*2.0
                           !Energy change caused by the flip of
                           spin(i1,i2).
!Execution of the Metropolis method----------------------
          if(delta_E<0)then
            spin(i1,i2)=-spin(i1,i2)  !Flip if energy decreases.
            else if(exp(-delta_E/T)>ran(iran))then
            spin(i1,i2)=-spin(i1,i2)  !Probabilistic flip for energy
            increase.
          end if
!--------------------Measurement ----------------------
Stat:    if(mod(mcprocess,measure*L*L)==0)then ! Measurement at some
                                                interval
          if(mcprocess>discard*L*L)then   ! Skip until the system
                                           equilibrates.
```

```
            ecurrent=0.0
        do i=1,L
        do j=1,L
         field=spin(ip(i),j)+spin(i,ip(j))+spin(im(i),j)
         +spin(i,im(j))
         ecurrent=ecurrent+spin(i,j)*field
        end do
        end do
        ecurrent=-ecurrent/2.0
            !Divide by two since each bond is counted twice.
        energy=energy+ecurrent
      end if
    end if  Stat
!---------------End of the Monte Carlo loop -----------
     end do Main_loop
!----------------------Avereage------------------------
     denominator=(mcs-discard)/real(measure)*real(L*L)
                !Number of data points divided by the system size
     energy=energy/denominator
                ! Simple average approximates the canonical average.
!-----------------------Output-------------------------
     write(6,200)energy
 200 format(' Average energy per spin :',f12.6)
     end
```

A.20.2 Code in C

Note that 'genrand()' is a random number generator between 0 and 1 and should be prepared appropriately.

```
/* ---------------------------------------------------------------
Monte Carlo simulation of the two-dimensional Ising model
Metropolis method.  Ferromagnetic interaction, J=1.
-------------------------------------------------------------*/
#include <stdio.h>
#include <stdlib.h>
#include <math.h>
#include <time.h>
#define ls 40           // linear size of the system

int main(void){
    int mcs=100000;        // total MC steps (per spin)
    int discard=500;       // # of steps to remove initial effects
    int measure=10;        // measurement interval
    int i,j;               // variables to control loops
    int i1,i2;             // site index for flip trial
```

```
int mcprocess;           // current Monte Carlo step
int spin[ls][ls];        // spin configuration
int ip[ls],im[ls];       // table of right and left neighbors
double t=2.0;            // temperature
double delta_E,energy;   // energy change and energy
double ecurrent;        // local energy
double field;           // local field
double denominator;     // measurement normalizer
double genrand(void);   // random number generator
//------------------Initialization--------------------
// Table of nearest-neighbor sites for periodic boundaries
for (i=0;i<ls;i++){
    ip[i]=i+1;         // right (upper) neighbor
    im[i]=i-1;         // left (lower) neighbor
}
ip[ls-1]=0;            // right (up) of ls-1 is 0.
im[0]=ls-1;            // left (bottom) of 0 is ls-1.
//   Initial configuration.  All up.
for (i=0;i<ls;i++){
    for (j=0;j<ls;j++){
        spin[i][j]=1;
    }
}
srand((unsigned)time(NULL));     // random number initialization
//------------------- Main loop ------------------------
energy=0.0;
for (mcprocess=1;mcprocess<=mcs*ls*ls;mcprocess++){
 i1=(int)(genrand()*ls);
     // Choose a site for flip trial. x coordinate.
 i2=(int)(genrand()*ls);
     // Same for y coordinate.
 field=spin[ip[i1]][i2]+spin[i1][ip[i2]]+spin[im[i1]][i2]
   +spin[i1][im[i2]];
     // Sum of spin states around spin[i1][i2]
 delta_E=field*spin[i1][i2]*2.0;
     // Energy change caused by the flip of spin[i1][i2].
// Execution of the Metropolis method---------------------
    if(delta_E<0){
        spin[i1][i2]*=-1;   // Flip if energy decreases.
    }
    else{
        if(exp(-delta_E/t)>genrand()) spin[i1][i2]*=-1;
    }        // Probabilistic flip for energy increase.
//-------------------Measurement ----------------------
    if(mcprocess%(measure*ls*ls)==0){
```

```
        if(mcprocess>(discard*ls*ls))//Skip until the system
                                        equilibrates.
        {
          ecurrent=0.0;
          for (i=0;i<ls;i++){
          for (j=0;j<ls;j++){
           field=spin[ip[i]][j]+spin[i][ip[j]]+spin[im[i]][j]
            +spin[i][im[j]];
           ecurrent+=spin[i][j]*field;
          }
          }
            ecurrent/=-2.0;
                //Divide by two since each bond is counted twice.
            energy+=ecurrent;
          }
      }
//-----------------End of the Monte Carlo loop ------------
  }
//----------------------Avereage-------------------------
  denominator=(mcs-discard)/(double)measure*ls*ls;
            //Number of data points divided by the system size
  energy/=denominator;
            // Simple average approximates the canonical average.
//-----------------------Output-------------------------
  printf("Average energy per spin :%f\n",energy);
  return 0;
}
```

Appendix B
Solutions to exercises

Chapter 1

1.1

Let $S_i = 1$ correspond to $\sigma_i = -1$ and $S_i = 2$ to $\sigma_i = 1$. Then, we have

$$\delta_{S_i,S_j} = \frac{1}{2} + \frac{1}{2}\sigma_i\sigma_j, \quad \delta_{S_i,1} = \frac{1}{2} - \frac{\sigma_i}{2}. \tag{B.1}$$

This relation shows that the two-state Potts model is equivalent to the Ising model. To be more explicit, we can rewrite eqn (1.19) as

$$H = -\frac{J}{2}\sum_{\langle ij \rangle} \sigma_i\sigma_j + \frac{h}{2}\sum_i \sigma_i + \text{const.} \tag{B.2}$$

Chapter 2

2.1

According to the mean-field approximation of eqn (2.5), the Hamiltonian is reduced to

$$H \approx \sum_i H_i = -(Jmz + h)\sum_i S_i, \tag{B.3}$$

where $S_i = -S, -S+1, \cdots, S-1, S$, with S an integer or a half-odd integer. The total partition function is $Z = \prod_i Z_i$, where

$$Z_i = \sum_{S_i=-S}^{S} e^{\beta(Jmz+h)S_i} = \frac{\sinh\left((S + \frac{1}{2})\beta(Jmz + h)\right)}{\sinh\frac{\beta}{2}(Jmz + h)} \tag{B.4}$$

is the single-spin partition function. The magnetization is defined as

$$m = \langle S_i \rangle = \frac{1}{Z_i}\sum_{S_i=-S}^{S} S_i\, e^{xS_i} = \frac{d\log Z_i}{dx} = \left(S + \frac{1}{2}\right)\coth\left((S + \frac{1}{2})x\right) - \frac{1}{2}\coth\left(\frac{x}{2}\right),$$

$$\tag{B.5}$$

with $x = \beta(Jmz + h)$. At $h = 0$, the slope of the right-hand side of eqn (B.5) at $m = 0$ is greater than 1 when

$$\beta z J \frac{S(S+1)}{3} > 1, \tag{B.6}$$

and therefore the critical temperature is

$$T_\mathrm{c} = zJ \frac{S(S+1)}{3}. \tag{B.7}$$

Note that the Ising model in this notation corresponds to $S = 1/2$, and if one wishes to recover eqn (2.6) from eqn (B.5), the following changes must be applied to eqn (B.5), $2m, J/4, h/2 \to m, J, h$.

2.2

Differentiation of both sides of the equation of state with respect to h and setting $h \to 0$ give the following equation,

$$\frac{\partial m}{\partial h} = \mathrm{sech}^2(\beta Jmz)\left(\beta Jz\frac{\partial m}{\partial h} + \beta\right). \tag{B.8}$$

When the system is very close to the transition point, m is very small and we are allowed to expand $\mathrm{sech}^2(\beta Jmz)$ in the above equation to second order in m. Then, we can write the susceptibility as

$$\chi = \frac{\beta\left(1 - (\beta Jmz)^2\right)}{1 - \beta Jz\left(1 - (\beta Jmz)^2\right)}. \tag{B.9}$$

Below the transition point, eqn (2.7) gives

$$(\beta Jmz)^2 = \frac{3(\beta Jz - 1)}{\beta Jz} = 3\left(1 - \frac{T}{T_\mathrm{c}}\right), \tag{B.10}$$

and therefore

$$\chi = \frac{\beta\left(1 - 3(1 - T/T_\mathrm{c})\right)}{1 - \beta Jz\left(1 - 3(1 - T/T_\mathrm{c})\right)}. \tag{B.11}$$

The second term in the numerator, the term with the prefactor 3, is much smaller than the first term and can be ignored. By rewriting the denominator we have

$$\chi = \frac{1}{2(T_\mathrm{c} - T)}, \tag{B.12}$$

from which $\gamma' = 1$ is concluded.

2.3

From eqn (2.4) the total partition function is

$$Z = \mathrm{e}^{-\beta N_\mathrm{B} Jm^2}\left(2\cosh\left(\beta(Jmz + h)\right)\right)^N, \tag{B.13}$$

and the average energy per spin is, if we remember that $m = \tanh(\beta Jmz)$,

$$\varepsilon(T) = -\frac{1}{N} \left.\frac{\partial \log Z}{\partial \beta}\right|_{h=0} = \frac{Jz}{2}m^2 - \tanh(\beta Jmz)Jzm = -\frac{Jz}{2}m^2. \tag{B.14}$$

On the other hand, the specific heat ($h = 0$) is defined as, using $T_c = Jz$,

$$c(T) = \frac{\partial \varepsilon(T)}{\partial T} = Jz\beta^2 m \frac{\partial m}{\partial \beta} = \left(\frac{T_c}{T}\right)^2 \frac{m^2}{\cosh^2((T_c/T)m) - T_c/T}. \tag{B.15}$$

Note that $\varepsilon(T) = 0, c(T) = 0$, for $T \geq T_c$, and $\varepsilon(T) = 3(T - T_c)/2$, $c(T) = 3/2$, for $T \to T_c^-$. In other words, the specific heat is discontinuous at $T = T_c$.

2.4

We set $\boldsymbol{S}_i = \boldsymbol{m} + \delta \boldsymbol{S}_i$ and use the approximation that neglects quadratic terms in $\delta \boldsymbol{S}_i$ as in Section 2.1 to obtain

$$H = N_{\mathrm{B}} Jm^2 - (Jmz + h) \sum_i S_i^z. \tag{B.16}$$

Here, it has been assumed that $\boldsymbol{m}$ has only a z-component. The partition function is then written as

$$Z = \mathrm{e}^{-N_{\mathrm{B}}\beta Jm^2} \left[\int \mathrm{d}\boldsymbol{S}\, \mathrm{e}^{\beta(Jmz+h)S^z}\right]^N. \tag{B.17}$$

The integral is performed over the unit sphere $|\boldsymbol{S}| = 1$. Since the magnetization m is the average of S^z, we have

$$m = \frac{\int \left(\prod_i \mathrm{d}\boldsymbol{S}_i\right) S_i^z \mathrm{e}^{-\beta H}}{\int \left(\prod_i \mathrm{d}\boldsymbol{S}_i\right) \mathrm{e}^{-\beta H}} = \frac{\int \mathrm{d}\boldsymbol{S}\, S^z \mathrm{e}^{\beta(Jmz+h)S^z}}{\int \mathrm{d}\boldsymbol{S}\, \mathrm{e}^{\beta(Jmz+h)S^z}}$$

$$= \frac{\partial}{\partial(\beta h)} \log\left(\int \mathrm{d}\boldsymbol{S}\, \mathrm{e}^{\beta(Jmz+h)S^z}\right). \tag{B.18}$$

The integral here can be evaluated as follows using polar coordinates in three dimensions. S^z is the projection of the unit vector onto the z-axis and is $\cos\theta$. Thus,

$$\int \mathrm{d}\boldsymbol{S}\, \mathrm{e}^{\beta(Jmz+h)S^z} = \int_0^\pi \sin\theta\mathrm{d}\theta \int_0^{2\pi} \mathrm{d}\phi\, \mathrm{e}^{\beta(Jmz+h)\cos\theta}$$

$$= 2\pi \int_{-1}^1 \mathrm{d}\mu\, \mathrm{e}^{\beta(Jmz+h)\mu} = \frac{4\pi \sinh\beta(Jmz+h)}{\beta(Jmz+h)}. \tag{B.19}$$

The log-derivative of this equation with respect to βh

$$m = \coth\beta(Jmz+h) - \frac{1}{\beta(Jmz+h)} \tag{B.20}$$

is the self-consistent equation. We expand the right-hand side in powers of m with $h = 0$ in order to determine the critical point and critical exponent,

$$m \approx \frac{\beta Jmz}{3} - \frac{(\beta Jmz)^3}{45}. \tag{B.21}$$

The critical point is found to be $T_c = Jz/3$ from the condition that the coefficient of the linear term on the right-hand side is unity. Since the coefficient of the third-order term of the right-hand side is negative, we conclude that the critical exponent β is $1/2$ by using the same reasoning as in Section 2.2.

2.5

The magnetization vanishes above the transition point $m = 0$, and hence from eqn (B.9) we have $\chi = 1/(T - T_c)$. Comparison of this result with eqn (B.12) for the low-temperature side of the transition point reveals $1/2$ for the ratio of critical amplitudes, a universal value independent of the details of the system. The Landau theory also gives $1/2$ from eqns (2.22) and (2.23) for the susceptibility above and below the transition point.

2.6

As is well known in the van der Waals theory, the critical point $(T \to T_c)$ is defined from the equation of state as the inflexion point of the function $P(v) = T/(v - b) - a/v^2$

$$\frac{\partial P(v)}{\partial v} = \frac{\partial^2 P(v)}{\partial v^2} = 0, \tag{B.22}$$

or equivalently

$$\frac{T_c}{(v_c - b)^2} = \frac{2a}{v_c^3}, \quad \frac{T_c}{(v_c - b)^3} = \frac{3a}{v_c^4}, \tag{B.23}$$

which gives the critical volume $v_c = 3b$ and the critical temperature $T_c = 8a/(27b)$. By replacing these critical values in the equation of state, we obtain $P_c = a/(27b^2)$. The ratio between these critical parameters is a universal number

$$\frac{P_c v_c}{T_c} = \frac{3}{8}. \tag{B.24}$$

Let us compute some critical exponents directly from the equation of state. Rewrite the latter in terms of the reduced variables $p = (P - P_c)/P_c, t = (T - T_c)/T_c$ and $\mathcal{V} = (v - v_c)/v_c$ as

$$p + 1 = \frac{8(t + 1)}{3(\mathcal{V} + 1) - 1} - \frac{3}{(\mathcal{V} + 1)^2}. \tag{B.25}$$

By setting $t = 0$ (the critical isotherm) and expanding near $\mathcal{V} = 0$, one finds

$$p = -\frac{3}{2}\mathcal{V}^3 \propto -\mathcal{V}^\delta, \tag{B.26}$$

which means $\delta = 3$. Similarly, one can compute the isothermal compressibility

$$\kappa_T = -\frac{1}{v}\frac{\partial v}{\partial P}\bigg|_T = \left(\frac{Tv}{(v-b)^2} - \frac{2a}{v^2}\right)^{-1} \approx \frac{4b}{3}(T - T_c)^{-1}, \tag{B.27}$$

implying $\gamma = 1$. As we will see in Chapter 3, to determine the critical behavior we only need to compute two exponents since the others are related by scaling relations. Therefore, we conclude that the critical behavior of the van der Waals fluid is of the mean-field type.

2.7

First, notice that the free energy is not $\mathbb{Z}_2$-symmetric, i.e. $f(m) \neq f(-m)$. To determine the equilibrium value of the magnetization we differentiate the Landau free energy and set the result to zero, $\partial f(m)/\partial m = 2am + 4bm^3 + 3cm^2 = 0$. The solutions are

$$m = 0 \ , \ m_\pm = \frac{-3c \pm \sqrt{9c^2 - 32ab}}{8b}. \tag{B.28}$$

A non-vanishing physical solution must satisfy $9c^2 \geq 32ab$. If $a = kt$, this implies that $t \leq \bar{t} = 9c^2/(32kb)(>0)$. This means that the non-vanishing physical solution is still valid for temperatures larger than T_c ($t = 0$). When $t > \bar{t}$, $m = 0$ is the equilibrium solution. At $0 < t = t_1 = c^2/(4kb) < \bar{t}$, when $c < 0$ ($c > 0$), the solutions $m = 0$ and $m_+(m_-)$ have the same free energy, i.e. both are stable. For $t < t_1$, the non-vanishing solution is the equilibrium one (global minimum of the free energy). Therefore, at $t = t_1$ the magnetization (order parameter) jumps discontinuously, indicating a first-order transition. The addition of a cubic term to the free energy clearly leads to a first-order transition within the Landau framework. If the jump in magnetization at the transition point is not small, then the Landau expansion is not necessarily accurate. It is important to remember that fluctuation effects over the mean-field may change the order of the transition.

2.8

The equation indicating that the free energy is equal to 0 (the value at the origin) should have non-vanishing multiple solutions when the magnetization jumps from 0 to a non-vanishing value. The non-vanishing solution to the equation $am^2/2 + bm^4/4 + cm^6/6 = 0$ is $m^2 = (-3b \pm \sqrt{9b^2 - 48ac})/4c$. Hence, the condition for a multiple solution is $9b^2 - 48ac = 0$ or $a = 3b^2/16c$.

2.9

The application of the mean-field approximation to the Hamiltonian (1.18) with $h = 0$ leads to

$$H = N_B Jm^2 - Jzm\sum_i S_i - D\sum_i S_i^2. \tag{B.29}$$

Since each spin is independent in this equation, we can take the sum over $S_i = -1, 0, 1$ at each site, and the free energy per spin is derived as

$$f = \frac{Jm^2 z}{2} - T \log \left(e^{Kzm+\beta D} + 1 + e^{-Kzm+\beta D} \right). \tag{B.30}$$

Here, $K = \beta J$, and we have used $N_B = zN/2$. Expansion of the logarithm to sixth order in m gives, with $e^{\beta D}$ written as u,

$$f = \frac{Jm^2 z}{2} - T \left(\log(1 + 2u) + \frac{u(Kz)^2}{1 + 2u} m^2 + \frac{u(1 - 4u)(Kz)^4}{12(1 + 2u)^2} m^4 \right.$$
$$\left. + \frac{u(1 - 26u + 64u^2)(Kz)^6}{360(1 + 2u)^3} m^6 \right). \tag{B.31}$$

The coefficient of the fourth-order term changes sign at $u = 1/4$. This condition, $e^{\beta D} = 1/4$, has a solution if D is negative. It is straightforward to check that the coefficient of the sixth-order term is positive when $u = 1/4$.

2.10

Since $m_0 = m_1$, we equate eqn (2.63) to (2.64),

$$e^{\beta h} \left(2 \cosh(K + \beta h + \beta h_1) \right)^z - e^{-\beta h} \left(2 \cosh(-K + \beta h + \beta h_1) \right)^z$$
$$= e^{\beta h} \left(2 \cosh(K + \beta h + \beta h_1) \right)^z \tanh(K + \beta h + \beta h_1)$$
$$+ e^{-\beta h} \left(2 \cosh(-K + \beta h + \beta h_1) \right)^z \tanh(-K + \beta h + \beta h_1). \tag{B.32}$$

By collecting the terms with $e^{\beta h}$ on the left-hand side and those with $e^{-\beta h}$ on the right-hand side, we have

$$e^{\beta h} \left(\cosh(K + \beta h + \beta h_1) \right)^z \left(1 - \tanh(K + \beta h + \beta h_1) \right)$$
$$= e^{-\beta h} \left(\cosh(-K + \beta h + \beta h_1) \right)^z \left(1 + \tanh(-K + \beta h + \beta h_1) \right), \tag{B.33}$$

which is equivalent to

$$\left(\cosh(K + \beta h + \beta h_1) \right)^{z-1} e^{-K-\beta h_1} = \left(\cosh(-K + \beta h + \beta h_1) \right)^{z-1} e^{-K+\beta h_1}. \tag{B.34}$$

This is the desired relation.

2.11

The second term of the right-hand side of the self-consistent equation (2.66) is much smaller than the first term and can be neglected. Then, the self-consistent equation in the presence of an external field is

$$\frac{2\beta h_1}{z - 1} = 2 \tanh K \cdot (\beta h_1 + \beta h), \tag{B.35}$$

from which we have

$$\frac{h_1}{h} = \frac{\tanh K}{1/(z - 1) - \tanh K} \propto \frac{1}{T - T_c}. \tag{B.36}$$

We next study the relation between this ratio h_1/h and the susceptibility with $m = m_0 = m_1$ in mind. The expansion of eqn (2.61) for small h and h_1

$$Z_\pm = (2\cosh K)^z \left(1 \pm \beta h \pm z \sinh K \cdot (\beta h + \beta h_1)\right) \tag{B.37}$$

is inserted into eqn (2.63) to give

$$m_0 = \beta h + z \sinh K \cdot (\beta h + \beta h_1). \tag{B.38}$$

Differentiation of both sides with respect to h leads to

$$\chi = \frac{\partial m_0}{\partial h} = \beta + \beta z \sinh K \cdot \left(1 + \frac{\partial h_1}{\partial h}\right). \tag{B.39}$$

Since the final term on the right-hand side diverges as $1/(T - T_c)$ as in eqn (B.36), we conclude that the susceptibility also diverges with the same rate, and so $\gamma = 1$.

To evaluate δ it is useful to rewrite eqn (2.66) with an external field added as

$$2\tanh K \cdot \beta h = \frac{2\sinh K}{3\cosh^3 K}(\beta h_1 + \beta h)^3, \tag{B.40}$$

where the condition has been used that the system is exactly at the transition point. Combination of this and the relation

$$\beta h + \beta h_1 = \frac{m - \beta h}{z \sinh K} \tag{B.41}$$

that results from eqn (B.38) yields

$$2\tanh K \cdot \beta h = \frac{2\sinh K}{3\cosh^3 K}\left(\frac{m - \beta h}{z \sinh K}\right)^3. \tag{B.42}$$

This equation is not satisfied in the small-h limit unless m is proportional to $h^{1/3}$.

2.12

The integral is immediately separated by (a) as

$$g(\boldsymbol{r}) = \int_0^\infty \mathrm{d}u\, \mathrm{e}^{-ua^2} \prod_{i=1}^d \int_{-\infty}^\infty \mathrm{d}q_i\, \mathrm{e}^{-uq_i^2 + iq_i r_i}. \tag{B.43}$$

Gaussian integration of this expression easily leads to eqn (2.90). This integral is written in terms of the modified Bessel function of the second kind as

$$g(\boldsymbol{r}) = \pi^{d/2} a^{d-2}\, 2\left(\frac{2}{ar}\right)^{d/2-1} K_{d/2-1}(ar). \tag{B.44}$$

The asymptotic form of the modified Bessel function of the second kind in the limit of large r gives $g(\boldsymbol{r}) \propto r^{-(d-1)/2}\mathrm{e}^{-ar}$. The original problem (2.85) has kt/b for a^2 in eqn (2.88), and we have $\xi = \sqrt{b/kt}$ in the exponent of $\mathrm{e}^{-ar} = \mathrm{e}^{-r/\xi}$.

Chapter 3

3.1

Equation (3.15), to be written as Z_{local}, is

$$Z_{\text{local}} = e^{K(S_1+S_2+S_3+S_4)} + e^{-K(S_1+S_2+S_3+S_4)}. \tag{B.45}$$

Since the simultaneous inversion of all four spins $S_i \to -S_i$ ($\forall i$) does not change Z_{local}, it should be expressed in terms of even products of $S_1, \cdots, S_4$. The logarithm $\log Z_{\text{local}}$ should also have the same property, and hence we can write

$$Z_{\text{local}} = A \exp\{K_1'(S_1S_2 + S_1S_3 + S_1S_4$$
$$+ S_2S_3 + S_2S_4 + S_3S_4) + K_2'S_1S_2S_3S_4)\}. \tag{B.46}$$

Other fourth-order terms like $S_1^2 S_2^2$ and higher-order terms such as $S_1 S_2^2 S_3^2 S_4$ reduce to the above form because of the identity $S_i^2 = 1$. As eqn (B.45) includes $S_1, \cdots, S_4$ in a symmetric way, eqn (B.46) is also written in a form symmetric under the exchange of any pair of spins.

We next have to fix the values of A, K_1', K_2' as functions of K by comparison of eqns (B.45) and (B.46). This problem turns out to be relatively easy if we equate those two equations at specific values of $S_1, \cdots, S_4$. Since we need only three equations to determine the three parameters A, K_1', K_2', we write the relation that eqn (B.45) is equal to (B.46) for three cases, the first for all S_i to be unity, the second $S_1 = -1$ and all others 1, and the third $S_1 = S_2 = -1$ and $S_3 = S_4 = 1$,

$$2\cosh 4K = Ae^{6K_1'+K_2'}, \quad 2\cosh 2K = Ae^{-K_2'}, \quad 2 = Ae^{-2K_1'+K_2'}. \tag{B.47}$$

By taking the ratio of these relations, we find the explicit expressions for K_1' and K_2' as

$$K_1' = \frac{1}{8}\log\cosh 4K, \quad K_2' = \frac{1}{8}\log\cosh 4K - \frac{1}{2}\log\cosh 2K. \tag{B.48}$$

The nearest-neighbor spins after renormalization, $S_1S_2, S_2S_3, S_3S_4, S_4S_1$ (see Fig. B.1), acquire the same renormalized interaction K_1' from the adjacent block of spins, and thus the renormalized neighboring coupling is $K' = 2K_1'$. The

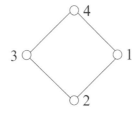

Fig. B.1 A single step of renormalization produces a system with interactions among four spins $S_1, \cdots, S_4$.

next-nearest-neighbor interactions, $S_1 S_3$ and $S_2 S_4$, do not have such contributions and the interaction remains K_1', and so does the four-spin interaction K_2'.

3.2

The scaling law of the free energy is written as, with the lattice constant a included as a variable,

$$f(t, h, a) = b^{-d} f(b^{y_t} t, b^{y_h} h, b^{y_a} a). \tag{B.49}$$

After renormalization by the scale b, the lattice constant should be multiplied by $1/b$, and the exponent is $y_a = -1$. Hence, a is irrelevant.

3.3

Since the correlation function does not decay as a power law, eqn (3.55) implies $y = y_h = d$.

3.4

Let us equate eqn (3.59) with eqn (3.60) and insert ± 1 into S_1 and S_3 to have

$$e^{2K+h} + e^{-2K-h} = A e^{K'+2h_1} \tag{B.50}$$

$$e^h + e^{-h} = A e^{-K'} \tag{B.51}$$

$$e^{-2K+h} + e^{2K-h} = A e^{K'-2h_1}. \tag{B.52}$$

The ratio of eqns (B.50) and (B.52) gives

$$e^{4h_1} = \frac{\cosh(2K+h)}{\cosh(2K-h)}. \tag{B.53}$$

Hence,

$$e^{2h'} = e^{2h+4h_1} = \frac{e^{2h}\cosh(2K+h)}{\cosh(2K-h)}. \tag{B.54}$$

This is eqn (3.63). Next, from the ratio of eqns (B.50) and (B.51),

$$e^{2K'+2h_1} = \frac{\cosh(2K+h)}{\cosh h}. \tag{B.55}$$

Further, from the ratio of eqns (B.52) and (B.51),

$$e^{2K'-2h_1} = \frac{\cosh(2K-h)}{\cosh h}. \tag{B.56}$$

Multiplication of these two equations leads to eqn (3.62),

$$e^{4K'} = \frac{\cosh(2K+h)\cosh(2K-h)}{\cosh^2 h}. \tag{B.57}$$

Lastly, from eqn (B.51), we have eqn (3.64),

$$A^4 = e^{4K'}(2\cosh h)^4 = 16\cosh^2 h \,\cosh(2K+h)\cosh(2K-h). \tag{B.58}$$

3.5

The scaling relation for the specific heat is derived from the scaling law of the free energy by differentiation of the latter twice with respect to t, the scaling field corresponding to the temperature. The discussions for the one-dimensional system had this t replaced by $x = e^{-4K}$. As mentioned in Section 3.6.3, the dependence on the original temperature variable in the limit $K \to \infty$ is recovered only after the correction of x^2. The field is also replaced by $y = e^{-2h}$, but the fixed point of our interest is at $h = 0$, which reduces the correction factor to unity, i.e. $y = 1$.

3.6

We just repeat the same procedure as in the case of $b = 2$. The quantity to calculate is

$$\sum_{S_2, S_3, \cdots, S_b} e^{K(S_1 S_2 + S_2 S_3 + \cdots + S_b S_{b+1})}. \tag{B.59}$$

To perform the sum over all spins simultaneouly will complicate the situation, and so we start with $S_2 = \pm 1$,

$$\sum_{S_2} e^{K(S_1 S_2 + S_2 S_3)}$$

$$= \cosh^2 K \sum_{S_2} (1 + S_1 S_2 \tanh K)(1 + S_2 S_3 \tanh K)$$

$$\propto 1 + S_1 S_3 \tanh^2 K. \tag{B.60}$$

The next sum to be taken is over S_3. Similarly to the above,

$$\sum_{S_3} (1 + S_1 S_3 \tanh^2 K)(1 + S_3 S_4 \tanh K) \propto 1 + S_1 S_4 \tanh^3 K. \tag{B.61}$$

The same process for S_4 gives

$$\sum_{S_4} (1 + S_1 S_4 \tanh^3 K)(1 + S_4 S_5 \tanh K) \propto 1 + S_1 S_5 \tanh^4 K. \tag{B.62}$$

It is now apparent that the result of consecutive summations over S_2 to S_b is $1 + S_1 S_{b+1} \tanh^b K \equiv 1 + S_1 S_{b+1} u'$, which gives the desired result $u' = u^b$.

3.7

We insert the scaling law of magnetization (3.76) and $a = kt$ to the Landau free energy $f = am^2 + bm^4 - hm$, $(a = kt)$ to have

$$f = t^2 \{ kg(ht^{-3/2})^2 + bg(ht^{-3/2})^4 - ht^{-3/2}g(ht^{-3/2}) \}. \tag{B.63}$$

This is of the form of the general scaling law (3.38) with the mean-field exponents, $d/y_t = 2 - \alpha = 2, y_h/y_t = \beta\delta = 3/2$.

3.8

It cannot be written. If we assume that m and h are small around the critical point and expand the hyperbolic tangent to third order in m, we obtain essentially the same equation as the equation of state of the Landau theory, eqn (3.74), and the scaling law is satisfied. A general m, not necessarily small, does not satisfy the scaling law because this law is valid only near the critical point.

3.9

We choose $b = t^{-1/y_t} = t^{-\nu}$ in the finite-size scaling (3.96),

$$f(t, h, L^{-1}) = t^{\nu d} f(1, t^{-y_h/y_t} h, t^{-\nu} L^{-1}). \tag{B.64}$$

Comparison with eqn (3.103) suggests that L^{-1} corresponds to D and hence the crossover exponent of L^{-1} is ν.

Chapter 4

4.1

The fixed-point equation is eqn (4.29) with $K = K' = K^*$, which is solved to give $e^{4K^*} = 1 + 2\sqrt{2}$, or numerically $K^* = 0.336$. We next linearize the parameters around the fixed point as $K' = K^* + \epsilon', K = K^* + \epsilon$. Insertion of these relations into eqn (4.29) and expansion to first order in ϵ give

$$\epsilon' = \frac{2(1 + e^{4K^*})\big(4(1 + 4K^*)e^{4K^*} + e^{8K^*} + 3\big)}{(e^{4K^*} + 3)^3} \cdot \epsilon. \tag{B.65}$$

The coefficient of ϵ on the right-hand side is the renormalization eigenvalue $\lambda = b^{1/\nu}$. From $e^{4K^*} = 1 + 2\sqrt{2}$, the eigenvalue is numerically $\lambda = 1.624$, and consequently $\nu = 1.13$ from $b = \sqrt{3}$.

4.2

The same dimensional analysis as in Section 4.2.1 will give the desired result under the assumption that the sixth-order term is renormalized as $v \to b^{y_v} v$. Invariance of the Hamiltonian, in particular the sixth-order term, leads to $y_v = 6 - 2d$. Since the quartic term has $y_u = 4 - d$, the exponent of the sixth-order term is smaller for $d > 4$ and is more irrelevant.

4.3

The scaling law for magnetization at $t = 0$ around the Gaussian fixed point is

$$m(u, h) = b^{1-d/2} m(b^{4-d} u, b^{1+d/2} h). \tag{B.66}$$

If we choose b such that the h dependence disappears from the right-hand side, we have

$$m(u, h) = h^{(d-2)/(d+2)} m(h^{(2d-8)/(d+2)} u, 1). \tag{B.67}$$

The u dependence of $m(u, h)$ can be revealed by the mean-field theory. The equation of state of the Landau theory for the Gaussian model at $t = 0$, $4um^3 - h = 0$, suggests $m \propto u^{-1/3}$, and hence m on the right-hand side of eqn (B.67) depends on the first argument with power $-1/3$. Consequently,

$$m(u, h) = h^{(d-2)/(d+2)} \cdot \left(h^{(2d-8)/(d+2)}u\right)^{-1/3} \propto h^{1/3}. \tag{B.68}$$

Chapter 5

5.1

If the interaction $\mathbf{S}_i \cdot \mathbf{S}_j$ is decomposed into components, $S_i^{(1)} S_j^{(1)} + S_i^{(2)} S_j^{(2)} + \cdots + S_i^{(n)} S_j^{(n)}$, the process of the Hubbard–Stratonovich transformation in the text can be applied separately to each component. The differences from the single-component case are, first, that the variable $\boldsymbol{\sigma}$ has n components $(\sigma_i^{(1)}, \sigma_i^{(2)}, \cdots, \sigma_i^{(n)})$ for each site i and, secondly, that the summation in eqn (5.17) is replaced by the integral

$$g(\{\boldsymbol{\sigma}^{(j)}\}) = \int \left(\prod_{i=1}^{N} \mathrm{d}\mathbf{S}_i\right) \delta(\mathbf{S}_i^2 - 1) \exp\left(-\sum_{j=1}^{n} \boldsymbol{\sigma}^{(j)} \cdot \mathbf{S}\right). \tag{B.69}$$

The result of this integral is independent of the direction of the vector $\boldsymbol{\sigma}^{(j)}$ but depends only on its magnitude $(\boldsymbol{\sigma}^{(j)})^2$ because of the rotational symmetry of the integration weight $\delta(\mathbf{S}_i^2 - 1)$. Therefore the effective Hamiltonian in terms of ϕ should be rotationally invariant, i.e. a function of the magnitude of the ϕ-field, $\sum_{j=1}^{n} \phi_j(\mathbf{r})^2$. The expansion of this isotropic potential gives the second and third terms on the right-hand side of eqn (5.25). The first term proportional to the gradient emerges just as in the case of a single component.

5.2

One can use the ansatz $\phi(\mathbf{r}) = \phi_0 \cos(\mathbf{q} \cdot \mathbf{r})$ to compute the free energy in the two cases $\mathbf{q} = 0$, i.e. no modulation, and $\mathbf{q} \neq 0$. The result of this simple calculation for the effective Hamiltonian per volume is

$$f^{\mathrm{h}} = t\phi_0^2 + u\phi_0^4 \tag{B.70}$$

$$f^{\mathrm{inh}} = \frac{1}{2}\left(cq^2 + Dq^4 + t\right)\phi_0^2 + \frac{3u}{8}\phi_0^4, \tag{B.71}$$

where we have used the fact that the averages over the periodic functions are

$$\frac{1}{V}\int \mathrm{d}^d\mathbf{r} \sin^2(\mathbf{q} \cdot \mathbf{r}) = \frac{1}{V}\int \mathrm{d}^d\mathbf{r} \cos^2(\mathbf{q} \cdot \mathbf{r}) = \frac{1}{2}, \ \frac{1}{V}\int \mathrm{d}^d\mathbf{r} \cos^4(\mathbf{q} \cdot \mathbf{r}) = \frac{3}{8}. \tag{B.72}$$

If $c > 0$, it is energetically favorable to have a homogeneous phase, i.e. $\mathbf{q} = 0$. Therefore, if $t < 0$ then the stable solution is the ordered phase $\phi_0 \neq 0$, while for $t > 0$ the disordered phase $\phi_0 = 0$ is more stable. The phase transition is second order.

On the other hand, when $c < 0$, one needs to consider all the three phases and determine which one has minimal free energy in the (Δ, T) plane. Let us start by looking for the modulation that minimizes f^{inh}. By taking the derivative of $cq^2 + Dq^4$, we find that the modulation is given by

$$q_0 = \sqrt{\frac{|c|}{2D}}, \tag{B.73}$$

and the minimal free energy for the modulated phase is

$$f^{\text{inh}} = \frac{1}{2}\left(t - \frac{c^2}{4D}\right)\phi_0^2 + \frac{3u}{8}\phi_0^4. \tag{B.74}$$

When the coefficient of the quadratic term is positive, i.e. $t > c^2/(4D)$, the paramagnetic phase is favorable, while for $0 < t < c^2/(4D)$ the modulated phase is stable. The transition between the disordered and spatially modulated phases is second order.

It remains to analyze the case $t < 0$ and $c < 0$. The competing phases are the spatially homogeneous and inhomogeneous states. At the minimum, as shown by the minimization of eqns (B.70) and (B.74) with respect to ϕ_0, the two free energies are given by

$$f_0^{\text{h}} = -\frac{t^2}{4u}, \quad f_0^{\text{inh}} = -\frac{1}{6u}\left(|t| + \frac{c^2}{4D}\right)^2, \tag{B.75}$$

which indicates that for $t < \bar{t}$ the stable phase is homogeneous, while for $t > \bar{t}$ it is modulated for some $\bar{t}$. This boundary value is determined by equating the two free energies $f_0^{\text{h}}(\bar{t}) = f_0^{\text{inh}}(\bar{t})$ and is given by

$$\bar{t} = -\frac{c^2}{4D}\left(\sqrt{\frac{3}{2}} - 1\right)^{-1}, \tag{B.76}$$

and represents a first-order transition. The point $t = 0, c = 0$ is a Lifshitz point. The resulting phase diagram is shown in Fig. B.2.

One can define general (d, n) Lifshitz points, where n is the number of components of the order parameter field $\phi(\boldsymbol{r})$. The present problem describes a $(d, 1)$ Lifshitz point.

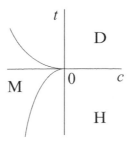

Fig. B.2 Phase diagram with a Lifshitz point at the origin. D, H and M stand for the disordered, homogeneous and modulated (inhomogeneous) phases, respectively.

Chapter 6

6.1

Translation is reproduced by $a = 1, c = 0, d = 1$ and finite b. Rotation has $b = 0$, $c = 0, d = 1$ and complex $a = e^{i\theta}$ with real θ. Similarly, dilation corresponds to $b = 0, c = 0, d = 1$ and real positive a. The special conformal transformation of eqn (6.5) is rewritten by using a complex number z as

$$f(z) = \frac{\dfrac{z}{|z|^2} + a}{\left|\dfrac{z}{|z|^2} + a\right|^2}. \tag{B.77}$$

By multiplying the denominator and numerator by $|z|^4$, we find

$$f(z) = \frac{|z|^2(z + a|z|^2)}{|z + a|z|^2|^2} = \frac{z(1 + a\bar{z})}{|1 + a\bar{z}|^2} = \frac{z}{1 + \bar{a}z}, \tag{B.78}$$

where $\bar{z}$ stands for complex conjugate. This is of the form of eqn (6.10).

6.2

The case $\epsilon = \bar{\epsilon} = 1$ is, according to eqns (6.11) and (6.12),

$$f(z) = z + \epsilon_{-1} \cdot 1, \quad \bar{f}(\bar{z}) = \bar{z} + \bar{\epsilon}_{-1} \cdot 1, \tag{B.79}$$

where we have arbitrarily introduced infinitesimal constants ϵ_{-1} and $\bar{\epsilon}_{-1}$. This is a translation. Next, when $\epsilon = z, \bar{\epsilon} = \bar{z}$, the transformation reads

$$f(z) = z + \epsilon_0 z, \quad \bar{f}(\bar{z}) = \bar{z} + \bar{\epsilon}_0 \bar{z}. \tag{B.80}$$

This is a dilation if ϵ_0 and $\bar{\epsilon}_0$ are real. It is a dilation and a rotation for complex ϵ_0 and $\bar{\epsilon}_0$ since the latter includes the change of the phase. Lastly, we discuss $\epsilon_0 = z^2$. The antiholomorphic part will have the same behavior and is omitted for simplicity. The transformation is

$$f(z) = z + \epsilon_1 z^2 = x + c_1(x^2 - y^2) - 2c_2xy + i(y + c_2(x^2 - y^2) + 2c_1xy), \tag{B.81}$$

where we have written $z = x + iy$, $\epsilon_1 = c_1 + ic_2$ with real x, y, c_1, c_2. The special conformal transformation of eqn (6.4) is rewritten for infinitesimal $\boldsymbol{a}$ as

$$\boldsymbol{r} \to \boldsymbol{r} + \boldsymbol{a}\, r^2 - 2(\boldsymbol{a} \cdot \boldsymbol{r})\, \boldsymbol{r}. \tag{B.82}$$

By writing $\boldsymbol{r} = (x, y)$ and $\boldsymbol{a} = (-c_1, c_2)$, the two components of this transformation are

$$x \to x + c_1(x^2 - y^2) - 2c_2xy, \quad y \to y + c_2(x^2 - y^2) + 2c_1xy, \tag{B.83}$$

which agree with eqn (B.81).

6.3

Only the holomorphic part is written explicitly here for simplicity. The application of the differential operator $(z_1^2\partial_1 + 2h_1z_1 + z_2^2\partial_2 + 2h_2z_2)$ to the right-hand side of eqn (6.30) yields an expression with the denominator $z_{12}^{h_1+h_2+1}\bar{z}_{12}^{\bar{h}_1+\bar{h}_2}$ and the numerator

$$-(h_1 + h_2)z_1^2 + 2h_1 z_1(z_1 - z_2) + (h_1 + h_2)z_2^2 + 2h_2 z_2(z_1 - z_2) = (z_1 - z_2)^2(h_1 - h_2). \tag{B.84}$$

This can vanish for $z_1 \neq z_2$ only when $h_1 = h_2$.

6.4

Using $f(z) = z + \epsilon(z)$ and taking only the leading order of ϵ, we easily see that the Schwarzian is $\{f, z\} = \partial^3 \epsilon$. Then, the right-hand side of eqn (6.49) minus $T(z)$ becomes

$$\left((1 + \partial\epsilon)^2 T(z + \epsilon) + \frac{c}{12}\partial^3\epsilon\right) - T(z) = 2(\partial\epsilon)T(z) + \epsilon\partial T(z) + \frac{c}{12}\partial^3\epsilon. \tag{B.85}$$

6.5

Straightforward applications of the definition of the Schwarz derivative show that both sides of eqn (6.51) are equal to

$$\frac{A}{2\left(\dfrac{\partial u}{\partial f}\right)^2\left(\dfrac{\partial f}{\partial z}\right)^2}, \tag{B.86}$$

where

$$A = 2\frac{\partial u}{\partial f}\frac{\partial^3 u}{\partial f^3}\left(\frac{\partial f}{\partial z}\right)^4 - 3\left(\frac{\partial^2 u}{\partial f^2}\right)^2\left(\frac{\partial f}{\partial z}\right)^4$$
$$+ 2\left(\frac{\partial u}{\partial f}\right)^2\frac{\partial f}{\partial z}\frac{\partial^3 f}{\partial z^3} - 3\left(\frac{\partial u}{\partial f}\right)^2\left(\frac{\partial^2 f}{\partial z^2}\right)^2. \tag{B.87}$$

The second half of the problem is solved by successive applications of two transformations,

$$T(z) \rightarrow \left(\frac{\partial f}{\partial z}\right)^2 T(f) + \frac{c}{12}\{f, z\}$$

$$\rightarrow \left(\frac{\partial f}{\partial z}\right)^2\left(\left(\frac{\partial u}{\partial f}\right)^2 T(u) + \frac{c}{12}\{u, f\}\right) + \frac{c}{12}\{f, z\}$$

$$= \left(\frac{\partial u}{\partial z}\right)^2 T(u) + \frac{c}{12}\{u, f\}\left(\frac{\partial f}{\partial z}\right)^2 + \frac{c}{12}\{f, z\}$$

$$= \left(\frac{\partial u}{\partial z}\right)^2 T(u) + \frac{c}{12}\{u, z\}. \tag{B.88}$$

6.6

The projective mapping of eqn (6.10) yields

$$\frac{\partial_z^3 f}{\partial_z f} = \frac{6c^2}{(cz + d)^2}, \quad \frac{\partial_z^2 f}{\partial_z f} = -\frac{2c}{cz + d}, \tag{B.89}$$

from which the Schwarzian is easily seen to vanish.

6.7

If the correlation function is written in terms of the conformal generators

$$\langle 0|T(w)T(z)|0\rangle = \sum_{m,n=-\infty}^{\infty} w^{-m-2}z^{-n-2}\langle 0|L_m L_n|0\rangle, \tag{B.90}$$

we notice that n actually runs only for $n \leq -2$ and m runs for $m \geq 1$ because of eqns (6.72), (6.94) and (6.95). We thus have

$$\langle 0|T(w)T(z)|0\rangle = \sum_{m=1}^{\infty}\sum_{n=2}^{\infty} w^{-m-2}z^{n-2}\langle 0|L_m L_{-n}|0\rangle. \tag{B.91}$$

Using the Virasoro algebra

$$[L_m, L_{-n}] = (m+n)L_{m-n} + \frac{c}{12}m(m^2-1)\delta_{m-n,0}, \tag{B.92}$$

and the fact that from $[L_0, L_k] = -kL_k$ it results $\langle 0|L_k|0\rangle = 0$, we rewrite the above expression as

$$\langle 0|T(w)T(z)|0\rangle = \frac{c}{12}\sum_{n=2}^{\infty} w^{-n-2}z^{n-2}n(n^2-1) = \frac{c}{12wz^3}\sum_{n=3}^{\infty} n(n-1)(n-2)\left(\frac{z}{w}\right)^n, \tag{B.93}$$

where we have rewritten $n \to n-1$ in going from the second to the third expression. We can confirm that this expression is equal to $c/2(w-z)^4$ by taking the third-order derivative of the series expression

$$\frac{1}{w-z} = \frac{1}{w}\sum_{n=0}^{\infty}\left(\frac{z}{w}\right)^n \tag{B.94}$$

by z.

6.8

The following consequences of the Virasoro algebra

$$[L_2, L_{-2}] = 4L_0 + \frac{c}{2}, \quad [L_1, L_{-1}] = 2L_0 \tag{B.95}$$

and the relations $L_n|0\rangle = 0$ $(n \geq -1)$ lead to the desired results.

Chapter 7

7.1

The calculations remain the same for an arbitrary d up to the second line of eqn (7.24). In the third line, $(4 - \sum_\delta e^{-iq\cdot\delta})$ is changed to $(2d - \sum_\delta e^{-iq\cdot\delta})$. The right-hand side of eqn (7.25) is now $2\cos q_1 + 2\cos q_2 + \cdots + 2\cos q_d$, and accordingly in the final expression of eqn (7.26) $q_x^2 + q_y^2$ is replaced by $q_1^2 + q_2^2 + \cdots + q_d^2 = q^2$. The integral (7.28) has $(2\pi)^d$ in place of $(2\pi)^2$ and the other factors and the integrand remain the same. This integral diverges as $h \to 0$ if $d \leq 2$, and $m \to 0$ follows as long as $T > 0$. When $d > 2$, the integral is finite, and no inconsistency arises with finite m.

7.2

We evaluate the equivalent of integral (7.31) for general d by using eqn (7.9). The r dependence of the result of integral (7.9) is $-r^{2-d}/(d-2)$ if $d \neq 2$. When $d > 2$, the r dependence vanishes in the large-r limit, and the correlation function (7.30) converges to a finite value, which means the existence of long-range order. For $d = 2$ we have already shown the power law decay of the correlation function. In the case of $d < 2$, the integral diverges to $+\infty$ as $r \to \infty$ and the correlation function vanishes. These analyses show that $d = 2$ is the lower critical dimension.

7.3

Shown in Fig. B.3.

7.4

Let us multiply $\phi(\boldsymbol{r})$ and $\phi(0)$ by p in eqn (7.30). Then, the correlation function is

$$\langle \cos \left(p(\phi(\boldsymbol{r}) - \phi(0)) \right) \rangle = \exp \left(-\frac{p^2}{2} \langle (\phi(\boldsymbol{r}) - \phi(0))^2 \rangle \right). \tag{B.96}$$

Thus, the final expression of the correlation function is eqn (7.32) with the power of r multiplied by p^2. This means that the scaling dimension x_p is $x_p = Tp^2/4\pi J$ according to eqn (3.80). Then, $y_p = d - x_p = 2 - Tp^2/4\pi J$, and the relevance condition $y_p > 0$ is $T < 8\pi J/p^2 \equiv T_p$. It should be noticed here that the temperature T_p should be lower than $T_{\mathrm{KT}}(= \pi J/2)$ in order for the above argument based on the spin-wave theory to make sense. We therefore have $\pi J/2 > 8\pi J/p^2$ or $p > 4 \equiv p_0$. In conclusion, the angle variable ϕ_i tends to take discrete values $2\pi k/p$ in the low-temperature range $0 < T < T_p(< T_{\mathrm{KT}})$ when p is larger than four.

7.5

We collect the x dependence of eqn (7.55) to the left-hand side and integrate,

$$\int \frac{\mathrm{d}x}{x^2 + ct} = \int \mathrm{d}l. \tag{B.97}$$

Fig. B.3 $n = 2, 2, -2, -2$ from left to right.

The integral is simplified by the change of variable $x = \sqrt{ct} \tan \theta$,

$$\int \frac{\mathrm{d}\theta}{\sqrt{ct}} = \int \mathrm{d}l = l + \mathsf{const}, \qquad (\text{B.98})$$

from which we find

$$l = l_0 + \frac{\theta}{\sqrt{ct}} = l_0 + \frac{1}{\sqrt{ct}} \arctan \frac{x}{\sqrt{ct}}. \qquad (\text{B.99})$$

Chapter 8

8.1

The zero-temperature limit $\beta \to \infty$ of the equation of state (8.9) has the solutions $m = 0$ and $m = 1$ for $h_0 < Jz$ and only $m = 0$ for $h_0 > Jz$. The zero-temperature limit of the free energy (8.8), on the other hand, is $F(0) = -Nh_0$ and $F(1) = -NJz/2$ for $m = 0$ and $m = 1$, respectively, because of $N_\mathrm{B} = zN/2$. Comparison of these two values reveals the transition point $h_0 = Jz/2$ at which $F(0)$ and $F(1)$ match.

8.2

The equation of state (8.9) for the Gaussian distribution reads

$$m = \frac{1}{\sqrt{2\pi}h_0} \int_{-\infty}^{\infty} e^{-h^2/2h_0^2} \tanh \beta(Jmz + h)\,\mathrm{d}h. \qquad (\text{B.100})$$

In the zero-temperature limit $\beta \to \infty$, the factor $\tanh \beta(\cdot)$ in the integrand is either $+1$ or -1 according to the sign of $Jmz + h$,

$$m = \frac{1}{\sqrt{2\pi}h_0} \int_{-Jmz}^{\infty} e^{-h^2/2h_0^2}\,\mathrm{d}h - \frac{1}{\sqrt{2\pi}h_0} \int_{-\infty}^{-Jmz} e^{-h^2/2h_0^2}\,\mathrm{d}h. \qquad (\text{B.101})$$

We rewrite the right-hand side in terms of the error function using the notation $h = \sqrt{2}\,h_0 x$,[1]

$$m = 1 - \frac{2}{\sqrt{\pi}} \int_{Jmz/\sqrt{2}h_0}^{\infty} e^{-x^2}\,\mathrm{d}x = \mathrm{Erf}\left(\frac{Jmz}{\sqrt{2}h_0}\right). \qquad (\text{B.102})$$

Expansion of the error function to third order of its argument gives

$$m = \frac{2}{\sqrt{\pi}}\left(\frac{Jmz}{\sqrt{2}h_0} - \frac{1}{3}\left(\frac{Jmz}{\sqrt{2}h_0}\right)^3\right). \qquad (\text{B.103})$$

Since the coefficient of the third-order term is negative, we conclude the existence of a second-order transition when the coefficient of the first-order term becomes unity, $h_{0,\mathrm{c}} = \sqrt{2}\,Jz/\sqrt{\pi}$.

---

[1] The error function is defined as $\mathrm{Erf}(x) = \frac{2}{\sqrt{\pi}} \int_0^x e^{-t^2}\,\mathrm{d}t$.

8.3

It suffices to closely follow the discussions of Section 4.2.1. We replace $\phi(r)$ by the field of the spin glass order parameter $q(r)$ and $u\phi(r)^4$ by $vq(r)^3$. After the change of scale this third-order term is multiplied by $b^{-d+y_v+3(d-y_h)}$. Invariance then implies $-d + y_v + 3(d - y_h) = 0$, from which we find $y_v = 3 - d/2$ using the result of invariance of the first term, $y_h = d/2 + 1$. This argument is valid irrespective of the presence of replica indices in q as $q_{\alpha\beta}$. Only the existence of the third-order term matters.

8.4

We follow the processes for β and γ. In the following expression,

$$M_0 = \sum_s n_s(p) \approx \int ds\, s^{-\tau} f\big((p - p_c)s^\sigma\big), \tag{B.104}$$

the integration variable is changed as $z = (p_c - p)s^\sigma$,

$$M_0 \propto (p_c - p)^{(\tau-1)/\sigma} \int dz\, f(-z) z^{-1+(1-\tau)/\sigma}, \tag{B.105}$$

from which $2 - \alpha = (\tau - 1)/\sigma$ is derived.

8.5

1. When site 0 and r belong to different clusters, variables $\delta_{S_0,1} - q^{-1}$ and $\delta_{S_r,1} - q^{-1}$ are independent of each other. Since all values of S_i contribute with the same weight in a cluster, $\sum_{S_0=1}^q (\delta_{S_0,1} - q^{-1}) = \sum_{S_r=1}^q (\delta_{S_r,1} - q^{-1}) = 0$.

2. Two sites are in the same cluster. Let us call the summation over spin variables within a cluster the 'spin summation' and consider the spin summation of terms appearing in the expansion of $(\delta_{S_0,1} - q^{-1})(\delta_{S_r,1} - q^{-1})$ by noting $S_0 = S_r$ within a cluster. The spin summation of $\delta_{S_0,1}\delta_{S_r,1}$ is 1, those of $q^{-1}\delta_{S_0,1}$ and $q^{-1}\delta_{S_r,1}$ are both q^{-1}, and that of q^{-2} is q^{-1}. Then, the spin summation of $(\delta_{S_0,1} - q^{-1})(\delta_{S_r,1} - q^{-1})$ is $1 - q^{-1}$, and consequently the expectation value is obtained by dividing this by q as $(q - 1)/q^2$.

3. For $q = 1 + \epsilon$, $(q - 1)/q^2 = \epsilon + \mathcal{O}(\epsilon^2)$. The solutions to the previous two problems indicate that coefficient of the term ϵ in the expansion of the correlation function of the Potts model represents the probability that 0 and r are in the same cluster, which is the correlation function of percolation.

Chapter 9

9.1

For a free boundary, the partition function is written as

$$Z_N^{(F)} = \sum_{\{S_i\}} e^{K\{\delta(S_1,S_2)+\delta(S_2,S_3)+\cdots+\delta(S_{N-1},S_N)\}}. \tag{B.106}$$

We follow the method for the solution to the Ising model and first sum over the variable S_N as

$$Z_N^{(F)} = \sum_{S_1,\cdots,S_{N-1}} e^{K\{\delta(S_1,S_2)+\delta(S_2,S_3)+\cdots+\delta(S_{N-2},S_{N-1})\}}$$

$$\cdot \sum_{S_N=0,1,2} e^{K\delta(S_{N-1},S_N)} = Z_{N-1}^{(F)} \cdot (e^K + 2). \tag{B.107}$$

This is a recursion relation for the partition function, which we repeatedly use to obtain the partition function as $Z_N^{(F)} = 3(e^K + 2)^{N-1}$. The free energy per spin in the thermodynamic limit is

$$\beta f = - \lim_{N\to\infty} \frac{1}{N} \log Z_N^{(F)} = - \log(e^K + 2). \tag{B.108}$$

For a periodic boundary, the transfer-matrix method applies. The transfer matrix for the three-state Potts model has the elements $T(S_i, S_{i+1}) = e^{K\delta(S_i,S_{i+1})}$. More explicitly,

$$T = \begin{pmatrix} e^K & 1 & 1 \\ 1 & e^K & 1 \\ 1 & 1 & e^K \end{pmatrix}. \tag{B.109}$$

The eigenvalues are $e^K + 2, e^K - 1, e^K - 1$ with the corresponding un-normalized eigenvectors ${}^t(1,1,1)$, ${}^t(1,e^{2\pi i/3},e^{4\pi i/3})$, ${}^t(1,e^{-2\pi i/3},e^{-4\pi i/3})$, respectively. The partition function is then

$$Z_N^{(P)} = (e^K + 2)^N + 2(e^K - 1)^N. \tag{B.110}$$

Since $e^K + 2 > e^K - 1$, the free energy per spin in the thermodynamic limit is

$$\beta f = - \lim_{N\to\infty} \frac{1}{N} \log Z_N^{(P)} = - \log(e^K + 2), \tag{B.111}$$

which agrees with the free-boundary result (B.108).

9.2

The partition function is given by

$$Z_N^{(P)} = \mathrm{Tr}\, T^N, \tag{B.112}$$

where the transfer matrix is the 4×4 matrix

$$T = \begin{pmatrix} e^{2K_1+K_2} & 1 & 1 & e^{-2K_1+K_2} \\ 1 & e^{2K_1-K_2} & e^{-2K_1-K_2} & 1 \\ 1 & e^{-2K_1-K_2} & e^{2K_1-K_2} & 1 \\ e^{-2K_1+K_2} & 1 & 1 & e^{2K_1+K_2} \end{pmatrix}, \tag{B.113}$$

with matrix elements $T(S_{11}, S_{21}, S_{12}, S_{22}) = e^{K_1 S_{11} S_{12} + K_1 S_{21} S_{22} + K_2 (S_{11} S_{21} + S_{12} S_{22})/2}$, and basis vectors $|S_{1j} S_{2j}\rangle$

$$|11\rangle = \begin{pmatrix} 1 \\ 0 \\ 0 \\ 0 \end{pmatrix}, \ |1-1\rangle = \begin{pmatrix} 0 \\ 1 \\ 0 \\ 0 \end{pmatrix}, \ |-11\rangle = \begin{pmatrix} 0 \\ 0 \\ 1 \\ 0 \end{pmatrix}, \ |-1-1\rangle = \begin{pmatrix} 0 \\ 0 \\ 0 \\ 1 \end{pmatrix}. \quad (B.114)$$

The eigenvalues of the matrix T are:

$$\lambda_{1,2} = 2 \left(\cosh(2K_1) \cosh(K_2) \pm \sqrt{1 + \sinh^2(K_2) \cosh^2(2K_1)} \right), \quad (B.115)$$

$$\lambda_{3,4} = 2 \sinh(2K_1) e^{\pm K_2}, \quad (B.116)$$

and, therefore, the partition function is given by

$$Z_N^{(P)} = \lambda_1^N + \lambda_2^N + \lambda_3^N + \lambda_4^N. \quad (B.117)$$

Since λ_1 is the largest among the four eigenvalues, only this term remains in the thermodynamic limit.

To compute the correlation function $\langle S_{1j} S_{1j+r} \rangle$, one needs to observe that

$$\langle S_{1j-1} S_{2j-1} | T S_{1j} | S_{1j} S_{2j} \rangle = \langle S_{1j-1} S_{2j-1} | T \bar{\sigma} | S_{1j} S_{2j} \rangle, \quad (B.118)$$

where

$$\bar{\sigma} = \begin{pmatrix} 1 & 0 & 0 & 0 \\ 0 & 1 & 0 & 0 \\ 0 & 0 & -1 & 0 \\ 0 & 0 & 0 & -1 \end{pmatrix}, \quad (B.119)$$

because of the choice of the basis as in eqn (B.114). Then, the correlation function is given by

$$\langle S_{1j} S_{1j+r} \rangle Z_N^{(P)} = \mathrm{Tr} \left(T^{j-1} \bar{\sigma} T^r \bar{\sigma} T^{N-j-r+1} \right) = \mathrm{Tr} \left(\bar{\sigma} T^r \bar{\sigma} T^{N-r} \right). \quad (B.120)$$

9.3

The expression

$$H_h(u) \equiv H(u) + \frac{h^2}{2Ku^2}, \quad (B.121)$$

which appears in the saddle-point condition, is a monotone decreasing function, diverging as $u \to 0$ in any dimension if $h \neq 0$ and approaching 0 as $u \to \infty$. Consequently, the saddle-point condition $H_h(u) = 2K$ has a solution for any K. Since any derivatives of $H_h(u)$ do not diverge for $u > 0$, this function is not singular. This implies that the solution u of the saddle-point condition is not singular as a function of K. Thus, the spherical model has no phase transition when $h \neq 0$.

9.4

Differentiation of eqn (9.65) with respect to h gives $-m$. Hence,

$$m = \frac{h}{2Ku}. \tag{B.122}$$

The relation between u and h is determined by eqn (9.66). Since the solution u of eqn (9.66) vanishes as $h \to 0$, u should be small for small h. Then, eqn (9.66) is written as

$$H(0) - cu^{d/2-1} + \frac{h^2}{2Ku^2} = 2K. \tag{B.123}$$

Since the second term on the left-hand side $u^{d/2-1}$ is smaller than the other term, we neglect it and use $H(0) = 2K_{\mathrm{c}}$ to rewrite the above relation as

$$\frac{h^2}{2Ku^2} = 2K - 2K_{\mathrm{c}}. \tag{B.124}$$

This equation is solved for h/u and we insert the result into eqn (B.122) to have

$$m = \sqrt{1 - \frac{K_{\mathrm{c}}}{K}}. \tag{B.125}$$

9.5

Let us evaluate the integration part, named I, of the general equation (9.86). We write $\beta J = K$ and have

$$I = 2 \int_0^{\pi} \frac{\cos^2 q\, e^{-K\cos q}}{(1 + e^{-K\cos q})^2}\, dq$$

$$= 2 \int_0^{\pi/2} \frac{\cos^2 q\, e^{-K\cos q}}{(1 + e^{-K\cos q})^2}\, dq + 2 \int_{\pi/2}^{\pi} \frac{\cos^2 q\, e^{-K\cos q}}{(1 + e^{-K\cos q})^2}\, dq$$

$$= 2 \int_0^{\pi/2} \frac{\cos^2 q\, e^{-K\cos q}}{(1 + e^{-K\cos q})^2}\, dq + 2 \int_0^{\pi/2} \frac{\cos^2 q\, e^{K\cos q}}{(1 + e^{K\cos q})^2}\, dq$$

$$= 4 \int_0^{\pi/2} \frac{\cos^2 q\, e^{-K\cos q}}{(1 + e^{-K\cos q})^2}\, dq. \tag{B.126}$$

For large K, contributions from small $\cos q$ dominate the integral. We thus change the variable as $x = \pi/2 - q$ so that the behavior of the integrand around $q = \pi/2$, where $\cos q$ is small, is easier to see,

$$I = 4 \int_0^{\pi/2} \frac{\sin^2 x\, e^{-K\sin x}}{(1 + e^{-K\sin x})^2}\, dx. \tag{B.127}$$

Since the leading contribution as $K \gg 1$ comes from the range of small x, we approximate $\sin x \approx x$ to find

$$I \approx 4 \int_0^{\pi/2} \frac{x^2 \, e^{-Kx}}{(1 + e^{-Kx})^2} \, dx$$

$$= 4K^{-3} \int_0^{\pi K/2} \frac{t^2 e^{-t}}{(1 + e^{-t})^2} \, dt$$

$$\approx 4K^{-3} \int_0^{\infty} \frac{t^2 e^{-t}}{(1 + e^{-t})^2} \, dt = \frac{2\pi^2}{3K^3}. \tag{B.128}$$

Multiplication of this result by the factor in front of the integral of eqn (9.86) yields eqn (9.87).

9.6

When $h > J$, the denominator of the integrand of eqn (9.84) is unity in the low-temperature limit because the exponent appearing in the denominator is always negative. We thus find

$$E_0 = \frac{h}{2} - \frac{1}{2\pi} \int_{-\pi}^{\pi} (J \cos q + h) \, dq = -\frac{h}{2} \tag{B.129}$$

as the ground-state energy. For $h < J$, the denominator of the integrand is unity when the absolute value of q is smaller than q_0 defined by $J \cos q_0 = -h$. Otherwise, the exponential function in the denominator grows indefinitely and the integral vanishes. Thus,

$$E_0 = \frac{h}{2} - \frac{1}{2\pi} \int_{-q_0}^{q_0} (J \cos q + h) \, dq$$

$$= \frac{h}{2} - \frac{J \sin q_0}{\pi} - \frac{h q_0}{\pi}$$

$$= \frac{h}{2} - \frac{\sqrt{J^2 - h^2}}{\pi} - \frac{h}{\pi} \arccos\left(-\frac{h}{J}\right) \tag{B.130}$$

is the ground-state energy.

9.7

The mapping from the Fermionic to the Ising spin Hamiltonian is realized by the relation $S_j = 2n_j - 1$. We disregard the boundary terms in the resulting spin Hamiltonian since we are only interested in the thermodynamic limit. Hence, our starting Hamiltonian is eqn (9.91).

Consider now a map between spin variables and bond variables

$$S_j S_{j+1} = \tilde{S}_j. \tag{B.131}$$

Clearly there are $N-1$ variables $\tilde{S}_j = \pm 1$, while there are N spins S_j. Thus, let us keep S_1, and the $(N-1)$ $\tilde{S}_j$ as our new variables. In terms of these new variables the Hamiltonian reads

$$\beta H = -K_1 \sum_{j=1}^{N-1} \tilde{S}_j - K_2 \sum_{j=1}^{N-2} \tilde{S}_j \tilde{S}_{j+1} \quad (\tilde{S}_j = \pm 1), \tag{B.132}$$

which corresponds to an Ising chain with $N-1$ sites in a longitudinal field K_1. The partition function is

$$Z_N^{(\mathrm{F})} = \sum_{S_1} \sum_{\{\tilde{S}_j\}} \mathrm{e}^{-\beta H} = 2 \sum_{\{\tilde{S}_j\}} \mathrm{e}^{-\beta H}. \tag{B.133}$$

We will use the transfer-matrix technique of Section 9.1.2 to determine $Z_N^{(\mathrm{F})}$, but notice that now we are dealing with free boundary conditions as opposed to periodic boundary conditions. The trick consists of using the same periodic boundary condition technique but with the last bond treated differently, i.e. we set $K_2 = 0$ in the last bond. The bulk transfer matrix is

$$T = \begin{pmatrix} \mathrm{e}^{K_2+K_1} & \mathrm{e}^{-K_2} \\ \mathrm{e}^{-K_2} & \mathrm{e}^{K_2-K_1} \end{pmatrix} = U \begin{pmatrix} \lambda_+ & 0 \\ 0 & \lambda_- \end{pmatrix} U^{-1}, \tag{B.134}$$

with the 2×2 matrix U given by ($UU^{-1} = U^{-1}U = 1$)

$$U = \begin{pmatrix} \dfrac{\mathrm{e}^{-K_2}}{\sqrt{\mathrm{e}^{-2K_2} + (\lambda_+ - \mathrm{e}^{K_2+K_1})^2}} & \dfrac{\mathrm{e}^{-K_2}}{\sqrt{\mathrm{e}^{-2K_2} + (\lambda_- - \mathrm{e}^{K_2+K_1})^2}} \\ \dfrac{\lambda_+ - \mathrm{e}^{K_2+K_1}}{\sqrt{\mathrm{e}^{-2K_2} + (\lambda_+ - \mathrm{e}^{K_2+K_1})^2}} & \dfrac{\lambda_- - \mathrm{e}^{K_2+K_1}}{\sqrt{\mathrm{e}^{-2K_2} + (\lambda_- - \mathrm{e}^{K_2+K_1})^2}} \end{pmatrix}, \tag{B.135}$$

$$U^{-1} = \begin{pmatrix} -\dfrac{\lambda_- - \mathrm{e}^{K_2+K_1}}{\sqrt{\mathrm{e}^{-2K_2} + (\lambda_- - \mathrm{e}^{K_2+K_1})^2}} & \dfrac{\mathrm{e}^{-K_2}}{\sqrt{\mathrm{e}^{-2K_2} + (\lambda_- - \mathrm{e}^{K_2+K_1})^2}} \\ \dfrac{\lambda_+ - \mathrm{e}^{K_2+K_1}}{\sqrt{\mathrm{e}^{-2K_2} + (\lambda_+ - \mathrm{e}^{K_2+K_1})^2}} & -\dfrac{\mathrm{e}^{-K_2}}{\sqrt{\mathrm{e}^{-2K_2} + (\lambda_+ - \mathrm{e}^{K_2+K_1})^2}} \end{pmatrix}, \tag{B.136}$$

and eigenvalues (eqn (9.25))

$$\lambda_{\pm} = \frac{\mathrm{e}^{K_2+K_1} + \mathrm{e}^{K_2-K_1} \pm \sqrt{(\mathrm{e}^{K_2+K_1} + \mathrm{e}^{K_2-K_1})^2 - 4\,\mathrm{e}^{2K_2} + 4\,\mathrm{e}^{-2K_2}}}{2} \tag{B.137}$$

$$= \mathrm{e}^{K_2} \left(\cosh(K_1) \pm \sqrt{\cosh^2(K_1) - 2\mathrm{e}^{-2K_2} \sinh(2K_2)} \right). \tag{B.138}$$

Therefore, the partition function is given by

$$Z_N^{(\mathrm{F})} = 2\mathrm{Tr}\left(T^{N-2} \begin{pmatrix} \mathrm{e}^{K_1} & 1 \\ 1 & \mathrm{e}^{-K_1} \end{pmatrix} \right) = 2\mathrm{Tr}\left(\begin{pmatrix} \lambda_+^{N-2} & 0 \\ 0 & \lambda_-^{N-2} \end{pmatrix} U^{-1} \begin{pmatrix} \mathrm{e}^{K_1} & 1 \\ 1 & \mathrm{e}^{-K_1} \end{pmatrix} U \right). \tag{B.139}$$

In the thermodynamic limit $N \to \infty$, since $(\lambda_-/\lambda_+)^N \to 0$, we obtain

$$\lim_{N \to \infty} \frac{F}{N} = -T \log \lambda_+. \tag{B.140}$$

The correlation function in the same limit is translationally invariant and is given by

$$\langle S_j S_{j+1} \rangle = \langle \tilde{S}_j \rangle \to \frac{1}{N} \frac{\partial \log Z_N^{(\mathrm{F})}}{\partial K_1} \to \frac{\sinh(K_1)}{\sqrt{\cosh^2(K_1) - 2e^{-2K_2} \sinh(2K_2)}}. \tag{B.141}$$

9.8

We closely follow the method to solve the two-dimensional Ising model. First, we replace the variables as $K^* \to \beta h, K \to \beta J$ in eqns (9.94) and (9.95) and ignore the factor $g(K)$. Since the density matrix has the Hamiltonian operator in the exponent, we do not have to make a product, like the product of V_1 and V_2, and write the partition function simply as

$$Z = \mathrm{Tr} \, \exp \left(\beta J \sum_j \sigma_j^z \sigma_{j+1}^z + \beta h \sum_j \sigma_j^x \right). \tag{B.142}$$

The representation by Majorana fields and the Fourier transformation apply similarly to lead to

$$Z = \mathrm{Tr} \prod_{q \geq 0} T(q) \tag{B.143}$$

$$T(q) = \exp \left[2\mathrm{i}\beta J \left(e^{-\mathrm{i}q} C_1(q) C_2^\dagger(q) + e^{\mathrm{i}q} C_1^\dagger(q) C_2(q) \right) \right.$$
$$\left. - 2\mathrm{i}\beta h \left(C_1(q) C_2^\dagger(q) + C_1^\dagger(q) C_2(q) \right) \right]. \tag{B.144}$$

The eigenvalues of $T(q)$ in the two-dimensional space spanned by $|00\rangle$ and $|11\rangle$ are two degenerate unities. The projection of $T(q)$ onto the space spanned by $|01\rangle$ and $|10\rangle$ is, corresponding to eqns (9.116) and (9.117),

$$\tilde{T}(q) = \exp\left(2\beta h(\tau^z \cos q - \tau^x \sin q) - 2\beta J \tau^z \right). \tag{B.145}$$

The quantity in the outer brackets on the right-hand side is written as a matrix as

$$2\beta \begin{pmatrix} h \cos q - J & -h \sin q \\ -h \sin q & -h \cos q + J \end{pmatrix}, \tag{B.146}$$

whose eigenvalues are $\pm 2\beta \sqrt{h^2 + J^2 - 2hJ \cos q}$. Thus, the partition function is

$$Z = \prod_{q \geq 0} \left(2 + 2 \cosh 2\beta \sqrt{h^2 + J^2 - 2hJ \cos q} \right)$$

$$= \prod_{q \geq 0} \left(2 \cosh \beta \sqrt{h^2 + J^2 - 2hJ \cos q} \right)^2. \tag{B.147}$$

The free energy per spin is

$$-\beta f = \frac{1}{\pi} \int_0^\pi \log\left(2\cosh\beta\sqrt{h^2 + J^2 + 2hJ\cos q}\right) dq,$$ (B.148)

where the variable has been changed as $q \to \pi - q$. This is the exact solution for the free energy.

The zero-temperature limit $\beta \to \infty$ of the free energy f is the ground-state energy. For $h > 0$, the zero-temperature limit of the above solution is

$$E_0 = -\frac{1}{\pi} \int_0^\pi \sqrt{h^2 + J^2 + 2hJ\cos q}\ dq$$

$$= -\frac{2(h+J)}{\pi} \int_0^{\pi/2} \sqrt{1 - k_1^2 \sin^2\omega}\ d\omega$$

$$= -\frac{2(h+J)}{\pi} E(k_1), \quad k_1^2 = \frac{4hJ}{(h+J)^2},$$ (B.149)

where $E(k_1)$ is the complete elliptic integral of the second kind. This function is known to have a singularity at $k_1 = 1$ ($h = J$). To confirm it, we set $h/J = 1 + \epsilon$ and use $k_1^2 \approx 1 - \epsilon^2/4$. We notice in eqn (B.149) that the singularity could emerge when the integrand vanishes and so rewrite the right-hand side as

$$E_0 \approx -\frac{4J}{\pi} \int_0^{\pi/2} \sqrt{1 - \left(1 - \frac{\epsilon^2}{4}\right)\cos^2\omega}\ d\omega$$

$$\approx -\frac{4J}{\pi} \int_0^{\pi/2} \sqrt{1 - \left(1 - \frac{\epsilon^2}{4}\right)\left(1 - \frac{\omega^2}{2}\right)}\ d\omega$$

$$\approx -\frac{4J}{\sqrt{2}\pi} \int_0^{\pi/2} \sqrt{\omega^2 + \frac{\epsilon^2}{2}}\ d\omega.$$ (B.150)

This is the same expression as eqn (9.129) and behaves singularly as $-\epsilon^2 \log|\epsilon|$ around $\epsilon = 0$. The singularity appears as a function of the transverse field, not as a function of the temperature.

9.9

For the single-spin case,

$$Z = e^{\beta h} + e^{-\beta h} = e^{\beta h}(1 + z^2) = 0,$$ (B.151)

which means $z = \pm i$. The two-spin system has

$$Z = e^{2\beta h + K} + e^{-2\beta h + K} + 2e^{-K} = 0.$$ (B.152)

This gives $z = -\mathrm{e}^{-2K} \pm \mathrm{i}\sqrt{1 - \mathrm{e}^{-4K}}$, whose absolute value is unity, as expected.

9.10

The integral expression of magnetization (9.138) is approximated as

$$m = 4\beta h \int_{-\pi}^{\pi} \frac{g(\theta)}{4\beta^2 h^2 + \theta^2}\, \mathrm{d}\theta. \tag{B.153}$$

Inserting $g(\theta) = \theta^a$ and changing the integral variable as $\theta \to h\phi$, we find

$$m = 4\beta h^a \int_{-\pi/h}^{\pi/h} \frac{\phi^a}{4\beta^2 + \phi^2}\, \mathrm{d}\phi \longrightarrow 4\beta h^a \int_{-\infty}^{\infty} \frac{\phi^a}{4\beta^2 + \phi^2}\, \mathrm{d}\phi, \tag{B.154}$$

where the last expression is valid in the asymptotic limit of $h \to 0$. Since the integral part is now independent of h, the correct behavior $m \propto h^{1/\delta}$ results only if $a = 1/\delta$.

Chapter 10

10.1

The upper right and lower left graphs of Fig. 10.3 represent the sixth-order term of the high-temperature expansion, and no other contributions exist of this order. There are N ways to locate either one of these graphs on the square lattice because such a number is identical to the number of ways to locate the lower-left corner of a graph. Hence, the total number is $2N$, which is the coefficient of the sixth-order term according to eqn (10.18).

10.2

We need the original Boltzmann factor $u(\xi_i - \xi_j)$ and its Fourier transform $\lambda(\eta_{ij})$ to apply the theory of Section 10.3. The former is already written in the text for the Potts model, $u(0) = \mathrm{e}^K, u(1) = \cdots = u(q-1) = 1$. The latter is obtained using the inverse transformation (10.35) with $u(0) = u(q)$ taken into account as,

$$\lambda(\eta) = \sum_{\xi=1}^{q} \mathrm{e}^{-2\pi \mathrm{i}\xi\eta/q} u(\xi) = u(0) + \sum_{\xi=1}^{q-1} \mathrm{e}^{-2\pi \mathrm{i}\xi\eta/q} = \begin{cases} \mathrm{e}^K + q - 1 & (\eta = 0) \\ \mathrm{e}^K - 1 & (\eta \neq 0) \end{cases}. \tag{B.155}$$

Since the ratio of the two different Boltzmann factors is $u(1)/u(0) = \mathrm{e}^{-K}$ for the original system, the dual ratio is used to define the dual coupling as $\lambda(1)/\lambda(0) = \mathrm{e}^{-K^*}$,

$$\mathrm{e}^{-K^*} = \frac{\mathrm{e}^K - 1}{\mathrm{e}^K + q - 1}. \tag{B.156}$$

The uniqueness assumption of the transition point allows us to identify the transition point with the fixed point of the duality, $K = K^* = K_c$, in the case of a self-dual lattice like the square lattice, leading to $\mathrm{e}^{K_c} = 1 + \sqrt{q}$.

As for the duality relation of the partition function, we notice that the following relation holds, as in eqn (10.37),

$$Z(K) \equiv \sum_{\{\xi_i\}} e^{K \sum_{\langle ij \rangle} \delta(\xi_i - \xi_j)} = q^{-1-N} \sum_{\{\mu_i\}} e^{\sum_{\langle ij \rangle} (K^* \delta(\mu_i - \mu_j) + a)}, \qquad (B.157)$$

where a is determined by $\lambda(1) = e^K - 1 = e^a$. Therefore, we have

$$Z(K) = q^{-1-N} (e^K - 1)^{N_B^*} Z(K^*). \qquad (B.158)$$

This is the duality relation of the partition function.

10.3

The same discussion applies as in the Ising model given in Section 10.1 with the necessary modification being the evaluation of $(T^*)'_c$. From $e^{K_c} = 1 + \sqrt{3}$ and eqn (B.156) we find $(T^*)'_c$ to be -1 as in the Ising case. Accordingly, the relations for the critical exponents and critical amplitudes, $\alpha_+ = \alpha_-$ and $A_+ = A_-$, remain intact.

10.4

The argument developed in Section 10.1 applies. If the left-hand side of eqn (10.11) represents the singular part of the free energy for the triangular lattice, the right-hand side is for the hexagonal lattice. Hence, eqn (10.14) relates critical exponents and amplitudes of the Ising models on the triangular and hexagonal lattices. If we denote these quantities for the triangular lattice by $A_\pm^t, \alpha_\pm^t$ and those for the hexagonal lattice by $A_\pm^h, \alpha_\pm^h$, the obtained relations are $A_\pm^t = A_\mp^h$ and $\alpha_\pm^t = \alpha_\mp^h$.

10.5

Let us take a trace over S_0 in the Boltzmann factor B,

$$\sum_{S_0 = \pm 1} e^{K^* S_0 (S_1 + S_2 + S_3)}$$

$$= \sum_{S_0 = \pm 1} \cosh^3 K^* (1 + S_0 S_1 \tanh K^*)(1 + S_0 S_2 \tanh K^*)(1 + S_0 S_3 \tanh K^*)$$

$$= 2 \cosh^3 K^* \left(1 + (S_1 S_2 + S_2 S_3 + S_3 S_1) \tanh^2 K^* \right). \qquad (B.159)$$

We define $\tilde{K}$ such that the above expression is equal to $A e^{\tilde{K}(S_1 S_2 + S_2 S_3 + S_3 S_1)}$,

$$A e^{\tilde{K}(S_1 S_2 + S_2 S_3 + S_3 S_1)}$$

$$= A \cosh^3 \tilde{K} (1 + S_1 S_2 \tanh \tilde{K})(1 + S_2 S_3 \tanh \tilde{K})(1 + S_3 S_1 \tanh \tilde{K})$$

$$= A \cosh^3 \tilde{K} \left(1 + \tanh^3 \tilde{K} + (S_1 S_2 + S_2 S_3 + S_3 S_1)(\tanh \tilde{K} + \tanh^2 \tilde{K}) \right)$$

$$= A \cosh^3 \tilde{K} (1 + \tanh^3 \tilde{K})$$

$$\times \left(1 + \frac{\tanh \tilde{K} + \tanh^2 \tilde{K}}{1 + \tanh^3 \tilde{K}} (S_1 S_2 + S_2 S_3 + S_3 S_1) \right). \qquad (B.160)$$

Hence,

$$\tanh^2 K^* = \frac{\tanh \tilde{K} + \tanh^2 \tilde{K}}{1 + \tanh^3 \tilde{K}} = \frac{\tanh \tilde{K}}{1 - \tanh \tilde{K} + \tanh^2 \tilde{K}}, \tag{B.161}$$

is the desired relation. $\tilde{K}$ is a function of K through K^*.

The system with interaction $\tilde{K}$ is the Ising model on a triangular lattice with the same number of sites as in the original Ising model on the triangular lattice.[2] Thus, the partition function of the system we just obtained is identical to the partition function of the original system up to a trivial prefactor, $Z(K) \propto Z(\tilde{K})$. We eliminate K^* using eqns (10.3) (or its equivalent $e^{-2K} = \tanh K^*$) and (B.161) to establish a relation between K and $\tilde{K}$,

$$(e^{4K} - 1)(e^{4\tilde{K}} - 1) = 4. \tag{B.162}$$

This equation shows that $\tilde{K}$ is a monotone decreasing function of K. Therefore, we may identify the fixed point of this relation with the unique singularity of the free energy. From the fixed-point condition $K = \tilde{K}$, we finally obtain $e^{4K_c} = 3$ as the transition point.

10.6

When μ_i is a continuous variable, the Boltzmann factor $e^{-(\mu_i - \mu_j)^2/2T_r}$ has the same form as the spin-wave approximation discussed in Section 7.2. It has already been shown there that the expectation value of $(\mu_i - \mu_j)^2$ diverges in two dimensions as $\log r$ when r grows. This means that the height variable μ_i is uncorrelated with another height variable at a far-away place, and hence the surface is rough.

Chapter 11

11.1

The heat-bath method trivially satisfies the detailed balance condition. As for the Metropolis method, if $H(b) > H(a)$, $w(a \to b) = e^{-\beta(H(b)-H(a))}$ and $w(b \to a) = 1$. Then, their ratio is easily confirmed to satisfy the detailed balance. When $H(b) < H(a)$, $w(a \to b) = 1$ and $w(b \to a) = e^{-\beta(H(a)-H(b))}$, which is also seen to be compatible with the detailed balance condition.

11.2

Energies of spin configurations are $H(a) = H(d) = -J$ and $H(b) = H(c) = J$. With the notation $u = e^{-K}/(2\cosh K)$ and $v = e^K/(2\cosh K)$, where $K = \beta J$, non-vanishing transition probabilities are $w(a \to b) = w(a \to c) = w(d \to b) = w(d \to c) = u$ and $w(b \to a) = w(b \to d) = w(c \to a) = w(c \to d) = v$. The corresponding matrix is

$$\mathcal{L} = \begin{pmatrix} 1 - 2u\Delta t & v\Delta t & v\Delta t & 0 \\ u\Delta t & 1 - 2v\Delta t & 0 & u\Delta t \\ u\Delta t & 0 & 1 - 2v\Delta t & u\Delta t \\ 0 & v\Delta t & v\Delta t & 1 - 2u\Delta t \end{pmatrix}. \tag{B.163}$$

---

[2] There is a small correction due to boundary effects, which is irrelevant to the singularity and is therefore ignored here.

Notice that $\mathcal{L}_{ab} = w(b \rightarrow a)\Delta t$ $(b \neq a)$. The right eigenvalues are $1, 1 - 2u\Delta t$, $1 - 2v\Delta t, 1 - 2u\Delta t - 2v\Delta t$ and their corresponding un-normalized eigenvectors are

$$
\begin{pmatrix} v \\ u \\ u \\ v \end{pmatrix}, \quad \begin{pmatrix} -1 \\ 0 \\ 0 \\ 1 \end{pmatrix}, \quad \begin{pmatrix} 0 \\ -1 \\ 1 \\ 0 \end{pmatrix}, \quad \begin{pmatrix} 1 \\ -1 \\ -1 \\ 1 \end{pmatrix}. \tag{B.164}
$$

The first eigenvector is the equilibrium distribution after proper normalization. The components of each of the other eigenvectors sum to zero.

Index